Prehis

MW01491231

ALSO BY ALLEN A. DEBUS

Dinosaurs in Fantastic Fiction:
A Thematic Survey (McFarland, 2006)

Paleoimagery: The Evolution
of Dinosaurs in Art (McFarland, 2002)

Prehistoric Monsters

The Real and Imagined Creatures
of the Past That We Love to Fear

ALLEN A. DEBUS

McFarland & Company, Inc., Publishers
Jefferson, North Carolina, and London

LIBRARY OF CONGRESS CATALOGUING-IN-PUBLICATION DATA

Debus, Allen A.
    Prehistoric monsters : the real and imagined creatures of the
past that we love to fear / Allen A. Debus.
        p.   cm.
    Includes bibliographical references and index.

    ISBN 978-0-7864-4281-2
    softcover : 50# alkaline paper ∞

    1. Dinosaurs in mass media.   2. Dinosaurs—Mythology.
3. Popular culture.   I. Title.
P96.M6D43   2010
567.9—dc22   2009040779

British Library cataloguing data are available

Cover art: Bob Eggleton, *Gorgzillicus,* acrylic on canvas, 18" × 24",
1997. Background images: ©2009 Shutterstock.

Manufactured in the United States of America

McFarland & Company, Inc., Publishers
  Box 611, Jefferson, North Carolina 28640
  www.mcfarlandpub.com

To the memory of
Richard W. Debus (1957–2007),
brother and friend,
and
in memory of
Allen G. Debus (1926–2009),
father and mentor

Table of Contents

Preface

Most "monster books" nostalgically emphasize movieland monsters, fondly acknowledging pivotal acting talents of Boris Karloff, Bela Lugosi, Vincent Price, Lon Chaney, Jr., Christopher Lee and many others who have thrilled us on the silver and television screens, playing evil or ghoulish roles. Focusing on actors, movie plots and behind-the-scenes special effects wizardry has become the traditionalized way of exploring and introducing monsters in print ever since *Famous Monsters of Filmland* began publication during the late 1950s.

I've enjoyed many of these cult and genre movies since my early years of watching them on *Creature Features* and *The Early Show* on Chicagoland stations back in the 1960s. But while Frankenstein, Dracula, Wolfman, Morlocks, assorted invisible men, and even less sublime robot or "green slime" monsters provided many hours of enjoyable, low-brow entertainment then, the monster films I most admire were and have always been those involving prehistoric creatures. Perhaps it all started with *King Kong*, which I first saw televised circa 1959 or 1960 at a period in my young life when my Dad was instilling in me a love for paleontology, a fascinating field forcing me to ponder the meaning and absolutely mind-bending infinitude of geological time. Then a young handful of years later, Godzilla entered my life.

I'm not the first to connect "prehistoric" with "therefore monstrous" viscerally. This is a time-honored association, as captured in the title of the Rev. H. N. Hutchinson's popular book, *Extinct Monsters: A Popular Account of Some of the Larger Forms of Ancient Animal Life* (1892), one of the earliest of its kind. Here, the author introduced dinosaurs, prehistoric mammals and marine and flying saurians then known to science as genuine monsters of natural history.

Why this emphasis on prehistoric *monsters*? Isn't this an unusual path? True. Scholarly works on prehistoric life traditionally adopt an onerous, encyclopedic tone, difficult for non-specialists to digest and appreciate. An expected result is that few are savvy about genuine prehistoria other than those ever-

1

so-popular dinosaurs and pterodactyls. Many since childhood are attuned to the idea of all manner of prehistorians being categorized loosely as monsters, regardless of their authenticity, phylogeny, or form. Yet few may realize how these fascinating creatures gained such a distinctive pejorative — *monsters*. And so here I shall illuminate how and why our concept of *prehistoric monsters* (including among the commonly recognized species everything from gigantic extinct sloths, and dinosaurs through Godzilla and Reptilicus) originated.

This volume illustrates how mankind's appreciation of monsters that are also popularly (or scientifically) regarded as prehistoric originated, and may be regarded as a companion to both my *Dinosaurs in Fantastic Fiction* (2006), in its treatment of man's eternal struggle (really a love-hate relationship) with giant dino-monsters, and *Paleoimagery* (2002), with its considerations, relevant here, as to how paleoart influenced man's visual perceptions of prehistoria translated into pop culture. The Appendix fuses featured monsters of the movies, science fiction and horror with their most famous genuine counterparts thought most monstrous.

In moving from the earliest historical examples as recorded by ancient scholars and myth makers, on into years when the paleontological and evolutionary sciences matured, there's a bones of violence — stone cold savagery emphasis at large, meaning that fears of the prehistoric world often stemmed from projection of fossil-founded primitivism and prehistoric savagery into modernity. Consider all the prehistoric monsters that have, either anachronistically or cryptozoologically, plagued mankind. From the mythical Monster of Troy, winged dragons, Kong, Rhedosaurus, the Gillman, and *Jurassic Park III*'s splendid *Spinosaurus*. All are interrelated not through supernatural phenomena or paranormal studies but through the noble study of fossils from real or imaginative places with ominous names like Big Bone Lick, the Black Lagoon, Hell Creek, Skull and Odo Islands, and the lands of Curupuri and where Time forgot.

This book germinated in October 2004 when, simply to amuse myself, I began listing the "greatest prehistoric monsters of all time." The scratch list of names evolved into a more informative, four-part series of short articles published in Dennis Druktenis' movie fanzine *Scary Monsters* (January-June 2006, issues nos. 57–58, and the 2006 *Monster Memories Yearbook*). At the time, who could anticipate what form that excursion could take — so greatly expanded and further defined, herein?

Few know that as a college undergraduate I originally declared geology as a major. Although this changed to chemistry a year later, during undergraduate, graduate and even post-graduate studies I accumulated credits in geology, geochemistry, mineralogyk, clay-mineralogy and paleontology — collectively nearly enough for a double major. My interrelated trilogy of books for McFarland must therefore represent, however tangentially, a means for revisiting my first academic love.

Much of this writing took place during the months following the loss of my brother, Richard, who was a geologist. This is a book he would have enjoyed. (Sara, here it is!)

Thanks go to the following individuals who assisted, contributed to, or otherwise in some fashion facilitated completion of this volume: Jack Arata, Martin Arlt, Lynne Clos, Diane Debus, Allen G. Debus, Brunilda L. Debus, Lisa Debus, Sara Debus, Karl Debus-Lopez, Kristen L. and Ryan Dennis, Dennis Druktenis, Bob Eggleton, Mike Fredericks, Don Glut, Robert Hood, Bruce Horton, J. D. Lees, William Schoell, Tanner Wray.

Introduction: Prehistoric Monster Reflections

Modernity and the prehistoric are, in a sense, like matter and anti-matter. To avoid catastrophe, they should never come together. In his *The Great Orm of Loch Ness*, F. W. Holiday notes that the word monster "derives from the Latin word *monstrum*, an omen, since the rare appearance of these very strange and sinister creatures was thought to betoken unusual events."[1] Holiday's definition clarifies our view of prehistoric monsters (as the Loch Ness monster is so commonly regarded), because more so than supernatural vampires and werewolves of our dreamscape, the thought of prehistoric monsters threatening mankind reveals our worst nightmare: catastrophe — distorted Nature horribly, perhaps irrevocably out of balance. So prehistoric monsters project somber messages, presaging global and ecological disaster — they're often cast as doomsday dinosaurs, signifying Man's folly — making them highly relevant today!

"Monster" has another meaning, this time entrenched in the Frankensteinian joining of bodily parts, resurrected, "that should have remained separate (and dead)."[2] Thus, the long-necked marine reptile, *Plesiosaurus*, had turtle- and snake-like affinities. In early light, the *Iguanodon* melded features of the rhinoceros, crocodiles, and lizards. Pterodactyls were viewed as fiendish bat-winged vampires. "Monstrosity was inscribed into their names: *Ichthyosaurus* ('fish-lizard'), *Pterodactyle* ('wing-finger'), *Pterichthys* ('winged fish')."[3] Like fictional Dr. Frankenstein, Albert Koch composed fossil material into larger-than-life skeletal displays of impossible prehistoric animals. Perhaps it is not so curious that fossilists of the early 19th century, such as Gideon Mantell and Thomas Hawkins, had a strange fascination with Mary Shelley's then new novel, *Frankenstein; or, the Modern Prometheus*.

Prehistoric monsters are a breed apart from other monsters of popular culture (e.g., vampires, aliens, werewolves, and zombies), with origins entrenched in discovery of genuine lost worlds existing at the proverbial dawn of time. While most famous monsters of fiction and film infest modernity, monsters judged prehistoric usually (but not always!) terrorize primeval settings. Take Kong's Skull Mountain for example, a misty, swampy jungle island setting where time — meaning evolution — has seemingly stood still, a place where gigantic monsters incessantly battle for supremacy. After all, when they aren't crushing or devouring people, or within civilization's confines overturning tanks and smashing buildings, isn't that what prehistoric monsters specialize in: practicing pseudo–Darwinian survival of the fittest behavior by combating *each other*? At least, considering Kong's 1933 fight with *Tyrannosaurus*, or turning to Toho's prolific series of *daikaiju eiga* films, Gigantis' dispatching of Anguirus, or even Kong's later, much publicized match with Godzilla, movie producers would have us believe so. While savage nature and struggle for existence have become traditional themes commonly noted in popular prehistoric animal reconstructions, ultimately, in order to properly rank as both *prehistoric* and (therefore) *monstrous*, prehistoric monsters must also prove menacing to mankind.

Of course most monsters have a common basis in reality. True, folklore surrounding ancient and medieval perceptions of vampires and werewolves also was often based in a reality as known to contemporary naturalists and scholars. However, facts supporting, say, a vampire legend would be quickly quashed when re-interpreted on basis of modern scientific knowledge (assuming we had all those facts to ponder). Those legends would indeed be falsified by modern forensic conclusions addressing those same facts allegedly suggesting the existence of vampires. Yet, although it's a certain bet there are no vampires alive (or undead), vampires persist in modern pop culture because, well, people enjoy a good (non-threatening) scare. Conversely, there *was* a genuine prehistoric age and there really *were* animals alive then which would have seemed truly frightening, as well as menacing, had we coexisted with them. And in some cases our primeval ancestors did encounter these prehistorians because, yes, man was formerly prehistoric and, in a popular vein, monstrous too.

A practical dimension to prehistoric nature enters from the modern perspective. For when zoologist W. Douglas Burden was asked whether the famed Komodo dragons are prehistoric, he replied, "I confess to finding the question difficult to answer. In the literal sense, every living organism is prehistoric, for are we not all, man included, the outcome of millions of years of evolution? If however, the word has come to mean great age with little change, it is correctly applied to the carnivorous lizards of Komodo."[4] Therefore, for present purposes, the term "prehistoric monsters" shall refer to either genuine fossilized creatures known to science that may seem monstrous (e.g., "big, fierce, extinct"),[5] or, although derived from prehistoric stock, in a fantastic or fictional sense, living in the present. While such a moniker —"prehistoric"— applies to disparate forms, from the Silurian Period's huge sea scorpions (Eurypterids) to, perhaps, the extant Komodo dragons, instead we will focus upon real vertebrate animals that are extinct, or fictional vertebrates that should be extinct (or never truly existed).

Mankind has resurrected its most famous, yet often fictitious prehistoric monsters from real fossilized remains of *natural* (not supernatural) creatures known only from the ashes of time. Furthermore, we note that mankind's conception of the most dominating of such creatures has evolved historically. Popular science fictional prehistoric monster manifestations are founded in science. However, they are more so extrapolated or composited from pre-packaged artistic restorations of real prehistoric animals rather than detailed diagnoses of skeletal anatomy.

How is it that genuine animals such as fin-back reptiles, pterodactyls and dinosaurs could then be regarded as "monsters"? The answer is woven into how people of past historical times came to regard them culturally. Perceptions of these animals in cultural history is a result of public exposure to visual paleoart, imagetext, literary writings, and restorations and portrayals witnessed in imaginary time tours (artificial primeval islands "inhabited" by sculpted dino-monsters, Hollywood, or otherwise), presenting prehistoric animals in monstrous ways. By the mid–19th century, most of geology's prehistoric time periods had been validated on the basis of fossils and global stratigraphical relationships; an interested public craved more concerning what these mysterious past worlds contained.

Interest in imaginary monsters of natural history transformed over many decades and centuries into intrigue over genuine monsters of prehistory and then later, tangentially, into fictional, often filmic creatures of horror, science fictional dinosaurian spinoffs. While for today's prehistoric monsters, movie producers usually turn to paleontology, unwittingly, millennia ago, myth makers did the same.[6] Hence there is a strange sense of continuity linking monsters known in ancient times—centaurs, giants and dragons—to, say, a recent Sci Fi Channel production, *Pterodactyl* (first aired in August 2005).

Shortly after scientifically minded men learned the rudiments of prehistoric animal reconstruction, came the idea of the prehistoric monster *battle*, as Thomas Jefferson idly conceived it during the 1790s.[7] Although initially not regarded as a true paleontological "monster," the riddle of the huge hairy mammoth became a proving ground in early enlightenment over the nature of fossil objects. Through the 19th century, science reveled in its ability to re-create the appearance of monstrous prehistoric animals in sculptures and illustrations, chiefly prehistoric mammals such as the American mastodon and Mesozoic marine sauria. A conceptual quantum leap from science into literary mass culture was made by Jules Verne through 1860s editions of his novel *Journey to the Center of the Earth*, dramatizing encounters between modern man and disturbing prehistoric monsters then known to science.

Public consciousness of prehistoric monsters cannot be heightened without popularization. And popularization of geology and its prehistoric monsters was also greatly facilitated in the early to mid–19th century by visionary earth scientists who embedded poetry within their scientific writings (or who themselves wrote "poetically"), as they theatrically unveiled each of the worlds existing before Man. Accordingly, geological texts of the time were peppered with poetic verses and stanzas spliced and inserted into otherwise scientific passages in order to reinforce the meaning of the latter. Such literary devices and allusions to classical writings that educated individuals studied at leisure staged the proper mood and atmosphere for unveiling geological science. Or, in other words, a spoonful of sugar helps the medicine go down!

Authors and artists tasked to do the unveiling, therefore, resorted to the familiar in order to facilitate comprehension of the unfamiliar, mind stimulating "diversions." And, in a fashion that would seem largely lost on us today, but seemed most familiar to gentlemen of the times—the reading public and others who enjoyed the theater — was literature composed by men such as the Romantic poets Lord Byron, Percy Bysshe Shelley, or works like Dante's *Inferno* and John Milton's *Paradise Lost*, or even the Roman poet Virgil. Furthermore, gentlemen readers of the time would have gravitated to scripturally inspired visions of antediluvian times, choice draconian allusions,[8] and descriptions of Hell itself (although "retooled" to represent the world's primeval times as yet unprepared for Man, or the Earth's fiery central regions), and (analogous to the current station of the film industry today) theatrical performances. These "stories" became "textual correlatives of the pictorial 'scenes from deep time.'"[9]

A literary admixture of science and poetry was appropriate because, to many, geology was akin to "romance," except its denizens and landscapes were prophetically real rather than imaginary thus heightening its practicality. And while geology's gallery of monsters could be conveyed in a variety of means, as Ralph O'Connor suggests in referring to the "poetics of geology," the 19th century reading public were a more *imaginative* lot who, through repeated mental exertions, could glean much more from applying their mind's eye to the printed word than many can today. Accordingly, by the mid–1830s through the 1850s, geology had aspired to become the most popular science of all.[10] In unveiling the monsters out of prehistory in their haunting paleoworlds, many writers of the period crafted paleo-themed "purple passages," anticipating the "techniques of later science fiction writers."[11] Relying on melded visual and literary arts, largely through the strange appeal of its paleo-monsters, geology was marketed successively and vigorously in the years following Charles Lyell's *Principles of Geology*, William Buckland's *Geology and Mineralogy*, Mantell's *Wonders of Geology*, and certainly throughout the period when Scotsman Hugh Miller dominated the popular press with his poetical visions of the past.[12]

Accordingly, from early verbal imaginings, descriptions, and "poetic" geological re-creations, sometimes accompanied by visual printed restorations, came attempts to more fully embody and capture the majesty and sublimity of prehistoric worlds in foreboding isolated places. Some of the earliest artificial settings to become thus conventionalized were Benjamin Waterhouse Hawkins' (1809–1889) "Frankensteinic" Crystal Palace Secondary and Tertiary islands, Jules Verne's sublime subterranean literary realm in *Journey* (1864), or other imaginary chronal refugia permitting "cyber" contact with assorted beasts from deep time.[13] A later example is Sinclair's Cretaceous oasis at the 1933/34 Chicago World's Fair. Only those illuminated by geology's light and reason, possessing the necessary knowledge, field credentials and bravado, could figuratively penetrate time and elucidate the nature of these recreated monster islands properly to the public.

Charles Lyell (1797–1875), William Buckland (1784–1856), Gideon Mantell (1790–1852) and others (e.g., Robert Chambers [1802–1871] and Louis Figuier [1819–1894]) sought to educate classes of interested gentlemen through popularization; their entertaining geological writings were often sprinkled with imaginative musings or rhapsody. Increasingly, from the 1830s through the 1880s, other popular writers joined the fray, publishing books and other forms of popular scientific journalism; material digested from the researches of principal investigators dramatically melding speculative literary passages with visual effects (i.e., printed restorations and reconstructions).[14] Such presentations intended to bedazzle fascinated readers with accounts of Prehistory's "theatrical" unveiling. For instance, Mantell's *Wonders of Geology* went through several late 1830s editions; artist John Martin's (1789–1854) lurid frontispiece therein seems more thematically derived from St. George's draconian exploits rather than Georges Cuvier's (1769–1832) sterile objectivity. Consequently, Mantell's production relying on the attraction of "extinct monsters was an early example of how scientific accuracy came to be sacrificed to the demands of commercially driven showbiz and popular hype."[15]

By the early 1900s, two highly significant dinosaurs were added to the escalating pantheon of prehistoric monsters—carnivorous *Tyrannosaurus* and three-horned *Triceratops*, memorialized together in paintings by Charles R. Knight (1874–1953). Once science introduced these recreated Cretaceous monsters to an adoring public, they were quickly whisked into pop culture, science fiction and horror. Modern man could also speculatively be introduced into the prehistoric formula — now facing Prehistory's two most formidable monsters engaged in mortal combat. Knight's beautiful paintings seemed so realistic and compelling. Gazing into the picture frame was like moving into the geological past through a time portal, without Wellsian time travel machinery. In a larger sense, much of what came thereafter involving prehistoric monsters—evident in classic horror, monster and science fiction literature — is derived from Knight's timeless and expertly executed original

"Rex & Tops" symbolic imagery, although reflecting contemporary cultural matters. A final, most curious and emblematic piece of the formula, of course, was the revelation that Man himself had been prehistoric.

While in Britain, Mantell shared colorful visions of his Wealden "country of the *Iguanodon*," in France, Pierre Boitard (1789–1859) published an intriguing popular article, "Fossil Man," in an 1838 issue of *Magasin Universel*. Boitard incorporated Susemihl's restoration of such an anthropoid walking in upright bipedal gait, carrying a stone ax and wearing a cloak made of animal fur, although depicted with "strongly *simian* (and negroid?) bodily features." This image proved unsettling because it captured what many feared most about the burgeoning paleontological sciences, namely that scientists were truly on the verge of revealing Man's "transformist" ancestry from ape-like, bestial ancestors.[16] The most recent geological lost worlds uncovered by science, therefore, were becoming more harrowing in nature and alarming to the human psyche, heightening their intrigue, and counterintuitively perhaps, their charm. By the mid–1830s, men of science realized as did their devoted readership that, tantalizingly, if primates had existed in the fossil past, *humans may have as well*. As a train wreck out of control, it was hard for the public to turn away from the welter of scientific, pseudoscientific and theological interpretations about the fossil past and its reconstructions that were out there on view. Our dark and forbidding ancestry as well as our natural plight, perhaps, became symbolized in the triumphant yet tragic prehistoric ape — Kong.

Today, when it comes to selecting monsters deemed prehistoric for a monster story, dinosaurs of one kind or another comprise the cast of usual suspects. Their ubiquitous popularity especially since the 1930s may be attributed to a number of irreducible factors. Yet as stated by a child psychologist, children's interest in dinosaurs stems from the fact that, in popular culture, their (non-avian) forms are generally "big, fierce, and extinct." There's also a possibility that these "giant creatures personified the uncontrollable force of looming authority. They were symbolic parents. Fascinating and frightening, like parents. And kids loved them as they loved their parents."[17] Also, in growing up dinosaur, that special bond many adults felt as kids for monster movies and, particularly, that unique, extinct breed of creatures they so loved in their youth is rekindled.

Yes—big, fierce, extinct: how we so love them!

ONE

Inventing the
Prehistoric Monster

In the interior of Africa, some great monster of the past may still roam the forests. A mammoth may be wandering now over the frozen wastes of our Arctic regions: last lonesome representative of his great race. A dinosaur in South America, a glyptodon in some unexplored area of the earth. That enormous bowl of water that covers nearly all our globe may conceal animals undreamed of, who would have the temerity to deny at least the possibility.

In this terrible age of disbelief and gullibility, people will swallow any tale of monsters of the past, but unless we find the bones of a centaur, no one will credit that myth. What have the scientists done but replace dragons, mermen and sphinxes with a new line of beasts? The people found the transposition easy. Where once they thought of dragons, they will have mammoths, and other extinct beasts to occupy the same mental pews: these never change.[1]

The story of mankind's enthrallment with prehistoric monsters begins with folklore and ends in mythology.

At first glance, *Jurassic Park*'s universe of spectacular, CGI-enlivened dino-monsters (*T. rex*, raptors, *Spinosaurus*, compys, and so on) would appear to have few connections to Ray Harryhausen's heart-stopping, stop-motion animated mythological and folkloric monsters, as featured in films such as *Jason and the Argonauts*, with its bat-winged harpies and seven-headed hydra, or *The 7th Voyage of Sinbad*, with its Cyclops, two-headed winged roc, and dragon. But there *are* connections, metaphorically linking the search for fossil vertebrates from antiquity to the present day. Intriguingly, monsters as known to the ancients and men of medieval times might seem uncannily familiar to modern paleontologists. Indeed, monsters of old in a sense served as progenitors, centuries later spawning a protean host of pop-cultural movieland dino-monsters and other reconstructed prehistoria. For millennia since their extinction, dino-monsters and associated prehistoria have fascinated and haunted mankind, in many ideological forms.

Which monsters of early historical times have been established on the basis of genuine fossils? How did certain monsters later become recognized and idealized as prehistoric? And how did their symbolic, savage struggles become traditionalized, first in nature, through various forms of paleo-art and eventually on film? One purpose of this book will be to demonstrate how interest in imaginary monsters of natural history transformed into intrigue over genuine monsters of prehistory. It wasn't the objective of 17th or 18th century scientist-naturalists to create a popular prehistoric monster. However, in elucidating the natural history of old, mysterious, poorly understood bones, and in applying scientific reasoning to fossils coupled with their knowledge of earth history, prehistoric creatures later derived as popular monsters did precipitate from early attempts at scholarship.

A startling, yet unheralded byproduct of the dinosaur renaissance (~1965 through the present) is the cultivation of dinosaur dictionary and encyclopedia volumes—paleo-zoologies utterly jam-packed with information—of which the connections to a time-honored tradition stemming from ancient times through medieval ages may seem obscure; their association has not been properly acknowledged. For thoughtful men like Aristotle (384–322 B.C.E.) and Pliny the Elder (23–79) compiled early bestiaries and natural histories, encyclopedias presenting information gleaned from various, often unreliable, sources about animals and plants. Pliny's presentation style proved eminently influential, as centuries later we note scholars such as Conrad Gesner (1516–1565), Ulisse Aldrovandi (1522–1605) and Edward Topsell (1572–1625) publishing manuscripts and folio volumes concerning creatures known to contemporary science. However, "While on occasion accounts of monsters were questioned ... every scrap of information that could be found was presented."[2] Thus we find in Topsell's work, for example, entries for doubtful creatures such as the Lamia, Mantichore, Unicorn and Dragons situated among descriptions of real vertebrates. Unknown to these men, substantiation for at least some of the animals cited in their works was founded upon fossil vertebrate skeletons having been discovered and noted since (at least) the time of the Trojan War, circa 1250 B.C.E.

The recorded tallest human, Robert Wadlow (born in 1918), grew to eight feet, eleven inches and weighed 365 pounds by the time of his death at age 22. Yet his height falls far short of the gargantuan sizes assigned to mythical gods, monsters and heroes by the ancient Greeks and Romans, who, as explained by Adrienne Mayor, attributed their lofty dimensions empirically upon real fossil bones that were becoming unearthed and increasingly encountered by learned, observant or curious men as civilization spread throughout the Mediterranean region over two thousand years ago.[3]

The largest giant on record representing medieval bestiary was Godzilla-sized, topping out at a staggering 400 feet tall. This earth-shattering creature was found on Sicily in about 1371; its discovery was documented by Italian writer and poet Giovanni Boccaccio (1313–1375), who thought the bones were of the Cyclops Polyphemus. Huge teeth allegedly belonging to the giant were exhibited in a church at Trepani. The skeleton, which was found in a sitting position in a cave, shortly crumbled to dust, leaving only its skull and teeth to posterity. Centuries later, German scholar Athanasius Kircher (1601–1680) commented in his *Mundus subterraneus* (1665) upon Boccaccio's claims.[4] Here, among a series of plates illuminating the Earth's mysterious interior, is a curious figure showing (two-eyed) Polyphemus compared relatively in size to famous historical giants, including the 13 foot tall biblical Goliath, the 14 foot tall Orestes, and 120 foot tall ogre Antaeus, found (and reburied) in Morocco in 81 B.C.E. No doubt, such height estimates founded on fossil bones were scaled comparatively to ordinary human thigh bones and teeth. Citing classical sources, Kircher corrected Boccaccio, stating Polyphemus's size was exaggerated. Instead, Kircher, who also published in his *Mundus subterraneus* figures of dragons thought to be lurking within the mysterious Earth, suggested that Polyphemus was only 30 feet tall.

In a high school Latin class we translated Virgil's (70–19 B.C.E.) epic poem, *The Aeneid*, where in Book II we learned the awful tale of Laocoon, the mythical Trojan priest who uttered prophetically, "I fear the Greeks even when they bear gifts." Laocoon had every right to be suspicious, for his countrymen were about to move the infamous Trojan Horse into their citadel, resulting in their ruination, as enemy Greek warriors were hidden inside ready to spring into attack under cover of darkness. While Laocoon railed against the tricky Greeks, vengeful gods—who so often meddled in human affairs then—sent two serpents with gleaming coils slithering from the sea toward Laocoon and his two sons. After Laocoon and his sons were crushed and injected with deadly venom, fearful Trojans took this as a sign that the gigantic Horse carving should indeed be carried into their city. The tales of the Trojan War and of Polyphemus the Cyclops and other fabulous mythological creatures are told in Homer's *The Iliad* and *The Odyssey* (inspirational for Virgil's *Aeneid*). Cyclops, serpents, and a host of other giants, monsters and larger-than-life heroes (as well as goddesses and heroines) are a common occurrence in Greek and Roman mythology. But until Adrienne Mayor conducted her researches, few realized that certain idealized, seemingly fantasy creatures may have been founded upon occurrences of *real* fossil bones.

Over two thousand years ago, sightings of big, stony bones, whether found *in situ* or unearthed in enormous manmade coffins—in which fossilized remains had been re-buried by still more ancient men—were explained in terms of contemporary logic, interpretations of natural phenomena and what men then understood from established history. So if bones were found in places where an ancient Greek hero had fallen in battle, or where a great monster had been defeated, then it seemed entirely plausible to believe such remains actually once belonged to these individuals. Fossil relics were also donated to temples, where they were placed on display, thus helping to define "an ancient city's self-identity and vision of the past, its trading habits, and status."[5] The retelling of giants and monsters in legend and lore, tales of the Neades, the sinister monsters of Samos and Joppa, Geryon, Campe, Typhon, Antaeus, jeweled dragons of India and the monster of Troy — discovered remains of which had been assigned to such as these — helped men of classical times marvel and "relive the ancient times,"[6] a practice which scientifically minded men of modernity still enjoy.

Not all giants were monsters, though. Besides monstrous giants, skeletons attributed to an admixture of real and mythological heroic figures— Pelops, Achilles, Ajax, King Teutobochus (yes, the original *T. rex*!— a Germanic king slain in battle in 105 B.C.E.), Orestes and winged horse Pegasus, for example, all of whom were evidently in the range of from 10 to 33 feet tall, and many others— were also disinterred from rocky tombs. And yet it is likely proof of their former existence, as well as substantiation of the myths such remains perpetuated, was founded instead upon teeth, tusks and bones of Tertiary and Pleistocene mastodons, mammoths, deinotheres and other fossil elephants, giraffids, rhinos, cave bears and whales! A large tooth allegedly assigned in 1613 to King Teutobachus, for example, was recently examined by French paleontologist Leonard Ginsburg: it's a deinothere tooth! Men of ancient times saw the bones but understandably interpreted them in terms of mythological monsters or historical heroes and men, different from today's perspectives and search image, because they lacked clear, anatomical/skeletal visuals of (both) modern fauna and prehistoric vertebrates, as well as knowledge of both geological time and Life's eternally evolving history.[7]

Mythological monsters abound throughout Greek mythology, and their struggles were titanic, both at the beginning of time, when Cyclops and Titans were created, but even much later, during the time of Hercules. However, as Edith Hamilton claimed in 1942,

> They were monsters. Just as we believe that the earth was once inhabited by strange gigantic creatures, so did the Greeks. They did not, however, think of them as huge lizards and mammoths, but as somewhat like men and yet unhuman. They had the shattering, overwhelming strength of earthquake and hurricane and volcano. In the tales about them they do not seem really alive, but rather to belong to a world where as yet there was no life, only tremendous movements of irresistible forces lifting up the mountains and scooping out the seas. The Greeks apparently had some such feeling because in their stories, although they represent these creatures as living beings, they make them unlike any form of life known to man.[8]

And perhaps they did because such creatures were *not* entirely imaginative.[9]

Six decades later, Adrienne Mayor's revelation of the Monster of Troy (although not the same as Laocoon's serpentine killers) is particularly illustrative of what ancients saw in nature and transformed into mythological proof. According to legend, the Monster of Troy was an awful creature that appeared on the Trojan coast, as related by Homer in the 8th century B.C.E. This monster antagonized local farmers, but before the king's daughter could be sacrificed to it in appeasement, Hercules (Heracles) came to the rescue, killing the Monster. A painting of the legendary event has been preserved on a vase dating from between 560 and 540 B.C.E., now held in the collection of Boston's Museum of Fine Arts, in which Hercules (Heracles) is depicted shooting arrows at the monster while the king's daughter, Hesione, throws rocks. But, as Mayor noted, the Monster appears to be a convincingly portrayed fossilized skull replete with jagged teeth, protruding in-situ from a rock outcrop! "Monsters of Greek myth were perceived in the popular imagination and portrayed by artists either as huge *beasts* or as giant *humans*.... But the unparalleled depiction of the Monster of Troy as a large fossil animal skull on the Boston vase points to a natural basis for the two branches of monster and giant images in art and literature. Here is powerful evidence that fossil remains of prehistoric animals influenced ancient ideas about primeval monsters!"[10] Mayor has also convincingly demonstrated that ancient portrayals and belief in the existence of winged Griffins most likely stemmed from ancient explorers' genuine encounters with fossilized *Protoceratops* skeletons in Mongolia.

No wonder several multi-headed monsters have been brought to life through Ray Harryhausen's talents. A variety of frightening monsters including multi-headed, multi-limbed forms abound throughout Greek mythology. Researcher Adrienne Mayor has noted that "According to Homeric-Hesiodic lore, strongmen of myth were often said to have multiple heads or limbs."[11] I wonder whether this is because men had noted the abundant vertebrate fossil deposits scattered throughout the Mediterranean region, associating gigantic bones with their deified mythological heroes and foes, who had walked or plundered the Earth so long before. If the bone bed represented a condition of mass mortality (i.e., more than one co-deposited skeleton — say, two or more Mastodon individuals represented therein), then in their early efforts to reconstruct the giant or dragon, a composite (chimeric), multi-headed beast resulted.

Thus, in Edith Hamilton's *Mythology* (1942), besides the Cyclops, evil Centaurs, Griffins, winged Harpies, the Chimaera, Laestrygons, Minotaur, the Monster of Troy and an assortment of ogres and sea serpents defeated by the heroic Argonauts, Hercules, Perseus, Odysseus and Theseus, for example, we also read of multi-headed creatures — the nine-headed Hydra, scaly winged Gorgons with snaky, serpentine hair so hideous that a mere glance could turn a mortal to stone, the three-headed/bodied giant Geryon, and the Underworld God Pluto's three-headed dog Cerberus. Most imposing of all, however, was Typhon, "A flaming monster with a hundred heads, Who rose up against all the gods, Death whistled from his fearful jaws, His eyes flashed glaring fire."[12] Only Zeus could destroy this terrible monster with his lightning bolts.[13]

Dragons and sea serpents, such as the Monster of Troy, regularly appear in Greek mythology. Jason must first spawn and then defeat, singlehandedly, an army springing to life from sown dragon's teeth (which he is able to do with help from the witch, Medea). Perseus rescues maiden Andromeda, who, in Fay Wray–like fashion, has been chained on a rocky coastline in sacrifice to a monster. Perseus, who had already killed the Gorgon Medusa (in Greek mythology, a winged dragon with snaky hair) cuts off the serpent's head and marries Andromeda. A sea serpent attacks and mortally wounds Theseus' estranged son, Hippolytus. Hercules' second labor was to kill the nine-headed Hydra of Lerna. Later, after defeating Geryon in his tenth labor, Cerberus in the twelfth labor, and later the giant Antaeus, it is Hercules' turn to rescue Hesione, a maiden who had been offered in sacrifice by her father, King Laomedon, in sacrifice to the aforementioned Monster of Troy. Much later, Roman poet Ovid (43 B.C.E.–A.D. 16) elaborated the story of Cadmus and a dragon he slew in his *Metamorphoses*.

Hercules was also immortalized in a Renaissance-era Austrian statue completed in 1636, where he is depicted in Klagenfurt's town square swinging a club in battle against a winged dragon, which, as paleontologist Othenio Abel (1875–1946) later proved, was restored on the basis of a fossilized Woolly Rhinoceros skull discovered in 1355. Certainly, in medieval times, common men could still safely cling to their beliefs in dragons; surprisingly, belief in dragons and sea serpents would persist into the modern age of paleontology.

Enlightened in our modern comprehension of Earth history and organisms of the past represented through fossils, it is all too often presumed that men of other civilizations or who lived before the dawning of the age of modern geological investigations had little or no perception of what fossils represented. And yet Mayor has shown how wrong is such thinking. Beside the ancients who misconstrued the Cyclops, and perhaps other fabulous monsters such as the Chimaera, Sphinx and the Minotaur — as well as their titanic battles with gods and early godlike men — from fossils of the Mediterranean region, likewise, Mayor has chronicled Native Americans' fossil legends in terms of North American fossil occurrences.[14] For, independently of the classical Europeans, various tribes of pre–Columbian Native Americans conjured their own evil giants and nightmarish monsters as well, founded upon genuine fossils, including the Flying Heads, Stone Giants, Quisquiss, the Monster Bear, Qwilla water dragons, Unktehi, Quetzalcoatl, Thunder Bird, and even the fossilized animal which our ancestors referred to as the *Incognitum* (of which more will be stated in the next two chapters).[15] Description of the appearance and nature of these fanciful, yet frightful, beings associates real fossils now known to science with tribal understanding of the ancient days of planetary history and perceived cosmic conflict between heavens and earth, particularly terrestrial water bodies, where vertebrate skeletons are often found eroding out of riverbanks. Thus, Mayor connects remains of Mesozoic animals (i.e., dinosaurs, marine reptiles and pterodactyls), prehistoric elephants, cave bears and a host of other fossil vertebrates with such legends, of which there are many.

Of these, perhaps most (vaguely) familiar to paleo-enthusiasts would be Indian lore concerning Unktehi, the ruling Water Monster, and Thunder Bird. According to the Sioux, in primeval times these two entities were pitted in a bitter cosmic struggle. As Mayor related, "the first creatures were insects and reptiles, under the domain of Unktehi, the Water Monster. In the days before humans, there were many kinds of reptiles, those with legs and tails, those without legs, and those with armored shells, and with many varieties of each kind. All became cold-blooded, ravenous monsters, devouring every living thing on land and in water, and finally had to be destroyed for natural balance to be restored. In the first of the four great ages— the age of Rock — the Water Monsters were blasted into stone by Thunder Bird's bolts of lightning. The earthly bodies of Thunder beings, like the Unktehi, were also buried in the ground."[16] Not only was the primeval world perceived as violent — a theme independently recurring across many cultures through the centuries— but the postulated sequence of events accounts for fossil occurrences of Mesozoic reptiles evidently recognized by the Sioux as marine and possibly even winged creatures, such as the *Pteranodon*.

Although one would naturally think that after scientists had a fairly refined concept, or paradigm, of what prehistoria were, say, by the late 19th century, belief in the monsters of antiquity would diminish. But we find this is not entirely so, for in his *Mythical Monsters* (1886), British geologist Charles Gould (1834–1893) offered curious notions analogous to Mayor's of a century later with respect to dragons, sea serpents and unicorns. In his Introduction Gould stated, "I have ... but little hesitation in gravely proposing to submit that many of the so-called mythical animals, which throughout long ages and in all nations have been the fertile subjects of fiction and fable, come legitimately within the scope of plain matter-of-fact natural history, and that they may be considered, not as the outcome of exuberant fancy, but as creatures which really once existed, and of which, unfortunately, only imperfect and inaccurate descriptions have filtered down to us, probably very much refracted, through the mists of time."[17] Gould claimed that these animals once lived alongside prehistoric man and that when viewed "by the lights of the modern sciences of Geology, Evolution, and Philology ... these creatures are not chimeras but objects of rational study."[18] So after suggesting, due to its universal existence spanning legends and myths or histories of many cultures, that the Deluge is not a myth ("It must arise from the reminiscences

Dragons are in effect "ancestral" to the dinosaur in mythology and lore. In this figure, (top) we see the earliest dragon (a two headed variety) to appear in an alchemical text, Elias Ashmole's *Theatricum Chemicum Brittannicum* (London: Grismand for Nathaniel Brooke, 1652; collection of Allen G. Debus, photographed by the author); at bottom is a dragon appearing in Kircher's *Mundus Subterraneus* (Rome, 1665 ed.; collection of Allen G. Debus, photographed by the author).

of a real and terrible event, so powerfully impressing the imagination of the first ancestors ... as never to have been forgotten by their descendants"), he tackled the natural history of Dragons.[19]

Gould addressed a host of references to serpents in ancient and medieval literature, concluding for instance that a dragon mentioned by Pliny was most likely "some large boa or python." Both among the Roman army and the Celts, Teutonic tribes and Berserkers, draconic emblems and dragon symbolism proved significant. In mythic lore Thor, Siegfried and Beowulf slayed dragons. The fabled King Arthur was a revered chief, or Pen*dragon*.[20] By the Middle Ages, we even find a number of stalwart Christians earning their saintly credentials after slaying dragons. And both in China and elsewhere, fossil remains of prehistoric vertebrates were "supposed to be the genuine remains of either dragons or giants, according to the bent of the mind of the individual who stumbled on them."[21]

In 1640, the aforementioned Aldrovandus published a folio on dragons and serpents wherein he published plates of true dragons known to contemporary scholars. One such specimen, which he obtained in 1551, had two legs, was winged and had a snaky tail. However, in the case of another biped dragon specimen with lacy wings, allegedly killed in 1572, Gould suggested Aldrovandus was "probably imposed on by some waggish friend."[22] Dragons of dubious construction were even displayed in early museums, such as Aldrovandus'. Phineas T. Barnum was apparently preempted!

Besides giants, Athanasius Kircher also wrote of dragons, citing the tale of a man who fell into a cavern in Mount Pilate, where he discovered a dragon lair. Left unharmed by a pair of winged dragons, the man resided with his draconian hosts for half a year before clinging to the tail of one as it flew to the outer world. Kircher also recounted the tale of a knight named Gozione who battled a four-legged, winged dragon in 1349. Kircher viewed dragons most piously, stating, "Since monstrous animals of this kind for the most part select their lairs and breeding-places in subterraneous caverns, I have considered it proper to include them under the head of subterraneous beasts. I am aware that two kinds of this animal have been distinguished by authors, the one with, the other without, wings. No one either can or ought to doubt concerning the latter kind of creature, unless perchance he dares to contradict the Holy Scripture, for it would be an impious thing to say it when Daniel makes mention of the divine worship accorded to the dragon Bel by the Babylonians, and after the mention of the dragon made in other parts of the sacred writings."[23]

Reliance on Pliny's natural history works, perpetuated through the centuries, gradually subsided as naturalists added new knowledge gleaned from exploration of new lands.[24] By the 17th century, scholars generally became more interested in real things they could observe directly as opposed to mythical animals and beasts which so far hadn't been found in living nature. Indeed

there was just something in the air around that time which increasingly promoted naturalists' curiosity, investigations and intrigue over phenomena and
objects that could be documented, observed and tested, as opposed to things
not seen, not evident in nature or which could not be found. So interest in
Lamia, the Mantichore, and even dragons became shuffled to the wayside:
such animals were neglected and eventually forgotten. Belief in dragons
diminished by the late 18th century, only to be superseded by skeletal evidence of *real* giants found in Europe and the New World. Perceptions of a
pre-biblical *pre*-history and discovery of fossil bones coalesced into a new
dimension of the *science* of natural *history*.

Charles Gould noted by the late 19th century that, "Though dragons
have completely dropped out of all modern works on natural history, they
were still retained and regarded as quite orthodox until a little before the
time of Cuvier."[25] However, a century later Gould was unable to dispense with
the possibility that dragons at one time did exist, appealing to Darwinian
evolutionary dynamics as a pseudo-scientific basis for their mysterious natural history. Gould wrote, "are the composite creatures, partly bird and partly
reptilian, occasionally referred to, so entirely incredible? Is it not possible
that some of those intervening types which we know from the teaching of
Darwin, must have existed; which we know, from the researches of paleontology have existed; types intermediate to the ... most reptilian of birds, and
the ... most avian of reptiles— is it not possible that some of these may have
continued their existence down to a late date, and that the tradition of these
existing as the descendants or the analogues of the *Archaeopteryx*, and the
toothed birds of America may be embalmed in the pages in question? Is it
impossible?... Why, then may not a few cretaceous and early tertiary forms
have struggled on, through a happy combination of circumstances, to an
aged, and late existence in other lands."[26] Thus Gould inferred with "little
doubt" that the real ancestral (although possibly now extinct) dragon which
prehistoric man often encountered and which later became a source for legends was a terrestrial lizard, carnivorous and "furnished with wing-like
expansions of its integument." During an October 29, 1879, lecture, one of
Gould's contemporaries, famed artist Benjamin Waterhouse Hawkins (1807–
1889), considered the anatomical absurdities of mythical dragons, concluding that the creatures most closely resembling dragons were the extinct winged
pterosaurs.[27]

Two centuries before Gould and his dragon theories, we note scholars
already struggling with geo-historical concepts and the devilish prospect of
fossilized bones unknown in any living species. Quite apart from giants and
dragons, one man in particular, the influential German philosopher Got-
tfried Wilhelm Leibniz (1646–1716), unwittingly promoted the first modern
silly and pseudoscientific reconstruction of a prehistoric terrestrial animal.
Curiously, this happened during a period when men were increasingly accept-

ing the theory of an organic nature for fossils and also that fossils must represent remnants of an ancient, possibly diluvial (although not necessarily prehistorical), age. Yet we find curious examples of misguided scholarship still rampant throughout the Middle Ages. For on the basis of evidence of a "genuine" unicorn horn, in his *Mundus subterraneus* Kircher stated that "It is certain that the unicorn, or *Monoceros*, exists."[28]

Although a prodigy, renowned for discovering differential calculus independently of Sir Isaac Newton, as well as for advanced understanding of geological history, Leibniz's reputation suffers mightily from an image of a unicorn posthumously published in his book — *Protogaea* (1749)! Embarrassingly — as modern historians are quick to note, scholars had prior to that time (circa 1720) made great strides into understanding what fossils truly represented.[29] So Leibniz's unicorn, a seemingly backward step, is often retrospectively derided today.

There were many on the forefront of knowledge elucidating the true nature of fossils, several of which made relatively little major impact on the field of study. One in particular, the brilliant British physicist and inventor Robert Hooke (1635–1703), dabbled in many conceptual matters presented by fossils. Besides his important work on springs and watches, Hooke reproduced the shapes of the lunar craters by dropping bullets into a soft clay mixture or a boiling alabaster, experimentation still conducted today by astrophysicists although using far more sophisticated equipment. Despite his observations Hooke concluded that lunar craters resulted from earthquakes. Hooke's *Micrographia, or Some Physiological Descriptions of Minute Bodies Made by Magnifying Glasses* (1665), included a — for its time — definitive chapter on fossil wood. Here, using a microscope he observed thin sections of fossil wood; he also commented on the origins of fossilized shells, or "figured stones," such as he'd often seen along the Isle of Wight coast.[30] Many of his astute geological ideas were published posthumously in his *Lectures and discourses of earthquakes and subterraneous eruptions, explicating the causes of the rugged and uneven face of the Earth, and what reasons may be given for the frequent finding of shells and other sea and land petrified substances, scattered over the whole terrestrial superficies* (1705).

Hooke stated that to understand the Earth's history one should study the Earth itself, instead of the Bible, and to this end fossils were key. To explain the origins of fossils, which to him represented organic remains of former marine life, Hooke invoked a catastrophic shifting in the Earth's axis occurring in the remote past causing powerful earthquakes and resulting in wide-scale oceanic movements, sometimes submerging continents. Such catastrophes would have caused paleo-climatological changes and consequent extinctions, and permitted newer organismal forms to take advantage of successive environmental conditions. As Stephen Inwood recently stated, "Hooke was not a seventeenth century Darwin, but his suggestion that species developed and disappeared in response to changing climatic and environmental conditions has a very modern ring to it, and showed that he was prepared to suggest possibilities that very few of his contemporaries would consider ... [however] ... it was not until the middle of the eighteenth century that most geologists accepted the ideas that Hooke had proposed in the 1660s. And by this time, of course, they had forgotten all about Hooke."[31]

Besides his focus on the nature of common invertebrate shelly fossils, Hooke's fertile mind evidently also, albeit briefly, probed the puzzle of extinct vertebrate remains. Apparently, he ascribed to a consensus that in the past there really were giants in the earth. For we note a passage in his *Lectures and discourses of earthquakes*, stating "that there are now divers[e] species of creatures which never exceed at present a certain magnitude, which yet, in former ages of the world, were usually of a much greater and gygantick [*sic*] standard; suppose ten times as big as at present; we will grant also a supposition that several species may really not have been created of the very shapes they are now of, but that they have changed in great part their shape, as well as dwindled and degenerated into a dwarfish variety."[32] What types of prehistoric vertebrate fossils Hooke had witnessed or heard of is left to the imagination.

But Hooke was relatively silent on the matter of unicorns, where Leibniz could not refrain from his imaginings. Of course there were no real unicorns, many of which were simply imagined or magnified from fossilized mammoth and woolly rhino bones. And this is the pitfall Leibniz unwittingly fell into, as recounted by historian Claudine Cohen.[33] Two figures published in his book, *Protogaea, or A Dissertation on the Original Aspect of the Earth and the Vestiges of Its Very Ancient History in the Monuments of Nature*, strikingly contrast in their suggested meaning to modern viewers. The first presents fossilized shark teeth situated alongside the head and gaping maw of a modern shark, while the other is a bizarre skeletal reconstruction of a two-legged unicorn. This marine monster conceptually represents a bold step forward in paleontological comprehension of prehistoric vertebrates while the latter illustration takes us backward, ideologically. Both pictures are derived from previous works. The shark head with tooth was the handiwork of a 16th century engraver, Italian naturalist Michele Mercati (1514–1593).[34] Leibniz's unicorn stemmed from information gleaned from Otto von Guericke's (1602 – 1686) observations (von Guericke was burgomaster of Magdeburg in Germany and inventor of the vacuum pump).

First, the shark head had been printed previously, artfully used by Danish geologist Nicolas Steno (1638–1686), who would shortly lay out the basis for modern understanding of sedimentary superposition, built on the foundation of comprehending how certain solids could become encased within other solid matter. Steno's 1667 account, *Canis carchariae dissectum caput* (or, "The head of a shark dissected") quite rationally argued that fossil shark teeth (or "tongue-stones" as they were commonly called), thought by many of his contemporaries to be inorganic products of nature, actually formed not within the Earth — where they are now found — but inside the mouths of living sharks instead.[35] True, fossilized tongue-stones were twice as large as those found in the shark he'd dissected, but their distinctive shape gave them away. One only had to consider modern aquatic areas to glean how such teeth could have become separated from dead sharks, deposited on the sea floor, and encased by soft mud which later hardened into stone. Subterranean emanations would have uplifted the rocky mass containing petrified teeth enclosed within; the petrifactions would therefore later be observed in terrestrial settings by modern naturalists. For Leibniz to have relied on Mercati's shark head illustration, reproduced in Steno's thoughtful masterpiece, implies a mode of thought quite progressive for its time.

Yet, while Steno's early researches are often lauded as a pivotal launching point for modern geology, Leibniz's speculations on unicorns became a source of ridicule. Hence we must address the *second* paleontologically themed image in Leibniz's *Protogaea*. Cohen asks where and how did this monstrosity originate? Von Guericke had this to say of the "skeleton of a unicorn, the hind part of its body being lowered and its head raised up and back," found excavated from a quarry at Quedlinburg on Mount Zeunikenberg in 1663. As reiterated by Leibniz, Guericke "reports the discovery of the skeleton of a one-horned animal. As is usual with such brutes, its posterior parts were very low and its head raised. Its forehead bore a horn nearly five ells long, as thick as a man's thigh but gradually tapering. Because of the ignorance and carelessness of the diggers, the skeleton was broken and extracted in pieces. However, the horn, which was attached to the head, several ribs, and the backbone were brought to the abbess of the town." In 1925, Austrian vertebrate paleontologist Othenio Abel recognized that Leibniz's unicorn figure combined mammoth molars, scapula and vertebrae with woolly rhino elements.[36] One wonders what the poor abbess did with such a peculiar prize.

Like mythical dragons, unicorns belonged to medieval bestiary, lingering in Renaissance fancy. It is unknown whether von Guericke supplied the figure that was later printed in *Protogaea*, or if Liebniz himself attempted to illustrate the animal's skeleton based on von Guericke's written description (even after noting that the skeleton was broken up and probably incomplete). Leibniz's *Protogaea* was begun in draft form in 1691, with a summary printed two years following during the author's lifetime. Not only was the unicorn interpretation incorrect, but the reconstruction of a two-(front) legged animal with its backbone dragging on the ground behind could not be natural. One may suggest that had Leibniz lived another 33 years, he might have edited references to the unicorn — including the unicorn figure — out of *Protogaea*.[37] For by the 1720s, learned men generally didn't believe in such nonsense and the burgeoning paleontological sciences were turning to more definitive matters involving cases of real fossilized animals. However, beyond the illustrations, *Protogaea* documents how staunchly Leibniz defended his belief in unicorns during the 1690s.[38] Cohen concludes, "If Leibniz had wanted the skeleton of this improbable two-legged beast, with its huge molars and single forehead horn, to appear among the illustrations of his book — despite the requirements of rationality shown in his thoughtful contemplation of fossils and the history of the earth — it is because at the dawn of the Enlightenment, the fabulous bestiary still remained a credible source of zoology."[39]

If a scholar of Leibniz's caliber could be so readily beguiled by alleged unicorn remains, then to be sure there were other early eminent 18th century naturalists whose interpretation of fossil vertebrates would also seem out of place today.[40] Indeed, belief in giants and a curious theory of gigantism were still soberly debated then, not unlike how the ancients viewed giant bones which unless anthropomorphized seemed otherwise incomprehensible. While modern paleontologists have confirmed that the largest terrestrial vertebrate animals are extinct, during the 17th and 18th centuries the debate over giant bones became deeply entwined with the nature of fossil objects and theories of their formation. But if fossils were viewed as organic remains — traces of genuine formerly living animals — then what could these evidently enormous animals really have been?

The modern period of American intrigue over fossil bones began in 1705, with the discovery of a "giant's tooth" in the Hudson River valley near Albany, at Claverack manor. This Giant of Claverack was judged 60 to 70 feet tall, inspiring an unfinished folkloric poem by Edward Taylor.[41] The prospect of American giants caused some religious souls to dwell on Genesis 6:4, wherein it is stated that "There were giants on the Earth, in those dayes." To many, such bones confirmed Mosaic Scripture. Boston physician Cotton Mather, for example, suggested the bones represented organic remains of antediluvian giants exterminated in the Flood. By 1714 a drawing and information concerning the American human giant remains were comprehended by members of the Royal Society in England: surgeon Thomas Molyneux (1661–1733) refuted the idea that similar bones found in Ireland belonged to giants. Instead quite rationally he suggested the teeth and bones could be favorably compared with bones of dissected elephants. Molyneux wrote, "As for the Hint of their being *human* or *gigantick*, 'tis so groundless a Thought, and so contradictory to *comparative* Anatomy and all *Natural History*, it does not deserve our consideration."[42] As historian Paul Semonin noted, "Coming on the heels of Mather's theories, Molyneux's plea for scientific rigor in analyzing the fossil remains marked the beginning of the formal refutation of the "Doctrine of Monsters" by England's anatomists and surgeons."[43]

Hooke had illustrated such a grinder for his *Posthumous Works* published in 1705, although then fancied as the "petrify'd tooth of a sea animal."[44] However, in one of the earliest documented references, Hooke described the animal, which later became known as Mammoth, more accurately. In a 1697 lecture Hooke stated, "We have lately had several Accounts of Animal Substances of various kinds, that have been found buried in the superficial Parts of the Earth, that is not very far below the present Surface, as particularly the parts of the Head of an *Hippopotamus* at Charthram in Kent, that of the Bones of the *Mammatovoykost*, or of a strange Subterraneous Animals, as the Siberians fancy, which is commonly dug up in Siberia which Mr. Ludolphus judges to be the Teeth and Bones of Elephants."[45]

Decades later, British anatomist Sir Hans Sloane (1660–1753) collected sufficient remains of Mammoths in places ranging from North America to Siberia to fortify publication of his definitive 1728 article denouncing Mather's "Doctrine of Monsters," titled "An Account of Elephants Teeth and Bones found under Ground," which proved the remains belonged to former elephants.[46] No longer could such remains properly be regarded by learned men as traces of erstwhile human giants, dragons or mythical creatures, as espoused by the Doctrine. Accounts compiled by German physician Daniel Gottlieb Messerschmidt (1685 –1735) and subsequently German anatomist Dr. Johann P. Breyne (1680–1764), who in 1737 considered the remains of a mammoth cadaver preserved with soft tissues including skin and hair found in Siberia, underscored that the Mammoth of Russia (and other high latitude localities as well) was a known entity — an *elephant*. Huge although clearly not a monster, the riddle of the mammoth became an ideological, intellectual proving ground in early enlightenment over the true nature of fossil objects.

Ah — if only comprehending the mysteries of prehistory and its denizens were so straight-forward! The era of scientific intrigue in the matter of prehistoric vertebrates — often viewed by the public as monsters — had only just begun. While in coming decades theoretical notions of what constituted prehistory's vistas and panorama issued from Europe, two gigantic monsters sprang from American soil whose interpretation would considerably alter the existing picture. And in spite of how volubly the Doctrine of Monsters could be denounced, elephantine Mammoth wasn't by any means the only huge animal which lived before a perceived Deluge.[47] We'll return to the animal represented by that giant's tooth found in Claverack, for example, which Sloane did not formally address in his publication, in chapter three.

In coming decades, fauna of the world would be increasingly investigated by scholars who applied the scientific method to their extant and extinct quarry. In 1735, Swedish botanist Carolus Linnaeus (1707–1778) classified living species of animals and plants, inventing a nomenclature system still used today by biologists. Then, beginning in 1752, French naturalist Georges Louis Leclerc, Comte de Buffon (1707–1788), delved deeply into the ancient history of species, incorporating Mammoths into his cosmic system of planetary evolution. Their works paved the way for long-neglected Spanish nat-

Thanks to many discoveries and correct scientific interpretation, by the mid–19th century, the woolly mammoth was a well known fossil vertebrate. This is Henry Ward's famous sculptural restoration as it appeared during Chicago's Columbian Exposition of 1892–93. *The Dream City: A Portfolio of Photographic Views of the World's Columbian Exposition* (St. Louis: N.D. Thompson Publishing Co., 1983).

uralist and artist Juan Bautista Bru's (1740–1799) encyclopedic book or atlas, *Collection of Illustrations Representing the Animals and Monsters of the Royal Cabinet of Natural History of Madrid* (1784–1786). As we shall see in the next chapter, Bru also became the first to accurately reconstruct the skeleton of a fossil vertebrate, perceived by naturalists as monstrous in the new United States.[48]

Through the 20th century, man's fascination with anthropoid giants became perpetuated through scores of big horror special-effect films with titles such as *King Kong*, with its titular prehistoric giant ape and enormous dinosaurians, and *The Amazing Colossal Man*. Although dragons, as understood by the ancients and medieval scholars, have never been proven to exist, mankind perhaps received its first, most realistic glimpse of what a genuine dragon would be like in the living flesh during the 1920s in Fritz Lang's silent movie *Siegfried* (1924), which featured an impressive, 70-foot-long anima-

tronic dragon, as well as following American Museum naturalist W. Douglas Burden's 1926
capture of two live Komodo
dragon specimens from Indone-
sia, later displayed in the Bronx
Zoo.[49] The giants theme scientifically persisted into the
1930s through Henry Fairfield
Osborn's theory of evolutionary
orthogenesis, in which it was
claimed that through geological
time, mystical evolutionary
forces caused vertebrate animals
to grow enlarged features or
to attain overall sizes greater
than what would be considered
adaptational through Darwinian
means.

The cultural significance of
classical, ancient monsters is
highly analogous to the relevance of prehistoria known to
science in America's post–Revolutionary War period two centuries ago, or perhaps more
prominently to that of today's
favored saurian *daikaiju-eiga*
and the familiar cast of Japanese
suitmation dino-monsters, several of which issue fiery (drago-
dinosauroid)[50] exhalations on
screen.

But which of the mon-
strous-looking animals first

In Fritz Lang's 1924 epic film, *Siegfried*, part of
"Die Nibelungen saga," the hero confronts a
mechanized dragon "costume" that closely
resembles the Komodo Dragon.

known to science in fossilized states were also extant? Conversely, were any
(or all) of them *extinct*? Were any contemporaneous with Man? What geological markers separated Man from what happened before Man's coming? In
other words, were the fossil beasts increasingly coming to light by the early
19th century to be considered strangely *pre*-historic? Such a term made little
sense then, for hadn't mankind existed at least since Noah's time, an event
which to many represented the most important, sacrosanct geological horizon of all — separating an antediluvial time of undetermined length from

later historical periods? Questions surrounding Man's place in relation to a crudely denoted geological timescale, the Earth's apparent sequence of paleontological ages, and whether species ever truly became extinct were contentiously debated during the late 18th and early 19th centuries.[51]

Eventually, as knowledge — gleaned from study of fossil vertebrates found in characteristic stratigraphical horizons, early geological maps and fossiliferous localities representing the span of all documented paleontological ages — accumulated and crystallized, scientists' attentions centered on the most recent geological revolution, as recorded in superficial diluvial deposits which to some indicated the occurrence of a former Deluge. By the mid–19th century, primarily through elucidation of disturbing bones, dating from what would later be known as the most recent Ice Age, scientists unleashed a triumvirate of monsters, haunting man's psyche — Great-Claw, the American *Incognitum*, and the Beast lurking within our very souls.

Two

Great-Claw

How did the idea of prehistoric monsters become rooted and why did they become popular? Because mankind's creation of prehistoric monsters, one of most our ingenious inventions, represented an unparalleled product of the imagination, conveyed most strikingly in the form of visual imagery.

If we lacked modern understanding of what animals called dinosaurs, pterosaurs, ichthyosaurs or glyptodonts and mammoths were, what would we think as we gazed upon their stone-cold fossil remains for the very first time? What living flesh and blood beings can be conjured from such often indiscriminate traces? These were more or less the circumstances faced by the earliest men now referred to as paleontologists over 200 years ago. Except then the science was different.

Today we know considerably more facts about physics, astronomy, chemistry, biology and the geological sciences than did the learned men of two centuries ago. And, true, we have a far lesser tendency to allow a discipline known as natural theology to creep into our comprehension of a mysterious past time known as prehistory. Yet the men who first recognized the remains of fossil organisms as being somehow special or different in nature from extant life forms weren't any less smart than are the curious-minded men of today who discover and describe dinosaur bones and other fossils belonging to extinct animals and plants.

So how is it that certain names, that aren't English names, instantly conjure an image, *your* images, of *T. rex, Triceratops, Stegosaurus*? You can practically see these animals in your mind's eye. And why is that? After all, none of us has access to a real time machine, like in the science fiction movies. It's because of an incredible force which has permeated from our research science institutions, laboratories, museum cases and into pop-cultural media, comprising an uncanny array of depictions and graphically portrayed information—*paleoimagery*.

Paleoimagery captures many themes, metaphors and ideas (often rooted in science). While paleoimagery encompasses virtually any image that can be even remotely linked to a paleontological theme, prehistoric animal or idea, the term *paleoart* is reserved for those images bearing on the science of paleontology. Paleoimagery can be produced in sculptural, two-dimensional or even animated form. There are numerous examples of paleoimagery, such as animated sequences in fantasy or science fictional (sf) dinosaur movies, that are either non-scientific in nature or which hold little value for practicing scientists. Therefore, it is instructive to classify paleoart into categorical subsets, of which life restorations, skeletal reconstructions and evolutionary iconography are the usual examples.

Paleoart emerged as a subset of natural history illustration, or even in the context of portraying biblical history — the Deluge, as fossils were commonly viewed as relics of Noah's Flood. Early scholars of medieval and renaissance times initially relied on paleoimagery in recording the appearance of fossils as plates in books and manuscripts for communicating to others who were also interested in these natural wonders, even though they may have been misguided in their understanding of what these curiously shaped rocks and stones actually represented. But there are subtleties at large here because in illustrating fossils, or in transforming the illustration into an engraving for printing, the artist, or artisan, infuses the drawing with his or her own interpretations of what he or she sees. And so, without proper contemporary theoretical understanding, from today's perspective, these early attempts at illustrating fossils may seem contrived or even ludicrous. Yet it would be improper to scorn these early efforts—for under the circumstances the early scholars did the best job possible given what they knew. Eventually, most scientists unfailingly grasped that (non-human) fossils represented organic remains of extinct organisms that lived in prehistory.

The means of replicating original illustrations of fossils, skeletal reconstructions and even life restorations into the pages of books and manuscripts became dramatically refined. Thus, rather than expensively translating original drawings and paintings into woodcuts or steel engravings, or relying on stone lithography, publishers could rely on the half-tone process. In fact many of the late 19th century dinosaur images we are familiar with today in popular books of the time — such as Joseph Smit's (1836–1929) restorations featured in Henry Neville Hutchinson's *Extinct Monsters*, were in fact printed using the half-tone process. Chromo-lithographs became the preferred means for reproducing color restorations in popular books or even collectible cards showing dinosaurs and other prehistoric animals.[1]

During the mid–19th century, scientists and naturalists were gaining skill reconstructing facsimiles of bones or the actual bones of extinct animals into assembled skeletons for public edification and museum display, while one famous paleoartist, Benjamin Waterhouse Hawkins, sculpted the earliest life-sized restorations of how Victorian paleontologists thought several prehistoric vertebrates appeared in life.[2] Hawkins' dinosaur statues were known to be anatomically inaccurate as early as the 1880s. Today, this remains a prevalent paleoart theme — revising a former paradigm of how a particular prehistoric animal was thought to have appeared and moved in life based on the latest interpretations of new fossil discoveries.

Yet the main iconic themes, dinosaurian strange attractors and imagery of other fossil organisms, remain unchanged since Benjamin Waterhouse Hawkins' day. For, as noted previously in my book *Paleoimagery*,[3] preoccupation with the hugeness of certain prehistoric animals as well as their alleged savage nature, conjuring the image of that magnificent paleontological holy grail — the dinosaur/bird evolutionary link that may be forever missing, and our innate quality, evident in prehistoric man, for artistically portraying wildlife forms — however ancient they may be — sustains Paleoart's eternal flame of mystery.

Early understanding of fossil vertebrates, fraught with numerous misconceptions, was thorny and problematic. Often imaginary recreations resulted that would have seemed most monstrous to their esteemed conjurers. Accordingly, in this vein, here and in the subsequent chapter let us consider America's first genuine pairing of prehistoric monsters: Great-Claw followed by its ancient nemesis, the *Incognitum*.

So you've nearly completed your tour through the museum dinosaur displays, nearing the end of a long day on your feet with children in tow who are just itching to invade the gift shop for expensive souvenirs. The 60-foot long-necked sauropods, tooth-baring carnosaurs, winged dragons and horned and plated dinos of every persuasion stirred your interests through Mesozoic life exhibits, followed by those large-horned thunder-beast brontotheres of the Tertiary. Now here, surrounded by Ice Age relics, you recognize the venerable mastodon and, of course, fossils of the obligatory Mammoth. But what is this last bastion of the Primeval — a bipedal concatenation of bones standing there in the darkened corner hugging an artificial tree trunk? A huge cave bear? No — it's a glorious ground-sloth fossil, now generally neglected among all the other celebrated post–Crichton paleo-splendor. And here lies the tale of this tree-hugging fellow and its near fossil relatives, prehistoria which at one time bore much (*theoretical*) significance for the emerging paleontological and evolutionary sciences.

In a 1942 *Proceedings of the American Philosophical Society* paper titled "The Beginnings of Vertebrate Paleontology in North America,"[4] American paleontologist George Gaylord Simpson (1902–1984) defined six periods in the history of vertebrate paleontology. My story of the ground sloth is framed within what Simpson referred to as the Pioneer Scientific Period (circa 1799 to 1842), and the First and Second Classic Periods, respectively (circa 1842 to 1865 and circa 1865 to 1895). For our purposes, here, we'll consider the relevancy of giant ground-sloths to contemporary theoretical notions about life

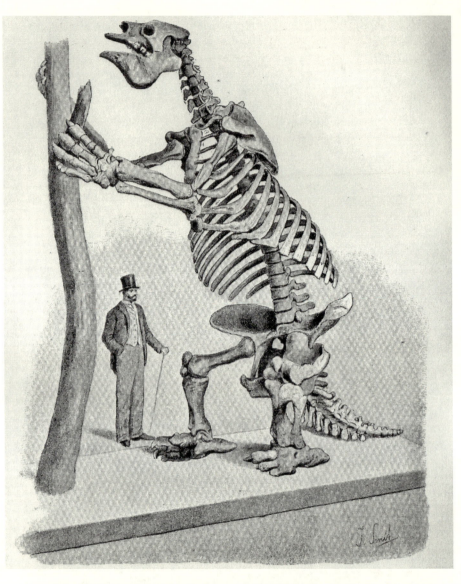

Joseph Smit's illustration of the giant ground sloth (*Megatherium*) as displayed in the British Museum during the late 19th century. *Megalonyx* bore a resemblance to *Megatherium*, although it was slightly smaller (from H. N. Hutchinson's *Extinct Monsters*, 1893 ed.).

of the past; how it vanished, evolved, plus apparent consequences for understanding human prehistory. And in the end, our story will stray into vertebrate paleontology's first Modern Period (since 1895 to circa 1930).

It was an earlier time in a newly founded nation. Fresh from fighting a political revolution with our French allies against the British, America's founding fathers became embroiled in a curious scientific debate with French scientists—prominently the illustrious Comte de Buffon—over the generative powers of American soil and New World climate. For, Buffon reasoned conceitedly, western animals seemed inferior in size and ferocity, relative to Old World (European, Asian and African) species. Perhaps Buffon's least favorite mammals of all were (extant) sloths—native to South America—described in his *Histoire naturelle* (*Natural History*), of which the first three of 36 volumes were published in 1749. Buffon stated, "Whereas nature appears to us live, vibrant, and enthusiastic in producing monkeys; so is she slow, constrained and restricted in sloths.... These sloths are the lowest term of existence in the order of animals with flesh and blood; one more defect would have made their existence impossible."[5]

To those of us familiar with modern geological principles, and imbued with the latest paleontological wisdom, America's old style brand of earth science would seem maddening indeed. For instance, in 1799, Charles Wilson Peale (1741–1827) lectured on his Theory of the Earth (a rationalization between biblical history and astronomy in which the Earth's orbit steadily decayed as the Sun's *electrical* attraction diminished, while the Earth-Sun distance increased steadily through time).[6] In 1788, Benjamin Franklin (1706–1790) accounted for occurrences of fossil objects with a theory of (gradual) pole axis shifting; *magnetic* phenomena resulting from the passing of a large (but not impacting) comet caused disturbances and deluges in the remote past.[7] And to parlay Buffon's chauvinistic claims, in 1799, then Vice President Thomas Jefferson (1743–1826) described in a scientific paper what appeared to be the bones of a gigantic lion found in a cave deposit in western Virginia in 1796. As we'll soon see, Jefferson was wrong too.

But as a backdrop to Jefferson, and Simpson's later historical periods, let's cross the Atlantic to France where anatomist Georges Cuvier (1769–1832) was avidly investigating a newfound prehistoric cousin to Buffon's lowly sloths. By 1812, Cuvier knew of at least two such South American ground-sloth specimens. The first very complete specimen to garner most scientific attention had been found in 1788 by a Dominican Friar named Manuel Torres on a Lujan River cliff, in Argentina southwest of Buenos Aires. This was excavated, packaged in seven boxes and shipped to Madrid's Royal Cabinet of Natural History, where it arrived in September. A second partial specimen had been discovered, this time in Paraguay, in 1795. (A Paraguayan specimen was discovered in 1832, and still another belonging to this genus arrived in 1841.)

Cuvier didn't have the actual bones of the extinct Madrid specimen to study, but rather a set of (as-yet) unpublished drawings made under the direction of the aforementioned Spanish naturalist and preparator/conservator Juan Bautista Bru y Ramon (1740–1799), referred to here as Bru.[8] These had been appropriated in 1795 by a French diplomat named Phillipe-Rose-Roume (governor of San Domingo), who was also a member of Institut de France. No doubt startled by the strange creature taking form out of the encrusting Pampas sediment before his eyes, Bru cleaned the bones and then, in unprecedented fashion, mounted the very first skeleton of a prehistoric vertebrate in a more or less naturalistic, quadrupedal pose. Bru's skeleton still resides on display in Madrid's Museo Nacional de Ciencas Naturales.

In a 1984 publication, Bru was referred to by paleontologist George Gaylord Simpson as "temperamental." When Bru departed from the Museum, he obstinately took his ground-sloth drawings, engravings and description of the skeleton. Simpson opined that the aging "Bru certainly knew that he had in his hands one of the greatest discoveries ever yet made and that priority of its publication would take him out of relative obscurity to absolute fame among scientists."[9] However, this wasn't to be, as the twenty-six-year-old rising-star Cuvier instead used Bru's prize handiwork to catapult himself to early fame. Simpson accusingly referred to Bru's proofs and engravings, which fell into Roume's hands, as loot.

The year 1796 turns out to have been a pivotal year in the early history of paleontology. Although in his lightning-fast assessment of 1796, Cuvier would beat the Spanish to the punch, Bru's papers on the bones were published posthumously in France in 1804 and 1812. Meanwhile, apparently outraged by these circumstances and in order to "defend the priority of Spain," Spanish naturalist Jose Garriga purchased Bru's materials, publishing his own paper on the *Megatherium*. (Garriga had been studying *Megatherium* for years and in fact had a paper in press on it at the time Roume asked him for a copy of the proofs.) In English translation this was titled "Description of the skeleton of a very corpulent and rare quadruped which is preserved in the Royal Cabinet of Natural History in Madrid." It has been suggested that Cuvier's rush to print triggered an international dispute, with Garriga claiming Spain had been cheated by the French. While Cuvier's scholarly (and today more widely known) description was accurate yet not quite so detailed, Garriga's more exacting description of 1796 was published shortly after Cuvier's. Garriga noted a number of mistakes in Cuvier's hasty work and described two additional, less complete specimens, which had also been unearthed. Garriga also rejected Cuvier's assigned name, *Megatherium*.

Martin J. S. Rudwick who translated many of Cuvier's papers in a book he edited, *Georges Cuvier, Fossil Bones, and Geological Catastrophes*, claimed, "Rather than feeling that Cuvier upstaged him, Bru may have valued the French naturalist's authoritative notice on the zoological affinities of the animal. Conversely, Cuvier knew almost nothing about its geological context, as text of his paper shows. Bru's plates were published in Spain later the same year, with his detailed description of the find, and a [Spanish] translation of Cuvier's paper; conversely, when Cuvier came to publish a more complete description he included a translation of Bru's work."[10] Simpson's perspective was that, "If one cannot help thinking that Cuvier had not been completely ethical in this affair, one also cannot help thinking that Garriga's inducing Bru to turn everything over to him and making this the basis of his own publication was not fully ethical either, even though he gave all the credit to Bru."[11] A redrawn version of Bru's engraving (with the animal facing the opposite direction) and a summary of Cuvier's 1796 paper were also published in a London journal, *The Monthly Magazine*,[12] a copy of which later fell into the hands of American naturalists investigating Jefferson's controversial "Great-Claw" Virginian lion.

Using Bru's Madrid proof, showing a quadrupedal skeletal reconstruction, as a visual aid, Cuvier recognized the new elephant-sized species as a relative of modern sloths, naming it *Megatherium americanum*, or "America's huge beast." This was his first published paper describing a fossilized vertebrate, publicizing the remarkable creature — iconic for its time among a limited circle of the intelligentsia.[13] (Cuvier erroneously referred to the Luxan creature as the "Paraguay" specimen in his 1796 description.)

Referring to Bru's engraving, showing the *Megatherium* in a quadrupedal stance, Cuvier stated the animal was 12 feet long (although there was no indication of a tail in the figure) and six feet high. Noting its "indicative characters," Cuvier astutely realized its lack of incisors allied *Megatherium* with other edentates — modern sloths, armadillos, anteaters and pangolins. However, no animal then known to science displayed the unusual skeletal features and proportions witnessed in this fossil form, particularly the enormous pelvis (actually the largest bone known in any animal, past or present) and stout limb bones, indicative of incredible strength. Another characteristic — front limbs longer than the back legs — "is a character singularly specific to the sloth genus." Cuvier also noted the huge, pointed claws, speculating that *Megatherium* may have possessed a short, tapir-like trunk. Cuvier concluded, "It adds to the numerous facts that tell us that the animals of the ancient world all differ from those we see on earth today; for it is scarcely probable that, if this animal still existed, such a remarkable species could hitherto have escaped the researches of naturalists."[14]

In a paper read before the National Institute on April 4, 1796, Cuvier described relationships between living and fossil elephants. Cuvier mentioned the great South American sloth again. "What has become of these ... enormous animals of which one no longer finds any [living] traces, and so many others of which the remains are found everywhere on earth and of which perhaps none still exist? ... the twelve-foot long animal, with no incisor teeth and with clawed digits, of which the skeleton has just been found in Paraguay: none has any living analogue." And then he added another line which in context of fossilized mammalian fauna had lingering impact on naturalists, "Why, lastly, does one find no petrified *human* bone?" (my italics).[15] Here, Cuvier's term "analogue" meant a fossil species *identical* to an extant form. Cuvier

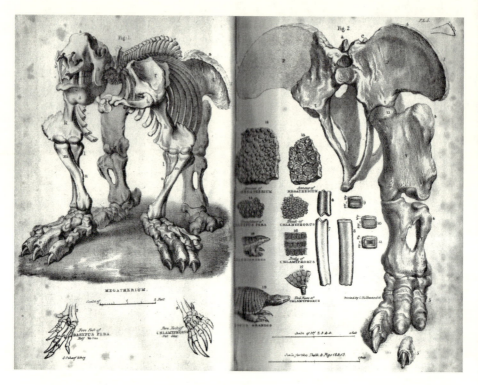

William Buckland included an in-depth discussion concerning the *Megatherium* in his *Geology and Mineralogy* (1837 ed.). Here (left) we see the beast reconstructed in quadrupedal stance, the way Bru originally mounted the "Paraguay" skeleton in Spain. At right are other bony elements then attributed to the ground sloth; the armor ossicles were later assigned to a glyptodont.

attributed the extinction of the fossil species which thrived in a "world previous to ours" to "some kind of catastrophe," a theory he and others would refine in decades to come.

What had been the fate of the mysterious fossilized beasts? There remained three conceptual possibilities for their disappearance from the globe (1) eradication due to extinction, (2) organic modification of fossil animals into modern varieties ("speciation"), or, (3) migration to other (i.e., now hidden) parts of the world. Migration didn't seem as likely a possibility for *terrestrial* forms, because surely by then they'd have already been discovered somewhere by explorers. And although evidence he and fellow scientists were gathering would eventually pave the way to understanding organic evolution through descent under the forces of natural selection, Cuvier clearly wasn't an early evolutionist. So this left *extinction* as Cuvier's chief explanation for occurrences of fossil vertebrates.

Cuvier perceived mighty revolutions or catastrophes which episodically exterminated fauna. These revolutions took the form of prolonged marine incursions, or sudden decline in global temperature. While at first it seemed as if a single revolution would suffice to explain the disappearance of mammoths in Siberia, South American megatheria and North American sloths, over time it became clear that there were other extinction boundaries in the fossil record as well, beside the most recent Quaternary event, which must have also been caused by catastrophes. Accordingly, mosasaurs and (later described) winged pterodactyls and other great sauria of the "Secondary Period" (i.e., Mesozoic Era), which we'll come to in Chapter Six, must have perished in a violent catastrophe even before the onset of huge mammalian forms. And so on, as further geological boundaries were identified, until by 1849 a total of 29 catastrophes had been invoked by European geologists to account for faunal "destructions" (i.e., extinction boundaries) and apparent progression through fresh creations, as noted in the fossil record.

In his *Reserches sur Les Ossemens Fossiles de Quadrupedes — Discours Preliminaire* (1812), Cuvier claimed that "it would be in vain to seek, among the forces that now act on the earth's surface, causes adequate to produce the revolutions and catastrophes the traces of which are shown in its crust ... life on earth has often been disturbed by terrible events; calamities which initially perhaps shook the entire crust of the earth to a great depth."[16] To Cuvier, proper interpretation of fossils was a key to constructing and comprehending the true theory of the Earth. To support the contentious fact of extinction, Cuvier's detailed anatomical analyses of fossilized bones proved such specimens were not identical to living forms, and thus were distinct species.

Furthermore, Cuvier was no cryptozoologist of his day, for, as a corollary, he predicted that vertebrates preserved in a fossilized state would *not* be found alive on the earth today, an outrageous theoretical notion for its time, and one which Thomas Jefferson would have disputed, as in the case of his North American lion. But by 1804 Cuvier would pronounce on Jefferson's Great-Claw beast as well, counting it among the ranks of extinct fauna after astutely comparing it with his *Megatherium*.

Curiously, shortly after *Megatherium* had arrived in Madrid, one William Carmichael (born circa 1739–1795), the American chargé d'affaires in Madrid, procured a preliminary sketch of the skeletal reconstruction completed only 4 months after arrival of the bones in Madrid. Jefferson received this drawing, apparently made by Bru shortly after receipt of the bones, with Carmichael's January 26, 1789, letter, on April 6, 1789, as he was returning to America from Paris. Jefferson even referred to the contents of Carmichael's package as "interesting papers."[17] Remarkably, Bru's drawing shows an essentially correct quadrupedal mount (lateral view) of the bones — which weren't physically assembled as such until 7 years later. Included with the sketch was Carmichael's note that the Spanish Academy of Natural History "will soon publish an account of this Animal."[18]

According to historian of science I. Bernard Cohen, Carmichael advised that until their description had been published, news of the animal "should not be made public." Cohen and others have remarked that had he not obliged Carmichael, and otherwise proceeded in unethical fashion, Jefferson might have beaten both Cuvier and Garriga to the punch, thus attaining prestige and "the honor of preceding the Spanish naturalists and Cuvier in identifying and naming the animal."[19] In science, as in nature, the race is to the swift.

Intriguingly, even though on April 6, 1789, Jefferson acknowledged his receipt of Bru's preliminary 1788 drawing showing a reasonable reconstruction of the skeleton, sent by Carmichael in writing, Jefferson evidently forgot about it. Perhaps almost a decade later Jefferson didn't associate the Spanish specimen with his North American *Megalonyx*; he may have already been wedded, if not blinded, to the lion analogy. And furthermore, the sketch of 1788 doesn't show a creature recognizable as any feline; there is one claw in Bru's figure which doesn't stand out so prominently in relation to the rest of the illustrated bones (unlike the scary-looking claw appearing in a plate published in Caspar Wistar's 1799 paper, which we'll come to shortly). Curiously, eight years later, Cuvier's reporting of the *Megatherium* would rankle Jefferson, just as the latter was prepared to announce *Megalonyx*'s existence.

Nearly a year following publication of Cuvier's *Megatherium* paper, on March 10, 1797, Thomas Jefferson's paper titled "A Memoir on the Discovery of Certain Bones of a Quadruped of the Clawed Kind, in the Western Parts of Virginia" was presented to a gathering of the American Philosophical Society in Philadelphia.[20] Several months before, in 1796, laborers digging for one Frederic Cromer had encountered the titular "bones" buried in a two to three foot deep cave deposit in Greenbriar (now in West Virginia). Jefferson first learned of the discovery in April 1796. Colonel John Stuart (1749–1823), aka "Stewart," took pains to collect the Greenbriar remains, which evidently belonged to a "Tremendous Animal of the Clawed kind," forwarding what he'd salvaged to Jefferson; later a Mr. Hopkins who also visited the caves added to the collection. The fossilized creature was fragmentary, as only foreleg material — a femur, radius, ulna, three claws and other miscellaneous foot bones — were retrieved. In Jefferson's July 3, 1776, letter to astronomer David Rittenhouse (1732–1796), then president of the American Philosophical Society and who had curiously died the week before, Jefferson informally proposed the names "Megalonyx" and Great-Claw for the new beast. Jefferson also proposed to write an account of the specimen for the Society, a report published two years later (1799) in *Am. Philos. Soc. Trans.*, Vol. 4.[21]

Jefferson seized upon the idea that the new animal was a leonine quadruped, immediately making analogies to extant carnivores — in particular, the African lion, as described in Buffon's *Natural History*. "I will venture to refer to him by the name of the Great-Claw or *Megalonyx*," Jefferson formalized, "to which he seems sufficiently entitled by the distinguished size of that member."[22] Jefferson included a table in his report directly comparing anatomical dimensions between lion and Great-Claw. Much of Jefferson's 1799 paper dealt with dimensional analogies exhibited between his North American lion and the African variety; perhaps his most readily apparent motive was to falsify claims of degeneration in the New World made by French savants.

Using error-prone, back-of-the-envelope type calculations in the 1799 paper, comparing dimensions of the African lion, by analogy, Jefferson concluded that Great-Claw weighed 803 pounds in life, but possibly as much as 2,000 pounds based on comparison to the largest lion specimen then known. (This is in comparison to the average weight of an adult lion — as known in Jefferson's time — 262 pounds.) However, because Great-Claw's bones were quite robust, thick and therefore evidently load-bearing, "Let us only say then ... that he was *more* than three times as large as the lion; that he stood as pre-eminently at the head of the column of clawed animals as the mammoth stood at that of the elephant, rhinoceros, and hippopotamus: *and that he may have been as formidable an antagonist to the mammoth* [my italics] as the lion to the elephant."[23] Note here that in his mind's eye, Jefferson had already pitted his formidable lion in combat against the mighty "mammoth" or *incognitum*. Ironically, even fact-focused Cuvier, who ordinarily dissuaded introduction of speculative ideas into scientific discussion, adopted Jefferson's vivid suggestion for an 1804 *Annales de Museum* paper, here claiming that cow-sized *Megalonyx* may have preyed upon the "Ohio Animal" (i.e., the mastodon).

Anyone possessed of even slight appreciation for modern paleoart senses our visceral reaction and attraction toward restorations of battling prehistoric monsters — *Tyrannosaurus* "versus" *Triceratops* chief among the paleocombatants, for example. Such restorations mirror man's subjective impressions of prehistoric violence and savagery. Although Jefferson didn't accept the "mammoth" and Great-Claw as *extinct* animals he certainly invented the germ of the idea behind this ever — and most popular form of imagetext — battling prehistoric monsters — as documented through suggestion in his 1799 paper.

Then Jefferson pondered his "difficult question," namely, "What is become of the great-claw?" Reaching a conclusion diametrically opposed to Cuvier with respect to the *Megatherium*, over the next seven printed pages, Jefferson explained rationale for why he believed Great-Claw was an extant mammal. Jefferson reasoned that the largest wild mammals are driven from areas occupied by man, and that they may still dwell within the vast unknown regions of the continental interior. "Our entire ignorance of the immense country to the West and North-West, and of its contents, does not authorize us to say what it does not contain."[24] Next, he described historical accounts of possible encounters with Great-Claw — citing both Indian traditions, art and rock carvings as well as more recent tales of explorers who heard loud, disturbing sounds—such as roaring in the woods at night. Could these be telltale signs of Great-Claw stalking modern man? It was enough to give anyone the shivers!

In 1958, Princeton historian Julian P. Boyd documented that such wild speculations weren't Jefferson's alone or even originally, but rather those of compatriot John Stuart, who, using powers of suggestion, melded the sloth remains with dramatic, conjectural views. For on April 11, 1796, Stuart wrote to Jefferson, stating "I do not remember ... to have seen any account in the History of our Country, or any other, of such an animal, which was probably of the *Lion kind*" (my italics). As Boyd stated, "This planted a germ that incubated rapidly."[25] For soon after, we find Jefferson proposing in his July 3rd letter to the aforementioned Rittenhouse that the remains belonged to an "animal of the family of the lion, tyger, panther &c but as prominent over the lion in size as the Mammoth is over the elephant." Furthermore, Jefferson's embracing the possibility that *Megalonyx* may still be living in the American wilderness can be traced to John Stuart's (as well a relative's—Archibald Stuart's) correspondence with Jefferson. So Jefferson erred in accepting the Stuarts' speculations on faith, using them as the crux of his scientific paper. It was on May 26, 1796, that Jefferson revealed his intentions to Archibald of enlisting the bones in the war of ideas over American "degeneracy" against the French.

Unlike Cuvier who believed in extinction, Jefferson was a contemporary cryptozoologist. In fact, when Captain William Clark (1770–1838) and Meriweather Lewis (1774–1809) ventured forth on their famous expedition to the American west ostensibly in search of the great northwest passage — supposedly connecting the Mississippi River to the Pacific — Jefferson instructed them to keep an eye out for unusual fauna "which may [be] deemed rare or extinct" (quite possibly for living specimens such as his Great-Claw and "mammoths"); their existence would discredit the theory of extinction. Lewis and Clark also intended to observe and collect fossil bones. While by 1804 to 1806 they achieved startling success in finding fossilized remains, obviously their quest for a living Great-Claw failed.[26] However, this wouldn't be the last time that scientifically minded men would suggest that creatures of the fascinating Great-Claw race lived beyond prehistory into historical times, or, romantically, might still be living in hidden quarters of the globe.

Jefferson reasoned Great-Claw's alleged existence as an *extant* species was governed by a contemporary, theologically founded "theory"—belief in an unbroken chain of being, or continuity of species. Jefferson wrote:

> the bones exist: therefore the animal has existed. The movements of nature are in a never ending circle. The animal species which has once been put into a train of motion, is still probably moving in that train. For if one link in nature's chain might be lost, another and another might be lost, till this whole

system of things should vanish by piece-meal; a conclusion not warranted by the local disappearance of one or two species of animals, and opposed by the thousands and thousands of instances of the renovating power constantly exercised by nature for the reproduction of all her subjects, animal, vegetable, and mineral. If this animal then has once existed, it is probable on this general view of the movements of nature that he still exists.[27]

Indeed, four months earlier, in a letter dated November 10, 1796, addressed to Colonel Stuart, Jefferson opined that he could not "help believing that this animal, as well as the mammoth, are still existing. The annihilation of any species of existence is so unexampled in any parts of the economy of nature which we see, that we have a right to conclude as to the parts we do not see, that the probabilities against such annihilation are stronger than those for it."[28]

Now Jefferson could exploit Great-Claw as incontestable evidence of America's natural procreative vivacity, or virility. Indeed, sixteen years before, naturalist Jefferson had not only vigorously challenged Buffon's sneering remarks about American fauna and its peoples in his privately printed *Notes on the State of Virginia* (orig. 1781, later 1784–1785, 1786, French eds.; and 1787–1788, London and Philadelphia eds.),[29] yet invoked evidence of fossils into his arguments as well. In concluding his 1799 paper, Jefferson noted how powerful a destroyer was *Megalonyx* (which also explained its supreme rarity in nature), while challenging (then deceased) Buffon's natural history of America, referring to his *Des Epoques de la Nature* (*Epochs of Nature*); the fifth volume (1778) contained a controversial theory of the Earth's origin and subsequent geological evolution, triggered by a comet impacting the Sun. Striking a conciliatory tone, Jefferson suggested — counter to Buffon — that although some of the largest natural history specimens known to science had been found *alive* in the Americas (e.g., including remains of the presumably extant Great-Claw), that "Are we then from all this to draw a conclusion, the *reverse* of M. de Buffon. That nature, has formed the larger animals of America, like its lakes, its rivers, and mountains, on a grander and prouder scale than in the other hemisphere? Not at all, we are to conclude that she has formed some things large and some things small, on both sides of the earth for reasons which she has not enabled us to penetrate."[30]

Recall that there were no cell phones or E-mail systems operating then. Several months might elapse before word of a startling discovery made on one side of the Atlantic finally reached the opposite shore. So, out of the blue, after his paper had already been prepared for its March 1797 oral presentation to the American Philosophical Society, lo and behold, Jefferson finally obtained a copy of the September 1796 issue of the London *Monthly Magazine and British Register*[31] issue — carrying news of Cuvier's 1796 announcement of sloth-like *Megatherium*. Therein Jefferson read of the Paraguay huge beast, already mounted in Madrid. Horrors! Had the recently inaugurated vice president made a huge mistake?

After purchasing the *Monthly Magazine* in a local bookstore, Jefferson was bitterly disappointed that Cuvier evidently had assumed apparent priority; that is, *if* the two animals turned out to be the *same* species. However, even if the two fossilized creatures were not identical, yet belonged to the same family, then Jefferson's perception of *Megalonyx* as a Mega-lion would still be wrong. For, as Jefferson woefully realized, naturalists might draw close analogies between the two fossilized animals. If so, then — based on completeness of the "Paraguay animal" — disappointingly his *Megalonyx* couldn't be classified as a carnivorous feline after all. This would be embarrassing, so Jefferson hastily revised his paper to the tune of "significant alterations," generalizing many conjectural details such as his, or really Stuart's, speculations about *Megalonyx*'s feline affinities—for example, substituting the phrase "animal of the clawed kind" for "family of the lion, tyger, panther, &c." Going down to the wire, Jefferson's addendum was finished on March 10, minutes before the meeting convened.

Of the March 10, 1797, presentation itself, historian of science John C. Greene has conjectured:

> One could wish to have been present at this historic meeting when Jefferson's description of the remains of the megalonyx was first presented to the world of science and discussed, no doubt with considerable animation by [Joseph] Priestly, [Caspar] Wistar ... and the rest of the distinguished company. Did Jefferson then verbally assert, as he did in the published memoir, that the great clawed animal must still be alive in some remote part of the earth? Did Wistar take exception to Jefferson's guess that the newly found *incognitum* was a huge lion, as he did in his remarks appended to the published account? The minutes only record that the memoir was referred to the Committee of Selection and Publication with an authorization to hire someone to delineate the megalonyx bones and that Mr. [Charles W.] Peale was requested to "cause those bones to be put in the best order, for the Society's use."[32]

Jefferson's quick, impromptu assessment of Cuvier's *Monthly Magazine* synopsis was published as a Postscript to his 1799 description. Jefferson diplomatically suggested that, while several resemblances in skeletal structure are evident, overall, *Megatherium*'s anatomy differed sufficiently from the partially known *Megalonyx*. "It may be better, in the mean time," Jefferson decided, "to keep up the difference of name," which turned out to be a very good suggestion because in 1804 Cuvier proved that *Megatherium* and *Megalonyx* were actually distinct creatures. In retrospect, Jefferson's aforementioned lapse of memory concerning receipt of drawings mailed by Carmichael undermined — what turned out to be — a poorly conceived imaginary reconstruction. *Megalonyx* was an animal of the clawed kind, although it certainly was not leonine or feline. Had Jefferson only recalled Bru's sketch of 1788 sent to him by William Carmichael, perhaps Jefferson might not have erred in his classification of Great-Claw. It is tempting to speculate whether, had Jefferson recollected and referred to Bru's sketch nine years later when it would have mattered most, perhaps he might have recognized *Megalonyx*'s affinities to the Madrid specimen. Then again, maybe not, because even five months following presentation of his American Philosophical Society paper, Jefferson wrote to John Stuart on August 15, 1797, after the former had obtained in May 1797 a copy of Jose Garriga's 1796 folio volume on the *Megatherium*. Even then Jefferson asserted, "I met with an account published in Spain of the skeleton of an enormous animal from Paraguay, of the clawed kind, but not of the lion class at all." Julian P. Boyd surmised that by the summer of 1796, Jefferson may have been overly involved with political affairs, and so "in the final analysis, Bru's [i.e., 1788] drawing of the megatherium was probably blotted from memory."[33]

In retrospect, Jefferson's erroneous conclusion — that Great-Claw was a super-lion — was noted by paleontologist Henry Fairfield Osborn (1857–1935) as "very sensible ... in view of the limited knowledge of his day," and later excused by Joseph Leidy (1823–1891) as "a mistake of much less importance than many made by the best naturalists."[34] But whereas Osborn and Leidy praised Jefferson for his paleontological contribution, George Gaylord Simpson lambasted Jefferson's early efforts as insufficient for scientific standards. "Jefferson ... was perfectly aware that he had no accomplishments as a student of paleontology," Simpson argued in 1942. "Jefferson's slight paleontological work must ... be regretfully dismissed as poor." Simpson further complained that Jefferson's paper "departs from inaccurate observations and proceeds by faulty methods to an erroneous conclusion.... The fact that the conclusion was wrong obviously does not, in itself, warrant characterizing the paper as unscientific, but the methods and viewpoint do."[35]

So it was left to American anatomist Caspar Wistar (1761–1818), to describe the remains in what Simpson considered a more favorable, scientific light. This paper, titled "A Description of the Bones deposited, by the President, in the Museum of the Society, and represented in the annexed plates," Simpson claimed, "is a model of cautious, accurate scientific description and inference, an achievement almost incredible in view of the paleontological naiveté of his associates and of the lack of comparative materials."[36] Wistar's 1799 paper was published in Vol. 4 of the *Transactions of the American Philosophical Society*, pp. 526–531, probably without being read at a meeting like Jefferson's. Here, Wistar remarked on the preserved bones, subtly disagreeing with Jefferson in that the left phalanx was not retracted, *unlike* the anatomical case in felines.

However, limited to Cuvier's brief description and a drawing (i.e., Bru's appearing in *Monthly Magazine*), Wistar's observations fortified Jefferson's opinion that the megalonix [*sic*] differed from the Paraguay *Megatherium*. For, Wistar stated, "these bones could not have belonged to that animal," citing particulars of the ulna, radius and claw bones. Wistar noted similarities

between megalonix [*sic*] and a modern sloth (*Bradypus*), while no doubt recognizing the preliminary nature of his insights. While Jefferson's paper was
printed without accompanying visual aids, Wistar included a significant paleoart product, a plate engraved by James Akin from a chalk drawing by Dr.
W. S. Jacobs. As Simpson remarked in 1942, "This illustration and the accompanying memoir marked the beginning of technical vertebrate paleontology
in America."[37]

Until 1820, Jefferson's name *Megalonyx* was used in the vernacular, rather
than as a scientific name selected in a valid Linnaean fashion. Then, French
zoologist Anselm Desmarest (1784–1838) honored our third president by
assigning remains of a species to *Megatherium jeffersonii*. Thereafter, American paleontologist Richard Harlan (1796–1843) formally renamed the animal *Megalonyx jeffersonii* in volume 1 of a work titled "Fauna Americana;
being a description of the mammiferous animals inhabiting North America"
(1825). Therefore, Harlan (and not Jefferson, Wistar or the others) is cited as
the first *technical* author of the genus and species *Megalonyx jeffersonii*.
Regardless of Simpson's protestations of over a century later, in an 1826 eulogy
for Jefferson, his great contemporary Cuvier praised the American naturalist for his love and aptitude for — as well as contributions to — science.

Today, a 30 percent complete skeleton of Jefferson's *Megalonyx* discovered in 1890 may be viewed at Ohio State Museum's Orton Geological
Museum, while an excellent life restoration is on display at the University of
Iowa's Museum of Natural History in Iowa City. In 1966 a 100 percent complete *Megalonyx* skeleton was found in Darke County, Ohio, now on display
in Dayton.

In 1821, Russian paleontologist Christian H. Pander (1794–1865), with
German scientist and engraver named Eduard d'Alton (1803–1854), compared *Megatherium* with extant sloths in a lavish volume titled *Das Riesen-*

Publication of Jefferson's Great-Claw (*Megalonyx*) in Wistar's 1799 report is considered the origin of vertebrate paleontological illustration in America. The claw morphology gave false impressions to Jefferson and others that its owner might have been
a terrible "super lion" (illustration by research immunologist Kristen L. Dennis, 2008,
based on the original drawing in Wistar's report).

Faulthier (*Bradypus giganteus*). They regarded *Megatherium* as an "enormous earth-mole which obtained its nourishment beneath the earth's surface through continuous exertion of its colossal strength; and when perhaps by sinking of the ground to the sea-level, it was driven to live on the surface of the earth, its vast powers, lacking exercise, degenerated, and its size dwindled, until finally it became the weak and puny tree sloth of to-day." This work in comparative osteology (bearing "Lamarckian overtones") was later acknowledged by Charles Darwin (1809–1882) in the Preface to later editions of his *Origin of Species*. Later the German naturalist and apparent evolutionary-minded poet Johann Goethe (1749–1832) published an essay in which he suggested that tree-dwelling sloths most likely descended from extinct giant ground sloths. While these forms are closely related, actually, modern anteaters are closer cousins to the *Megatherium* than the latter is to tree-sloths. However, true relationships existing between modern sloth species and extinct forms wouldn't be elucidated for decades to come, that is until after Darwin returned to England from his voyage aboard the *Beagle*, carrying the germ of a fantastic idea[38] as well as the fossils of South American sloths.

British geologist William Buckland (1784–1856) tackled the anatomy of *Megatherium*, perhaps the first case of a prehistoric mammalian posing most vexing problems with respect to its curious functional morphology, in his *Geology and Mineralogy* (1837) contribution to the *Bridgewater Treatise*. Buckland treated *Megatherium* and other fossil animals as "proofs of the design of structure."[39] Despite its "apparent monstrosity of external form, accompanied by many strange peculiarities of internal structure, which have hitherto been but little understood," *Megatherium*'s anatomy was structured with perfection, or wise design, so it would lead its life in the most efficient way possible. That is, quadrupedal *Megatherium*'s stocky torso and immense musculature were perfectly (and wisely) adapted for digging roots for nourishment. Buckland then considered diagnostic facets of its skeletal anatomy and inferred musculature to prove his thesis. As one example, Buckland erroneously expounded on *Megatherium*'s alleged inch thick osseous cuirass (polygonally shaped bony armor) conceivably offering protection from sun, disease, predation and insect bites. However, the bony armor turned out to belong to a co-deposited glyptodont which Richard Owen named *Hoplophlorus* shortly following publication of Buckland's *Geology and Mineralogy*.

While the origin of this megathere-armadillo mistake is usually attributed to Charles Darwin, who found such fossils in association at Punta Alta, according to biographer Richard Darwin Keynes the error apparently originated with Cuvier in one of his later assessments of the Madrid specimen.[40] Richard Owen soon corrected the mistake, but to his credit Darwin did also consider the possibility that the "osseous polygonal plates" might belong instead to "an enormous armadillo," which turned out to be the actual case. But mistakes persist as one can glean from a poorly executed, cuirass-sheathed megathere situated in Pierre Boitard's (1789–1859) posthumously published 1861 novel, *Paris avant les hommes* (Paris Before Men).

Meanwhile, further news of giant North American sloths continued to startle the imaginations of early naturalists. In April 1823, a news item most unusual for its time appeared in Savannah's *The Georgian* newspaper, an account of the 1822 discovery of large fossilized bones eroding on Skiddaway Island [*sic*] along Georgia's seacoast. These had been discovered on a plantation owned by a Mr. Stark. Stark in turn notified Dr. Joseph Clay Habersham, who saw affinities between the fresh bones and those of the Mammoth, due to their great size.[41] Reassessment of the fossils by Dr. Samuel Latham Mitchell (1764–1831) and William Cooper conjured different conclusions, that instead (and respectively), the remains were of giant sloths either like Jefferson's *Megalonyx* or perhaps Cuvier's *Megatherium*. Mitchell's report was read to New York Lyceum attendees on November 17, 1823, while Cooper's paper allying Skidaway's bones to *Megatherium* was presented on April 19, 1824.

Two decades later, in 1842, more bones were exhumed from Skidaway. International interest and assistance offered by prominent individuals such as Scottish geologist Charles Lyell (1797–1875), who visited Skiddaway on two separate occasions— once in 1842, former United States President John Quincy Adams (1767–1848), as well as Great Britain's Professor Richard Owen (1810–1890), collectively encouraged Savannah businessman, naturalist, scholar and diplomat William B. Hodgson (1801–1871) to report on the discoveries and their locality — by then known as "Fossilossa"— in an 1843 paper read to the National Institute in Washington. Much of the substance of his lecture was published in Hodgson's *Memoir on the Megatherium and Other Extinct Gigantic Quadrupeds of the Coast of Georgia*, in 1846. (This was the second book to mention Skidaway's extraordinary remains, the first being Dr. Goodman's *Natural History of North America* [1831]. Lyell described the fossilized creatures and Skidaway deposit in two books, *Travels in North America* (vol. 1, 1845), and *A Second Visit to the United States of North America* (vol. 1, 1849).[42]

Hodgson noted how the deposit offered no evidence for "diluvial" action, nor signs of deceased animals having died in a "violent or sudden catastrophe." Furthermore, "the similarity of the fossil shells, found in the newer pliocene, underlying the fossil bones, with the existing species of the adjoining coast, shows that if a change of temperature sufficient to destroy the extinct gigantic mammalia, has occurred, there has been none since long prior to the historical period adequate to change, materially, the species of the marine mollusca of this coast." In other words, in contrast to Cuvier, it would appear that (like Charles Darwin) Hodgson was an early advocate of Lyell's actualistic gradualism, or 'uniformitarianism.' Rather than dying in a catastrophe, instead, as Habersham suggested, the "carcasses of the various animals" simply "floated, or fell, into the then lake or stream, and sinking to the sandy bottom, were gradually covered to their present depth by the alluvial deposit from the water."[43] In 1839, German naturalist Albert Koch (died circa 1866) found remains attributed to the *Mylodon* in Missouri.

Lyell, author of the deeply influential *Principles of Geology*, was certainly intrigued by *Megatherium*, then arguably the most popular prehistoric animal of its time across the Atlantic. Lyell had seen the Madrid specimen and when Darwin wrote to him, Lyell suggested he should be on the watch during his travels for further specimens. So perhaps Darwin was predisposed to take a special interest in the giant sloths once he fortuitously found them in Argentina between the fall of 1832 to 1833, fossils which really got him thinking about their disappearance, as well as the flip side of extinction —*evolution* of species.[44]

Citing the occurrence of a North American megathere discovered in 1838–1839 and found by fossil collector Hamilton Couper during excavation of the Brunswick Canal, Lyell pondered the nature of their extinction in 1849. Refuting a hypothesis that the most recent prehistoric fauna had been exterminated by Native Americans, Lyell concurred with Darwin, suggesting that causes of extinction are too intricately woven and subtle as to be able to identify *the* absolute factors leading to extinction of the great sloths such as *Megatherium* (as well as *Mylodon* and *Scelidotherium*). Clearly in the minds of contemporary geologists, any worthy theory of the earth had to satisfactorily account for America's giant ground-sloths.

Darwin elaborated further, stating that "If Buffon had known of the gigantic sloth and armadillo-like animals, and of the lost Pachydermata, he might have said with a greater semblance of truth that the creative force in America had lost its power, rather than it had never possessed great vigour,"[45] a statement which Thomas Jefferson would have (at least partially) welcomed. "Certainly, no fact in the long history of the world is so startling as the wide and repeated exterminations of its inhabitants," Darwin exclaimed. Furthermore, citing both the former occurrences of *Megalonyx* and *Megatherium*, ecologist-geologist Darwin advised that "We are, therefore, driven to the conclusion, that causes generally quite inappreciable by us, determine whether a given species shall be abundant or scanty in numbers."[46] And in cases where a species (first) becomes rare in nature due to, for example, loss of habitat or inaccessibility to food resources, its eventual extinction should be as unsurprising as the circumstances of a very sick man eventually dying. Although South American bone deposits were sometimes highly fossiliferous, Darwin noted this was neither necessarily an indication or proof of mass mortality nor of prehistoric catastrophe causing the assemblage.

By August 1846, one Dr. Dickerson of Mississippi had noted the apparent contemporaneity of *Megalonyx* and *Megatherium* from a Natchez deposit with humans—based on the observation of a co-stratified human fossil bone—an intriguing and controversial result.

By 1855, American paleontologist Joseph Leidy (1823–1891) published a seminal account of the "extinct sloth tribe of North America," in which he described all specimens of *Megalonyx* then known to science. Besides *M. jeffersonii*, other *Megalonyx* species had been named from newer fossils. Leidy suggested there were at most only two species, *M. jeffersonii* and *M. dissimilis*, the latter described in 1852. Leidy also compared *Megalonyx* to three species attributed to Richard Owen's genus *Mylodon,* and Cuvier's *Megatherium.*[47] (Leidy distinguished North American specimens with a distinct name—*Megatherium mirabile*—from the South American variety then assigned to the species *Megatherium cuvieri,* 1804.) The range of known *Megalonyx* specimens had expanded, with remains reported from Memphis; Big Bone Lick, Kentucky; Natchez Bluffs, Mississippi; South Carolina; and Alabama.

One specimen cited by Leidy was found eroding from a ferruginous sandy channel bank near the Ohio River by physician Dr. David Dale Owen (1807–1860), in 1850. D. D. Owen emphasized that the bones were of a "*very recent age,* at least as recent as the origin of most of the existing species of univalve shells now inhabiting the Ohio River and its tributaries." Some of the collected bones were lost in an 1854 ship wreck along the River near Mount Vernon. Yet, other remains of this Henderson, Kentucky specimen were at one time displayed in the Owen Museum, "known in its day as the largest museum west of the Allegheny Mountains." This skeleton was later exhibited in Bloomington's Indiana State Museum, but was eventually moved to Indianapolis. D. D. Owen's reference to the specimen's "very recent date" was verified in 1958 when petrified wood samples collected from the site in 1870 by paleobotanist Leo Lesquereux (1806–1889) were subjected to radiocarbon dating. The measured age of 9,400 years became accepted ("with some reservation") as the terminus in *Megalonyx*'s fossil record. Thus America's prehistoric Great-Claw outlasted prehistory![48]

As to the Skidaway sloths remains originally identified as *Megatherium,* today these skeletons are known as the twenty foot long "hermit sloth" *Eremotherium laurillardi.* In their *Journal of Vertebrate Paleontology* article paleontologists Castor Cartelle and Gerardo De Iuliis noted that this *single* Pan-American species— originated by Danish researcher Peter Wilhelm Lund (1801–1880)— ranged geographically from what is now southern Brazil through Central America and Mexico to the southeastern United States. It appears morphologically distinct from "Paraguay" beast *Megatherium* especially on the basis of forefoot, skull and jaw elements.[49] The genus name *Eremotherium* was assigned by a paleontologist to specimens formerly recognized as *Megatherium* by F. Spillman in 1948. Jefferson and the founding fathers would perhaps have welcomed the knowledge that *Eremotherium* is one of the largest of giant ground sloths known to science.

It is apparent that during the first half century following their nearly coincidental, if not intertwined discoveries, America's two most famous giant sloths had undergone quite a public transformation through various iterative restorations and reconstructions. Whereas, in *Megalonyx*'s case, Jefferson perceived a frightening super-lion, instead, Cuvier regarded "Paraguay's" beast as a huge terrestrial sloth possessing a short tapir-like trunk. D'Alton and Goethe viewed megatheres as aquatic animals or swamp dwellers. Alternatively, Lund claimed that *Megatherium* was arboreal, observing, "It is very certain that the forests in which these huge monsters gambolled could not be such as now clothe the Brazilean mountains, but it will be remembered ... that the trees we now see in this region are but the dwarfish descendants of loftier and nobler forests." And for a short time, giant South American ground-sloths were even thought (erroneously) to have been equipped with a protective bony carapace.[50]

Meanwhile, reconstructions of the bones on paper, in publications and as three-dimensional museum displays were attempted with — for their time — surprising accuracy. Bru's and Garriga's conventionalized quadrupedal pose corresponded closely with Buckland's later theory of the *Megatherium* whose great claws were tools adapted for digging up roots. While the earliest reconstructions showed *Megatherium* oriented quadrupedally, paleontologists next considered an alternative pose — bones mounted in a raised tripodal position (i.e., body balanced on hindquarters and tail). Thus from the 1850s onward, *Megatherium* became the earliest prehistoric monster in popular culture to become mounted in an essentially bipedal pose both in restorations and reconstructions.

How did this come about? Well, partly, remains of giant ground-sloths found in South American were found deposited in an upright position. But this pattern was attributed to them being engulfed in flood plains despite their attempts to escape by crawling up slippery mudbanks; further struggling only caused the bulky sloths to sink further and further into the encasing mud. Maybe such taphonomy got people thinking about the problem.

Following Buckland, Professor Richard Owen became Britain's expert on extinct American sloths, especially after receiving shipments of South American fossils excavated by Darwin during his *Beagle* voyage. From these remains, Owen published an elaborate memoir on the Mylodon, "Description of the skeleton of an extinct gigantic sloth *Mylodon robustus*, Owen, with observations on the osteology, natural affinities, and probable habits of the megatheroid quadrupeds in general" (1842). In the same year (1842) that Owen formalized the Dinosauria, he also described *Mylodon darwinii* and *M. robustus*. He also named genera, *Glossotherium* in 1840 and another, *Scelidotherium*. And between 1851 and 1860, Owen published a number of scholarly articles "On the *Megatherium*" in the *Philosophical Transactions of the Royal Society of London*, culminating in a beautifully illustrated 1861 memoir. As Owen stated in his *Mylodon* memoir, "the indications of an extensive

family of most singular quadrupeds, once spread over the American Continents, from the latitude of New York to Patagonia, cannot ... be pondered on
without exciting the strongest interest."[51]

While Lund viewed America's great extinct sloths as *arboreal* animals
because their ground motion would likely have been impeded by their enormous claws, by 1842, Owen could no longer believe such nonsense. For what
tree branch imaginable could bear elephantine weight? So instead of climb-
ing, Owen suggested, ground-sloths such as *Mylodon* and *Megatherium* used
their sturdy limbs and anchoring pelvis for pulling down or even uprooting
trees so they could masticate leaves. Owen stated, the giant sloth's probable
long tongue and proboscis coupled with "all the characteristics which coexist in the skeleton of the mylodon and megatherium, conduce and concur
to the production of the forces requisite for *uprooting and prostrating trees*.
The megatherians constituted an extensive tribe of leaf-devouring and tree-
destroying animals, of which the larger extinct species were rendered equal
to the herculean labors assigned to them in the economy of an ancient world.
In the task of thinning the American forests, the brute force of the mylodon
[and megatherium] has been superceded by the axe of the backwoodsman."[52]

As indicated in his *Journal of Researches* (1839), Charles Darwin, who
had discovered so many ancient sloth remains (including a relatively complete skeleton of the *Scelidotherium*) in an area of the Pampas— viewed as "one
wide sepulchre of these extinct gigantic quadrupeds"—completely agreed with
Owen's assessment. And in his 1855 memoir, Leidy noted "additional evidence
ingeniously inferred by Professor Owen, that the giant sloths were very liable
to accidents not unfrequently involving fracture of the bones, from which the
habit of uprooting trees, the boughs of which formed their food."[53] Here,
Leidy cited evidence of fractured ribs and shoulder blades as being consistent with Owen's view. Thus originated the image of *Megatherium* grasping
tree limbs with its clawed hands in raised stance.

Perhaps the earliest and most enduring restoration of a rearing, tripodal *Megatherium* is the first-of-its-kind concrete sculpture made by British
artist Benjamin Waterhouse Hawkins (1807–1889) as part of the geological
exhibit opening in 1854 at the Crystal Palace in Sydenham, England. This
model still stands, although because the tree which the massive sculpture was
hugging has grown, years ago its left arm snapped and later had to be replaced
with a fiberglass replica. In following Cuvier's hunch, the *Megatherium* sculpture has a noticeable proboscis. In January 1854, an *Illustrated London News*
journalist described Hawkins' and Owen's creation with awe, noting how
Megatherium, "Propped on this mighty tripod, grasping a tree with fore limbs
well adapted..., putting forth the powers of muscular and nervous system of
prodigious energy which diverged from the pelvis and acted even on the fore
limbs, he prostrated it in a moment."[54]

Although this model stands apart from the rest of Hawkins's more

For a time, paleontologists entertained the possibility that giant ground sloths such as *Megatherium* or *Mylodon* could have uprooted trees, as illustrated (at left) in F. Long's circa 1905 restoration, originally printed as a collectible card by the Kakao Company of Hamburg-Wandsbek, Germany. At right is Lankester's similar restoration of the *Megatherium* as seen in his *Extinct Animals* (1905 ed.; both figures collection of Allen G. Debus, photographed by the author).

famous prehistoric menagerie, such as his antiquated-looking *Hylaeosaurus, Iguanodon* and *Megalosaurus* dinosaurian cluster, according to paleontologist and historian of geological science William A. S. Sarjeant, Prince Albert, consort of Queen Victoria, had been so intrigued by Owen's descriptions of "antediluvian creatures, in particular the giant ground-sloth *Megatherium*,"[55] that he proposed the erection of a life-sized display of antediluvians; hence Hawkins' handiwork. So were it not for Owen's impressions of *Megatherium*, would Hawkins's celebrated chapter in early paleontological restoration never have been written? Hawkins completed further restorations and designs featuring tripodal, tree-hugging, tapir-trunked Megatheria (and *Megalonyx*) during the 1860s and 1870s as well; although by the 1870s, bipedal dinosaurs (particularly *Laelaps* and a sometimes tree-grasping *Hadrosaurus*) had visibly begun to eclipse the megatheres in prominence in these early diagrams.[56]

Ever since the *Megatherium* was first informally described in a circa 1832 lecture by Buckland, the first such case of an "extinct monster" being introduced to a popular unscientific audience, scientific skeletal reconstructions and restorations of Megatheres began appearing in a variety of media — museums and books, especially. Besides the Madrid specimen, other skeletal reconstructions were placed on display, such as at London's Hunterian Museum, Royal College of Surgeons, where a quadrupedally mounted *Megatherium* was exhibited by the mid–1830s. This specimen had been discovered by Woodbine Parish (1796–1882) at Buenos Aires Province's Salado River by 1832 adjacent to two other partial specimens. These were described by William Clift (1775–1849) in a *Transactions of the Geological Society* paper.[57] Since the Hunterian specimen was incomplete, casts of the missing bones were supplied from the more complete Madrid specimen.

By the mid–1850s, pop-cultural interest in giant ground sloths— particularly *Megatherium*—had peaked with Hawkins' rather forlorn-looking sculptural restoration still exhibited at Sydenham, and was beginning to wane. In 1857, several dedicated naturalists and scientist-enthusiasts formed an organization at the Smithsonian known as the "Megatherium Club." According to Todd Womack, the club was named after the Skidaway sloth remains. In their motto, anticipating that of *X-Files*' FBI detective character Mulder over a century later, the dauntless Megatheria claimed, "What we seek is the Truth."[58] By 1866, after the demise of founder William Stimpson and several other members who died while on exploring expeditions, the Club became "as extinct as its namesake."

Thereafter, particularly after Henry Ward's (1834–1906) fossil reproduction enterprise flourished during the1860s, it became more common to see 18 foot long casts of South America's *Megatherium* in museums both in Europe and the United States, as well as at industrial trade shows and expositions throughout America's Midwest. Whereas today a fossil cast of the complete *Eremotherium* skeleton might cost an individual hobbyist upward to $25,000, on page 14 of his 1866 *Catalogue of Casts of Fossils from the Principal Museums of Europe and America* Ward advertised a complete *Megatherium* for a mere $250 (i.e., "packed but not painted"); the price included a tree replica![59]

Giant ground-sloths also made sensational appearances in some of the early science fiction novels published in the1860s on into the first half of the 20th century, authored by notable writers such as Jules Verne, Edgar Rice Burroughs, H. G. Wells and Charles G. D. Roberts. Notably, paleontologist William Diller Matthew (1871–1930) wrote an excellent short story titled, "Scourge of the Santa Monica Mountains," published a 1916 issue of *American Museum Journal*,[60] concerning the demise of a giant ground-sloth which becomes mired in the La Brea asphalt and then attacked by a hungry sabertooth.

And with increasing prevalence and competency, beginning in the 1860s through the 1910s, visual artists such as Edouard Riou (1833–1900), Robert Bruce Horsfall (1869–1948), F. Long, Erwin S. Christman (1885–1921) and Joseph Smit (1836–1929) provided life restorations of giant ground-sloths for a number of popular and semi-popular books and articles. Then, by the mid–1920s, perhaps the greatest artist of extinct animal life—Charles R. Knight (1874–1953)—had most vividly restored giant ground-sloths (i.e., *Glossotherium*) for the Natural History Museum of Los Angeles County based on early 20th century excavations conducted at La Brea's famous Tar Pits (which began in March 1906). Knight also accurately painted South America's *Megatherium* for the American Museum of Natural History and Chicago's Field Museum, in poses conventionalized as in *Mylodon*'s case by Richard Owen eight decades before.[61]

A curious sloth-related incident took place during the 1880s when Mark

Twain satirized the claims of several witnesses who believed tracks discovered in a Carson City, Nevada, sandstone deposit were made by primeval man. In 1882, however, Yale Professor Othniel Marsh (1831–1899) showed they were instead made by the extinct sloth, *Mylodon*. This amusing instance exemplifies how, ever since Jefferson's time, scientists pondered the possibility whether any of the gigantic extinct animals represented as fossils remained alive somewhere on the globe. By the late 19th century, prominent men such as Lyell, Darwin, Alfred Russell Wallace (1823–1913) and E. Ray Lankester (1847–1929) each considered the problem from different angles, mindful of ground-sloth fossils found in curious association with human remains both in North and South America.

In the southern United States, Drs. Koch and Dickerson of Natchez had each, independently, found human bones and pottery fragments alongside remains of megatheres. In 1847, Lyell insisted the Natchez remains had fallen from an Indian grave situated on a cliff overlooking a river, into the fossil-bearing layer. For if the remains were coeval, most objectionably according to those skeptical of Charles Darwin's theories, "It would follow that the human race had survived the extinction of one group of terrestrial mammalia and seen another succeed and replace it."[62] Evolutionist Wallace, however — writing in 1887 — preferred the alternative, namely that the fossils were of equal antiquity.

Meanwhile, remains attributed to *Mylodon* came under scrutiny when an exquisitely preserved hairy pelt was found in a southwest Patagonian cave near the Chilean coast known as "Ultima Speranza." Besides the fur, remains found by a Dr. Nordenskjold in (circa) 1895 included bones of sloths and other species, some of which had been modified and sharpened into tools. The apparent freshness of the *Mylodon* pelt, combined with signs of human occupation of the cave, indicated to several late 19th and early 20th century naturalists that here, in historic times, a sloth had been held in domestic captivity by humans. In fact, decades before, Darwin had commented on the very slight petrification of a *Mylodon* skull he had delivered to Owen. And in 1928, an exceptionally well preserved *Nothrotherium* mummy was found at a 100 foot depth in a Texas lava fissure. The *Nothrotherium*'s skeleton was articulated and soft tissues were still attached to bones. Fur was found on the mummy's skin.

However, rather than these animals having died only a few centuries ago, the last of the sloths most likely perished *thousands* of years ago. The South American furry pelt has since been carbon-dated at 5,000 years in age and fossil dung found in the cave is twice as old. Their exquisite state of preservation may instead be attributed to the cave's dryness and other climatological factors. Nevertheless, ever since the early 20th century, some individuals continue to insist that giant ground-sloths (and other prehistoric animals as well!) may still live in the wilds of South America (or even in the

Florida Everglades). As one curious example, in 2004, Gavin Menzies suggested in his book *1421: The Year China Discovered America*, that the Chinese carried two living *Mylodons* back to China in their vessels during the 15th century![63]

The realm of prehistory is designated to inhabitants of a time greater than ten thousand years ago. So the last of the giant ground sloths may not have been quite so prehistoric as once thought, having surpassed the paleo-barrier.

The giant ground-sloths no longer are as prominent as other savage prehistoria which have come to replace them in public view through ensuing decades of discovery and restoration, since their arrival in 1788. Yet, their image persists and some day giant ground-sloths may even rise to pop-cultural prominence once more. Has their comeback been heralded by two Uruguay paleontologists (Richard A. Farina and R. E. Blanco) who in 1996 suggested that, far from docile, instead, *Megatherium* was "a gruesome carnivore which could stab saber-tooth cats with its long curved hand-claw daggers, or roll giant jeep-sized glyptodonts over, upside down onto their bony carapaces ... then commence to eating their soft belly flesh?"[64]

Megatherium and *Megalonyx* are two of the oldest-named prehistoric genera, and (with supporting star cast *Scelidotherium, Eremotherium* and the *Mylodon*) it is largely through considering the scientific meaning of their remains that the science of paleontology — in its larger sense — is what it has become today.

THREE

Furred Fury:
Exhumation of *Incognitum*,
America's First
Carnivorous Monster

Whereas the first American monster, later proven extinct, that was described in a journal article (i.e., Jefferson's Great-Claw, *Megalonyx*) wasn't reconstructed or restored artistically as a more or less recognizably complete animal for decades (thus diminishing its public impact), another mysterious beast known from fossils, America's *Incognitum*, proved a sensation, quite possibly because it *was* artistically restored and exhibited by avid naturalist-promoters to an adoring public following dramatic discovery of relatively complete specimens found in New York.

As previously mentioned, Jefferson had speculated that his Great Claw megalonyx "may have been ... formidable an antagonist to the *mammoth*." When he wrote those words, Jefferson referred to what was later called a mastodon, also known as the "Ohio animal,"[1] rather than a true mammoth. By 1797, the woolly mammoth was a fairly well understood creature, known principally from Siberian localities, while America's *Incognitum*— eventually known as mastodon — was not. The great fossil animal which Jefferson and perhaps John Stuart imaginatively pitted their Great Claw Megalonyx in battle against during the late 1790s was also known as an *Incognitum*, one of five established vertebrate — varieties, in this case, of the great "American *Incognitum*" (a term then also generally synonymous with "mammoth").

The Latin word "incognitum" means "unknown," certainly applicable to the formerly puzzling natural history of the mastodon. As Simpson[2] noted, by 1799, five animal incognita were known, comprising the Siberian mammoth, Great Claw megalonyx, possibly two genera of megatheria, and most mysterious of all, the "American *Incognitum*." Before discussing America's first popularized, "carnivorous" prehistoric monster further, the spate of confusing nomenclature surrounding the mastodon and mammoth, now known to be elephantine, should be clarified.

Historian Paul Semonin stated: "Like the giant of Claverack, the mammoth was an unknown creature whose discovery greatly puzzled European naturalists. The word *mammoth* soon became the popular term applied to fossil elephants in both Siberia and North America, introducing an element of confusion over the identity of the American *Incognitum*."[3] Woolly mammoth remains were first identified by those whose records would be passed along to modern scientists as early as 1695, while remains of what would be called the American mastodon first became of (misguided) scholarly intrigue by 1714. Today, the correct scientific name for the mastodon is *Mammut americanum*. According to Simpson, however, the generic name *Mammut* unfortunately "perpetuates the old confusion of the American mastodon and the Siberian mammoth and is inconvenient because of its apparently erroneous connotation. Many students therefore refuse to use it."[4] This is perhaps counterintuitive because late 17th century Siberian settlers who had found frozen and skeletal remains of woolly mammoths called them by a similar sounding word, "marmont," eventually resulting in a popular Anglicized name, "mammoth," first published in English in 1700–1701. By the late 18th century the word mammoth was associated with remains of the American *Incognitum* as well.[5] European and Siberian elephant fossilized remains could not be attributed to mastodon, however, because — based on fossil evidence — the American mastodon didn't inhabit those places.

In 1806, noting their peculiar nipple-shaped, bubbly-looking or knobby molar teeth, Cuvier referred to remains of the American *Incognitum* as "mastodontes"; hence, the origin of the term *mastodon* (meaning literally "breast tooth").[6] Earlier in 1785, referring to bones housed in the British Museum,

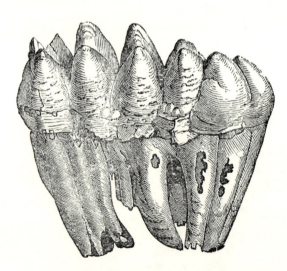

The pointy, conical projections of the mastodon molar differed considerably from those in mammoth molars (which are less pronounced and wrinkly). Therefore, some claimed, mastodon, or the American *Incognitum* as it was known for several decades, was a carnivorous beast, unlike the herbivorous mammoth (from Figuier's *Earth Before the Deluge*, 1867 ed.).

Christian Michaelis (1727–1804) published the term *Mammut americanus.* In 1799, German naturalist Friedrich Blumenbach (1752–1840) also formalized the American *Incognitum's* name using Linnaean nomenclature, this time with the similar, *Mammut ohioticum.*[7] He also applied the scientific name, *Elephas primigenius*, to Siberia's woolly mammoth. But even though Cuvier later coined the more familiar name *mastodon*, the term, *Mammut*, carries scientific priority and remains valid.[8] In 1973, however, the mammoth's phylogeny (evolutionary family tree) was formally revisited, with consequence that "woolly mammoths" are now referred to the genus, *Mammuthus.*[9]

So why was America's mastodon considered mysterious and unknown? Why wasn't it readily recognized as an extinct form of elephant, as was its evolutionary cousin, the mammoth? And why was it regarded as carnivorous? Readers enlightened by modern paleoimagery might proclaim, Shouldn't it all have been rather obvious? In sorting this puzzle, those nipple-shaped, monstrously sized molars were at the crux of the matter. If those distinctive teeth indeed belonged to the skeletons and tusks found in near association, then what did such an enormous animal eat? To naturalists of the time though, pointy teeth meant *carnivorous* diet. In fact, one interpretation made from those distinctive nipple or prong-shaped molars was that this kind of mammoth was actually a voracious carnivore!

But, here, let's first consider what Thomas Jefferson thought of the American *Incognitum*, a creature which he stated was "the largest of all terrestrial beings."[10] From Jefferson's correspondence, he was clearly gathering information about the beast and soliciting further skeletal remains from correspondents as early as 1781. Clearly he was highly intrigued by the American *Incognitum*. Jefferson's active interests and affiliations with America's *Incognitum* would persist through 1811, when he necessarily turned to other personal affairs. Although he fashioned his *Megalonyx* as a super-lion, by early 1797, he had reconsidered his erroneous carnivorous paradigm.

In the absence of North American archaeological ruins, relics, and classic monuments rivaling those found in Europe, Egypt or the British Isles, or a documented cultural history handed down from scholarly ancients, America's mastodon symbolically wedded our fresh perception of an albeit savage, yet distinctly *American* cultural heritage blended with the prospect for our future, ever-branching remotely toward majestic, sublime, western wildernesses. And so it was with patriotic undertones that Jefferson sought to combat and falsify Comte de Buffon's insulting theory of (American) biological degeneration. For according to Jefferson, Buffon had suggested that,

1.) ... the animals common to both the old and new worlds are smaller in the latter. 2.) That those peculiar to the new are on a smaller scale. 3.) That those which have been domesticated in both, have degenerated in America; and 4.) That on the whole it exhibits fewer species. And the reason he thinks is, that the heats of America are less; that more waters are spread over its surface by nature, and fewer of these drained off by the hand of man. In other words, that heat is friendly, and *moisture* adverse to the production and development of large quadrupeds.[11]

In 1787, Jefferson wryly shipped Buffon a skeleton and skin of the largest moose his hunters could find to show the French scholar how invigorating was the American habitat. (An adult moose's body—much larger than any European deer—is comparable in size to the Irish elk, a deer, by then generally thought to be *extinct* since no living animal had been found to possess such enormous antlers.) Furthermore, Jefferson remarked that the American mammoth was both larger than the Siberian variety as well as any modern elephant.

Over two decades before, in his comparative analysis of 1764, French naturalist Louis-Jean-Marie Daubenton (1716–1800) pictorially compared the shapes and sizes of thigh bones belonging to the Mastodon, the Siberian mammoth, and a modern elephant. While bone morphology does considerably contrast among represented specimens in his published drawing, Daubenton concluded that these animals were variants of the *same* species. Yet the common association of tusks with molars created a dilemma. While the tusks clearly belonged to elephants, Daubenton concluded, the large molars belonged to a large hippopotamus, a species then thought to be carnivorous. Brazenly, he even suggested that the American Indians had mistakenly associated the tusks with gigantic hippo teeth. In accepting Daubenton's conclusions, despite Jefferson's evident preoccupation with the size of the American mammoth's skeletal remains, Buffon countered Jefferson, stating that remains of the Siberian mammoth and the American variety (i.e., *Incognitum*) together represent the *same* species. Decades later, Jefferson viewed such reasoning skeptically. He reasoned that,

> Our quadrupeds have been mostly described by Linnaeus and Mons. de Buffon. Of these the mammoth, or big buffalo, as called by the Indians, must certainly have been the largest. Their tradition is that he was carnivorous, and still exists in the northern parts of America.... It is remarkable that the tusks and skeletons have been ascribed by the naturalists of Europe to the elephant, while the grinders have been given to the hippopotamus, or river horse. Yet it is acknowledged, that the tusks and skeletons are much larger than those of the elephant, and the grinders many times greater than those of the hippopotamus, and essentially different in form.[12]

> Wherever these grinders are found, there also we find the tusks and the skeleton.... It will not be said that the hippopotamus and the elephant came always to the same place, the former to deposit his grinders, and the latter his tusks and skeleton.... We must agree then that these remains ... are of one and the same animal.[13]

Echoing Jefferson's beliefs as to the size and whereabouts of this animal, in a lecture to the American Philosophical Society presented on April 3, 1789, the Rev. Dr. Nicholas Collin opined that "The vast Mahmot [*sic*] is perhaps yet stalking through the western wilderness; but if he is no more, let us carefully gather his remains, and even try to find a whole skeleton of this giant, to whom the elephant was but a calf."[14] But while there was certainly an array of questions to resolve surrounding the mysterious *Incognitum*, including matters such as to what extent was this animal like (or unlike) modern elephants — the likes of which few Americans had ever seen alive — or the Siberian Mammoth, and were these fossilized animals extinct, or living, above all, mastodon's eating habits proved most intriguing for early naturalists. For the prospect of a true, possibly living American monster — the largest anywhere in the world — not only hinged on *Incognitum*'s relative size but its presumed *voraciousness* as well.

Yet in fact this debate was already nearly two decades old by the time Jefferson found himself becoming embroiled in the matter. Scientists such as Daubenton, our own Benjamin Franklin, and British naturalists (botanist) Peter Collinson (1694–1768) and William Hunter (1718–1783) had already weighed in on the question between 1762 and 1768. Because men had never faced such a quandary before, it proved remarkably trickier to properly interpret the evidence at hand than would be the case today. Those peculiar teeth, which had became a source of intrigue for naturalists of the time, were at the root of the problem.

In comparison to mastodon molars, mammoth teeth have slightly raised, smooth yet wrinkly grinding surfaces, more or less similar to those of the two modern species of modern elephants (animals which, of course, are herbivorous). Jefferson was aware of these differences, claiming "The skeleton of the mammoth (for so the *incognitum* has been called) bespeaks an animal of six times the cubic volume of the elephant. The grinders are five times as large, are square, and the grinding surface studded with four or five rows of blunt points; whereas those of the elephant are broad and thin, and their grinding surface flat."[15] (Although large mastodon molars also somewhat resemble hippopotamus teeth.) Due to those enlarged, knobby protrusions, however, molars associated with skeletal and tusk remains attributed to the American *Incognitum* looked ominously carnivorous in nature. So was the American *Incognitum* therefore a carnivore? Some, like Jefferson — who melded Indian folklore with scientific opinions espoused by British anatomists — thought so, adding to the legend and lore of this distinctively *American* beast.[16] But then hadn't he erred in the case of his super-lion?

Benjamin Franklin claimed, in a letter dated August 5, 1767, addressed to George Croghan (1720–1782), that while "The tusks agree with those of the African and Asian elephant ... the grinders differ, being full of knobs, like the grinders of a carnivorous animal.... We know of no other animal with tusks like an elephant, to whom such grinders might belong."[17] Furthermore, Franklin suggested in order to account for the occurrence of elephantine fossils in both Siberia and North America, one may presume that "the earth had anciently been in another position, and the climates differently placed from what they are at present."[18]

But half a year later we read of Franklin's reservations over the plausibility of the sensational carnivorous conclusion. Writing to prominent French astronomer Abbé Chappe d'Auteroche on January 31, 1768, Franklin mentioned,

> Some of Our Naturalists here ... contend that these are not the Grinders of Elephants but of some carnivorous Animal unknown, because such Knobs or Prominences on the Face of the Tooth are not to be found on those of Elephants, and only, as they say, on those of carnivorous Animals. But it appears to me that Animals capable of carrying such large heavy Tusks, must themselves be large Creatures, too bulky to have the Activity necessary for pursuing and taking Prey, and that those Knobs might be as useful to grind the small Branches of Trees, as to chaw flesh.[19]

Franklin's flip-flopping on the matter reflected a scientific debate stirred in London by shipments of American *Incognitum* fossils. In his reply to Franklin, the Abbé Chappe opined that such creatures as woolly mammoths and the American *Incognitum* must have been washed there after drowning during Noah's Flood.

Peter Collinson addressed several fossilized teeth, tusks and a lower jaw in his 1767 *Philosophical Transactions* synopses of November 26 and December 10. These had been shipped in early 1767 to London by Croghan (the "King's superintendent of Indian affairs in America") from Kentucky's Big Bone Lick area, both to Lord Shelburne (who managed colonial affairs), as well as Benjamin Franklin then living in London. Collinson declared:

> It is very remarkable ... that none of the molares [*sic*], or grinding teeth of elephants are discovered with these tusks; but great numbers of very large pronged teeth of some vast animals ... which have no resemblance to the grinding-teeth, of any animal yet known. As no living elephants have ever been seen or heard of in all America, since the Europeans have known that country, nor any creature like them; and there being no probability of their having been brought from Africa or Asia; and it is impossible that elephants could inhabit the country where these bones and teeth are now found, by reason of the severity of the winters, it seems incomprehensible how they came there.[20]

Collinson further suggested that this great, yet still mysterious *browsing* animal used its heavy molars to bite off and eat thick tree branches for sustenance.[21] Was America's *Incognitum* a herbivore, then?

Interestingly, in his August 5, 1767, letter Franklin had re-aligned his thoughts on the diet of America's Ohio animal to conform with Dr. William Hunter's views. Hunter was a British physician and comparative anatomist who began examining the specimens Collinson scrutinized by the spring of 1767.

According to Semonin,[22] Hunter had already discussed remains of the Ohio animal — previously deposited in the royal family's Tower of London cabinet — with his younger brother, anatomist John Hunter (1728–1793), who in turn doubted whether the creature represented by the molars was an elephant. John Hunter "said the grinder was certainly not an elephant's. From the form of the knobs on the body of the grinder, and from the disposition of the enamel, which makes a crust on the outside only of the tooth, as in a human grinder, he was convinced that the animal was either carnivorous, or of a mixed kind."[23]

Further checking of the new collection sent by Croghan in 1767, as well as other bones and teeth from African elephants preserved in a London warehouse, only confirmed William Hunter's opinion that "the grinder tooth, brought from the Ohio, was not an elephant, but of some carnivorous animal, larger than an ordinary elephant. And I could not doubt that the tusk belonged to the same animal."[24] Then came the most dramatic conclusion of all, namely that "if this animal was indeed carnivorous, which Dr. H. believed cannot be doubted, though we may as philosophers regret it, as men we cannot but thank Heaven that its whole generation is probably extinct."[25]

In his February 25, 1768, presentation to the British Royal Society, William Hunter declared the alleged American elephant to be a "pseud-elephant, or animal incognitum, which naturalists were unacquainted with." Hunter also professed that the *incognitum* "would prove to be the supposed elephant of Siberia, and other parts of Europe; and that the real elephant would be found to have been in all ages a native of Asia and Africa today."[26] William Hunter's conclusions were reached through meticulous graphical comparison of bones, teeth and tusks as well as sound scientific reasoning. (Of course, his conclusion that the American *Incognitum* was a carnivore proved faulty. He erred further in suggesting that the Siberian Mammoth was synonymous with the American *Incognitum*.)[27]

The Hunters' collaboration resulted in a conclusion of carnivory for the American *Incognitum*, established shortly before the date of Franklin's aforementioned August 5, 1767, letter, a finding not rebutted by Collinson. However, another gentleman entertained the significance of carnivory in the animal realm before the Hunter brothers' pronouncement on mastodon. In 1764 English novelist and poet Oliver Goldsmith (1730–1774), a staunch defender of the theological Great Chain of Being, published thoughts on carnivore teeth, stating that carnivore teeth "serve ... as weapons of offence.... The rapacious animal is in every respect formed for war ... the beasts of the forest ... are formed for a life of hostility, and, as we shall see, possess [*sic*] of various methods to seize, conquer and destroy."[28] Later, in his *A History of the Earth and Animated Nature* published in 1774, Goldsmith echoed Hunter's conclusions concerning our mastodon *without* accepting his idea of probable extinction, stating "And if, indeed, such an animal exists, it is happy for man that it keeps at a distance, since what ravage might not be expected from a creature (endowed) with more than the strength of an elephant, and all the rapacity of a tiger."[29]

Although there was no consensus then as to its proper diet the monstrous reputation of the American *Incognitum* was mounting. Three decades later, on July 21 1797 (four months following Jefferson's announcement of the Megalonyx), Judge George Turner read his "Memoir on the extraneous fossils denominated mammoth bones; principally designed to show that they are the remains of more than one species of non-descript animal," to an audience of the American Philosophical Society. Noting differences between the tusks and teeth of the Siberian Mammoth and the American *Incognitum*, Turner "proved unequivocally that the Siberian and North American animals themselves were separate species."[30] Turner regretted that a complete skeleton of an *incognitum* had not yet been found, but that "The person who shall first procure the complete skeleton ... will render, not to his country alone, but to the world, a most invaluable present."[31]

Although Turner parted with Hunter's notion that mastodon and Siberian mammoth were the same species, based on his observations of fossils deposited in the Big Bone Lick region, he did concur with Hunter's carnivorous theory; in contrast the Siberian mammoth was decidedly a herbivore. Turner had witnessed bones of buffaloes along the Ohio River's salt lick marsh which had been fractured "most likely, by the teeth of the Mammoth [i.e., his term for the American mastodon, or *Incognitum*], while in the act of feeding on his prey. Now, may we not from these facts infer, that Nature had allotted to the Mammoth the beasts of the forest for his food?"[32] This led to a remarkable scenario! In order to catch buffaloes swiftly charging across the forest, paradoxically, the American *Incognitum* must have possessed great speed — unlikely in such a ponderous animal. So, relying on an Indian legend source, Turner instead dramatically speculated that *Incognitum* ambushed its prey.

> May it not be inferred, too, that as the largest and swiftest quadrupeds were appointed for his food, he necessarily was endowed with great strength and activity? That, as the immense volume of the creature would unfit him for coursing after his prey through thickets and woods, Nature had furnished him with the power of taking a mighty leap? — that this power of springing a great distance was requisite to the more effectual concealment of his bulky volume while lying in wait for his prey?[33]

Then, Turner, echoing Hunter's and Goldsmith's conclusions, remarked, "With the agility and ferocity of the tiger; with the body of unequaled magnitude and strength, it is possible the Mammoth [i.e., Mastodon] may have been at once the terror of the forest and of man! And may not the human race have made the extirpation of this terrible disturber a common cause." In other words, perhaps Native Americans killed the last of the American *Incognitum* in "self-defense."[34] Adding to the "ominous new images of warring species [of] the American landscape," at the time of its 1799 publication, was Turner's fevered speculation that the mastodon was a fierce clawed animal, like Jefferson's *Megalonyx*.[35]

Thus, America's *Incognitum* was taking on ferocious, bestial qualities, not unlike the monster's aura as expressed in Indian lore. In 1781, Jefferson related a Delaware tribal tradition offered in response to his query concerning the bones of the big buffalo.

> In ancient times a herd of these tremendous animals came to the Big Bone licks, and began an universal destruction of the bears, deer, elks, buffaloes, and other animals which had been created for the use of the Indians: that the Great Man above, looking down and seeing this, was so enraged, that he seized his lightning, descended on the earth, seated himself on a neighboring mountain, on a rock of which his seat and the print of his feet are still to be seen, and hurled his bolts among them till the whole were slaughtered, except the big bull, who presenting his forehead to the shafts, shook them off as they fell; but missing one at length, it wounded him in the side; whereon, springing round, he bounded over the Ohio, over the Wabash, the Illinois, and finally over the great lakes, *where he is living at this day*.[36]

Fanciful, yes, but as folklorist and historian Adrienne Mayor[37] contends in her fascinating 2005 account, *Fossil Legends of the First Americans*, such traditions offered explanations entirely comprehensible to Native Americans as to the nature of these remarkable creatures, the ghostly remains of our celebrated American monster.

As if in answer to Turner's prayers, by 1801 sufficiently complete remains of the mastodon *were* unearthed through the indefatigable spirit of Charles Willson Peale (1741–1827). Through his persistence and diligence, naturalists were then able to erect two *incognita* skeletons for exhibition, although to frame this episode properly the central significance history of the mastodon with respect to his famous museum in Philadelphia must be sketched.

Peale, who had resided in London during that pivotal paleontological year of 1767 where he befriended Franklin, with whom he no doubt discussed the Ohio animal's grinders, and who had served as a captain in the American Revolution, had obtained a collection of *Incognitum* fossils owned by Philadelphia physician Dr. John Morgan. Morgan was visited by German physician Christian Michaelis who had become awestruck with the fossils. Michaelis desired detailed illustrations of the bones to carry with him back to Germany and as there were no cameras or xerox machines available then, he sought the services of a reputable artist — Peale! Later, in 1785, Michaelis wrote about the Mastodon, rejecting Hunter's claim "that it had been a carnivore; but more surprisingly he argued that it had had neither tusks nor trunk, an inference based on his interpretation of a specimen of the upper jaw (which, as it turned out later, he had got back to front)."[38]

During 1783, Morgan's boxful of old bones attracted local interest. Soon people were stopping in to see the fossils and watch as Peale drew pictures of them. Peale's brother-in-law, Colonel Nathaniel Ramsey (1741–1817) suggested, "Charles, I wonder if you realize what you've got here. I would walk twenty miles to see this collection. Obviously, many others feel as I do. Why not, then, add these to your display of paintings? Obtain more of these oddities of nature — develop a museum and charge admission. It could not only become a source of income, but would it not also increase the sale of prints?"[39] However, while Morgan had no objections as to Peale's exhibition of the bones, by 1787 Morgan had sold his collection to Dutch anatomist Petrus Camper (1722–1789). So, dejectedly, Peale (himself, no "happy camper") was forced to part with his museum's central attraction. What could be done to restore this marvelous exhibit to its rightful place at America's hub, in Philadelphia?

Charles Willson Peale's famous mastodon exhumation has been recounted several times, authoritatively. A principal source for the record of how these remains were excavated and procured from two landowners is Rembrandt Peale's (1778–1860) thorough account, *Historical Disquisition on the Mammoth or, Great American Incognitum, an Extinct, Immense, Carnivorous Animal Whose Fossil Remains Have Been Found in North America* (London, 1803).[40] While Rembrandt's 1803 booklet has been judged as of no greater scientific value than Jefferson's 1797 formalization of Great-Claw, today it remains a principal contemporary source informatively presenting early ideas regarding reconstruction of our dawning (former) American symbol.

Perhaps we could frame discussion of the resurrection of America's first two mounted skeletal reconstructions of prehistoric vertebrates principally in context of three images, the first being a self-portrait by Charles W. Peale titled *The Artist in His Museum, 1822.* This painting, a magnificent metaphoric melding of American art, science and national symbolism, was used to commemorate the 150th anniversary of the Philadelphia Academy of Fine Arts, as Peale's painting was featured on a U.S. postage stamp in 1955. Here we see Peale standing at center raising the curtain of enlightenment into the grand hallway of his Philadelphia Museum. Situated on the floor, to his left at front, a pair of leg bones and a jawbone belonging to the mastodon are visible, lying next to an artist's palette with brushes.[41] Adjacent to the mastodon fossils, at lower left of the frame, is a wild turkey laying atop a taxidermist's toolbox. And the image of a bald eagle is displayed up high in a case opposite the mounted mastodon.[42] But — most mysteriously — what is that huge skeletal framework behind the curtain, ah, it must be the Mastodon — eleven feet high at the shoulders and fifteen feet long — nearly unveiled! Four pillar-like legs are visible, but curiously not the head or tusks. Presumably then, the tusks must have been placed in skull sockets in an *upraised* orientation, indicative of how "in flux" were perspectives on the creature's natural history at that time? We'll return to this curious matter shortly. Interestingly, the Mastodon was not exhibited in the room as shown in the famous 1822 painting. Instead, occupying that position in the "Long Room" of the museum were situated cases of small specimens, minerals and rocks. The mastodon skeleton was actually displayed in the adjacent Mammoth Room.

The second image of relevance here is Charles W. Peale's painting *Exhumation of the Mastodon* (1806–1808), commemorating his heroic 1801 excavation of the mastodon at John Masten's Newburgh farm in New York. According to Semonin and art historian Linda B. Miller, this painting repre-

In 1955, the United States issued a commemorative postage stamp honoring Charles Willson Peale. This was a rendition of Peale's fine art self-portrait, "The Artist in His Museum" (1822), showing a hallway in his Philadelphia museum where his mastodon was exhibited (behind the partly raised curtain) during the early 19th century.

sents "the birth of the monster itself."[43] Here, we see American ingenuity at work; the Peales are resourcefully relying on a pumping machine[44] devised for the purpose of drawing ground water from an open twelve-foot-deep marl pit excavation to facilitate the workers' search for mastodon bones. Only a diagram of the mastodon's leg bones is visible in this painting (i.e., as the monster hasn't been born yet or, rather, reconstructed through man's ingenuity). Viewers spy a tempest on the horizon, a developing thunderstorm at the right of the frame, which art scholars claim signifies the sublime, or a Romantic natural perspective toward American wilderness and its antiquity. As Semonin stated, "*The Exhumation of the Mastodon* depicted the marl pit at John Masten's farm as the womb of American ingenuity and industry as well as the mastodon's birthplace."[45] (Note, the lightning streak may also suggest Indian folklore and beliefs surrounding the big buffalo, or even Franklin's experiments with electricity — a striking association which would soon be made in Mary Shelley's rather famous novel of 1818 with its undying monster, *Frankenstein, or the Modern Prometheus*.) Through this painting, fortifying the importance and relevancy of nature studies, Peale united man with the natural world.[46]

The third image of significance here is Rembrandt Peale's 1803 study of the head of the American *Incognitum* with downturned tusks published in his *Philosophical Magazine* report, based on the famous jaw collected at Peter Millspaw's farm (reproduced on p. 336 in Semonin.)[47] This wasn't the first drawing showing a mastodon head reconstructed with tusks.[48] For instance, in 1801, Rembrandt had already illustrated an excellent and accurate skeletal reconstruction showing correctly positioned tusks. But his 1803 rendition is most likely the earliest scientific diagram known revealing the head of our carnivorous national emblem with tusks especially oriented to accentuate its alleged fierce disposition — looking like enormous fangs or sabers (as in the prehistoric feline *Smilodon*, which hadn't been discovered yet).

From the Peales' collection of mastodon bones made principally from three New York localities, two skeletons were reconstructed. Missing parts were carved in wood by artisan William Rush. Cork cartilage was used at bone junctures, increasing the animal's size (and majesty), although Caspar Wistar "took vigorous exception" to this practice.[49] The more complete skeleton from Masten's farm was missing the upper part of the skull; an elephant-shaped papier-mâché skull cap therefore crowned the head of the specimen exhibited at Peale's Philadelphia Museum. A horizontal red line on the specimen demarcated where real bone ended and reconstructed material began. Vertebrae at the shoulders were too highly arched, too bison-like rather than elephantine.

Rembrandt Peale, who had assisted with the New York excavations, documented his theory of the mastodon (referred to in his publications as a "Mammoth" for aforementioned reasons). While he believed the animal was extinct, father Charles thought it was extant. With one "Mammoth" skeleton on display at the Philadelphia Museum, the other was exhibited in New York City and then England; plans were made to also visit Paris. The European trip, with immense, carnivorous "Mammoth" in tow, was intended as a triumphant victory tour, demolishing Buffon's decades-old views of American degeneracy. Further opposing the French scientists Daubenton and Buffon, Rembrandt also offered keen insights into the mastodon's global occurrences relative to the Siberian mammoth's, species which he believed were distinct from one another.[50]

One museum patron during the 1802 New York leg of the exhibition was early American wood-engraver Alexander Anderson (1775–1870), who prepared a popular wood engraving of the skeleton, showing tusks curving upward (but erroneously jutting from the eye sockets). A similar version of this illustration was published in British ornithologist Thomas Bewick's (1753–1828) *A General History of the Quadrupeds: The Figures Engraved on Wood, Chiefly Copied from the Original of T. Bewick, by A. Anderson ... with an Appendix Containing Some American Animals Not Hitherto Described* (1804).[51] One early illustration of Peale's mastodon indicates that for at least some period of time, the skeleton displayed in Philadelphia's Hall of the American Philosophical Society may have been displayed *without* tusks, this outcome at the cautious urging of Caspar Wistar.[52] But Anderson also completed (at least) two additional, undated engravings, one showing tusks raised, and another with tusks curved downward to enhance the carnivorous aspect of its alleged natural history. Shortly after crossing the Atlantic, Rembrandt would angle the tusks downward.

Now ensconced in England, but to his father's chagrin,[53] Rembrandt commenced to writing about the beast. It is apparent that by 1802 Rembrandt was shifting his views of Mastodon's natural history.[54] One wonders if he may have been swayed in his opinions by any of the learned men of British society, possibly including Sir Joseph Banks (1743–1820), who took particular interest in Peale's reconstruction. Caught between the unforgiving rock of paleontological science and the hard place of trying to earn a wage overseas from pop-cultural exhibition of his monster, Rembrandt was spinning his views of the creature, rapidly. His reconstruction of the creature had become a moving target. Friedrich Blumenbach disagreed with Rembrandt's conclusions as to the mastodon's diet, while English anatomists protested the idea of its extinction and its curiously reconstructed natural history. For in 1802 Rembrandt wrote, "It's the opinion of many, that these tusks might have been reversed in the living animal, with their points downwards; but as we know not the kind of enemy it had to fear, we judged only by analogy in giving them the direction of the elephant."[55] Clearly by early 1803, however, Rembrandt's ideology had further transformed, as he subsequently wrote,

> I am decidedly of the opinion, since it cannot be contradicted by a single proof or fact, that the mammoth was exclusively carnivorous, ... that he made use no use of vegetable food, but either lived entirely on flesh or fish; and not improbably upon shell-fish, if as there are many reasons to suppose, he was of an amphibious nature.... But whether amphibious or not, in the inverted position of the tusks he could have torn an animal to pieces held beneath his foot, and could have struck down an animal of common size, without having his sight obstructed, as it certainly would have been in the other position.[56]

An amphibious Mastodon? Well, yes, as for now Rembrandt a walrus paradigm proved "a-Peale-ing" for his beast, fleshed-out much further in his next publication.[57]

Citing an area of the mammoth's under-jaw bearing a "most extraordinary roughness," Rembrandt stated this morphological feature resembles more the analogous bone in a walrus than in a modern elephant. Could it, perhaps, have possessed a trunk? The roughened skeletal area indicated that in life it must have supported "some unusual and immense appendage ... some powerful protuberant cartilagenous [sic] instrument for the purpose of taking up his prey, whether, like the Elephant, it was a nose elongated; or like the Walrus, it was a large and powerful lip; or like the ant-eater, it was a long and powerful tongue."[58] While dispensing with past suggestions that the carnivorous-looking teeth were hippo-like or pig-like, he stated Mammoth's teeth were unique in nature. Then Rembrandt addressed the reason for reorienting the tusks.

> When the skeleton was first erected, I was much at a loss how to dispose of the tusks; their sockets shewed [sic] that they grew out forwards, but did not indicate whether they were curved up or down. I chose, therefore, first to turn them upwards, not because they produced the same effect as in the Elephant; for it is evident they could not, from the different angles between the sockets for the tusks and the condyles of the neck (as before remarked); the horizontal position of which in the mammoth, together with the great curve of the tusks, would elevate them too high into the air, directing them backwards, twelve feet from the ground; so that they never could have been brought sufficiently near the ground for any kind of purpose. This position was evidently absurd; and there is infinitely more reason in supposing them to have been placed like those of the *Walrus*, and probably for a similar purpose.... The tusks which were ... very much worn at the extremities, and worn in so peculiar a manner as could not have happened in an elevated position; unless on the absurd supposition, that the animal *amused* himself with wearing and rendering them blunt, by rubbing them against high and perpendicular cliffs of rocks. This, in a state of nature, can never be supposed, whatever habits may be acquired when in a narrow confinement. There can be no doubt, then, of their having been *used* against the ground; and not improbably in rooting up shell-fish, or in climbing the banks of rivers and lakes.[59]

Peale was also back-pedaling from Turner's former position. "When it has been said of the mammoth that it must have been carnivorous, the word was not intended to convey the idea of his being a *beast of prey*, like the tiger, wolf, &c. but that his food must have been *animal,* because all vegetables (except fruit) require peculiar instruments to file, bruise, or grind them, totally unlike the teeth of the mammoth."[60] For, "such animals ... could not well escape him and would not require much artifice or speed to be caught."[61] And so after dining on shell-fish, turtles, fish or other lacustrine animals, the Mammoth could have also used its down-turned walrus-like tusks to grip mud and sediment, like crutches, as it ascended from steeply inclined river banks.

While after accessing the Continent the Peales had further intended to

metaphorically unleash their monster upon France, due to the mounting possibility of war soon to be waged between England and France, Rembrandt
never had that opportunity. However, a copy of his *Historical Disquisition on
the Mammoth* did reach Cuvier; by 1806, which would have the impact of facilitating the latter's more exacting scientific identification of the former *Incognitum* (to be discussed shortly). Nevertheless, as documented by an
illustration published in Frenchman Edouard de Montule's *Voyage en
Amerique* (1821), the mastodon's tusks would remain down-turned "for more
than a decade"[62] in the Philadelphia Museum, until around 1822.

Evidently another paradigm entered the picture following physician John
Godman's (1794–1830)— importantly, whose father-in-law was Rembrandt
Peale —1820s observations of the skeleton. Godman stated in 1826 that the
tusks had been mounted erroneously. "This position is certainly unnatural,
as Cuvier has clearly shown. Nothing therefore can justify us in placing these
tusks otherwise than in the elephant, unless we find a skull which has them
actually implanted in a different manner."[63] A lithograph prepared by Alfred
J. Miller in 1836, however, showing mastodon on display in Rembrandt Peale's
Baltimore Museum in 1836, reveals tusks in proper, elephantine position,
"although many Americans continued to see the mastodon as a ferocious
monster that dominated the natural world."[64]

It remains to discuss how and when America's carnivorous monster came
to be recognized as herbivorous, and then, briefly, its early pop-cultural
impact during this period.

To Georges Cuvier, creatures out of prehistory were less monstrous than
their terrible worlds, so often wrought in catastrophic upheaval. First, in a
pivotal 1796 lecture titled "Memoir on the Species of Elephants, Both Living
and Modern," presented to the French National Institute (and with its subsequent publication), Cuvier established that there were two *living* species of
elephants *and* a third extinct variety known only from fossils. This third genus
was Siberia's woolly mammoth, which by 1800 had become a reasonably well
known creature known from numerous assorted fossils, complete skeletons
and even frozen cadavers clothed with fur.[65] Cuvier suggested that the Siberian Mammoth had been suddenly extinguished by "some revolution of the
globe." Now there could be no doubt that the Woolly Mammoth was an
extinct elephant, which like modern Indian and African elephants was also
herbivorous.

Yet, Cuvier reserved his thoughts on America's *Incognitum*. Much like
when Jefferson erroneously pondered the natural history of the Megalonyx,
while Cuvier coincidentally — accurately — pronounced on the *Megatherium*,
Rembrandt Peale's erroneous speculations on the New York mammoth skeletons were being filtered and re-crystallized through Cuvier's meticulous
investigations. Then, by 1806, casting aside Rembrandt Peale's speculations,

Cuvier demystified our mastodon in a series of papers, one having the lengthy yet descriptive title "On the great mastodon, an animal very close to the elephant but whose molars are studded with large bumps, whose bones are found in various places on both continents, and especially near the banks of the Ohio in North America, incorrectly called *Mammoth* by the English and by the inhabitants of the United States."[66]

Like Rembrandt Peale had already concluded in his *Disquisition*, Cuvier reasoned that mastodon and Siberian mammoth were distinct from one another. Cuvier had reservedly refrained from entering furor concerning the mastodon's proper diet; the impression he cast in 1806, however, was that the Ohio animal was essentially an extinct elephantine form. Might one therefore not deduce that mastodon *wasn't* a carnivore? And so, years before its stately tusks had been reversed into correct orientation in hallowed museum hallways, the solution to the problem of mastodon's proper diet as well as its natural history took place without fanfare, although primary credit is usually granted to Cuvier. However, arguably, University of Pennsylvania American physician and botanist Dr. Benjamin Smith Barton (1766–1815), one of Cuvier's correspondents, may have solved the riddle of Mastodon's dietary habits, empirically in 1805.

One has to dig a little to find when Mastodon became herbivorous. A little publicized discovery is mentioned in Louis Figuier's (1819–1894) popular work on geology, *Earth Before the Deluge*, appearing as a French first edition in 1863. Here Figuier stated that, following discovery of the Peale's Mastodon skeletons, Barton discovered in-situ in a Withe [*sic*], Virginia salt lick,

> at a depth of six feet in the ground, and under a great bank of chalk, bones of the Mastodon were found sufficient to form a skeleton. One of the teeth found weighed about seventeen pounds; but the circumstance which made this discovery the more remarkable was, that in the middle of the bones, and enveloped in a kind of sac which was probably the stomach of the animal, a mass of vegetable matter was discovered, partly bruised, and composed of small leaves and branches, among which a species of rush has been recognized which is yet common in Virginia. We cannot doubt that these were the undigested remains of the food, which the animal had browsed just before its death.[67]

It was botanist Barton who supplied Cuvier with this information on October 14, 1805, via letter. Writing in an ingratiating tone, Barton excused himself for taking up Cuvier's valuable time with his missive, then proceeded to tell him of the amazing discovery. In his own words, Barton — recognizing the plants associated with remains of the "American Mammoth"— stated,

> Very lately, digging a well, near a salt-lick, in the county of Wythe, in Virginia, after penetrating about five feet and a half below the surface of the soil,

the workmen struck upon the *stomach* of one of those huge animals, best known, in the United States, by the name of the Mammoth. The contents of the viscus were carefully examined, and were to be "in a state of perfect preservation." They consisted of half-masticated reeds (a species of Arundo, or Arundinaria, still common in Virginia, and other parts of the United States), of twigs of trees, and of grass, or leaves.—"There could (says my informant) be no deception on the subject. The substances were designated by obvious characters, which could not be mistaken, and of which every one could judge: besides, the bones of the animal lay around, and added a silent, but sure, confirmation."[68]

There's no mention as to whether those characteristic mastodon *molars* were found associated with the skeleton or stomach contents though.

Barton continued, speculating that other soft part remains of the "American Mammoth" had been preserved and discovered on into recent times. Preservation had been favored by impregnation of the animal's body by saline solutions. Barton concluded, "As to myself, I have always leaned to the opinion that the Mammoth was an herbivorous animal." This, despite the fact that in "...a conversation which I had with that truly ingenious man [i.e., Mr. John Hunter], in the year 1787, on the subject of the mammoth, he observed to me, in a style rather authoritative, 'that the *Incognitum* had, certainly, been a carnivorous animal.'" Barton also believed that North America's fossil elephant ("*Elephas*") was the "exclusive domain of the Mammoth" (i.e., American mastodon).[69]

Coupling knowledge of Mastodon's skeletal anatomy and dietary remains, it must have been straight forward for Cuvier to view the former *Incognitum* as another extinct variety of elephant. It is documented that Barton also shipped a mastodon molar to Cuvier in 1806. (Cuvier also noted that mastodon teeth somewhat resembled human molars.)[70] The two naturalists finally met in Paris in April 1815, only months before Barton's death.

Simpson has noted that "one finds no less a man than Benjamin Smith Barton contributing to vertebrate paleontology not by any publications of his own but by gathering 'intelligence' and transmitting it to his paleontological correspondents, among whom was Cuvier."[71] Barton's letters were published in an 1814 volume, cited in Simpson. So perhaps primary credit for proper understanding of America's Mastodon should go to Rembrandt and Charles W. Peale, as well as Benjamin Smith Barton, with an armchair assist from Cuvier. (Or, recalling, Jefferson's and Buffon's feud, is this too chauvinistic? Let's bear in mind though that most naturalists of the time relied on workmen, informants, and other discoverers and correspondents for the fruits of what later became recognized as scientific knowledge.) When Jefferson retired from formalized studies of natural history in 1811, he passed the baton to Barton — the man who unmasked our symbolic monster, destroying the alleged carnivore.[72]

Cuvier illustrated his 1806 paper with a diagram drawn by British anatomist Everard Home (1756–1832) showing Peale's mounted mastodon, *sans* tusks, principally founded on the reconstruction delivered to England in 1802. Later, in 1852, a similar image of Peale's mastodon (although in reversed pose) appeared in American paleontologist's John Collins Warren's (1778–1856) *The Skeleton of the Mastodon Giganteus of North America*. Although he noted that Peale's mastodon had "disappeared several years earlier without any authentic account of its whereabouts,"[73] according to Paul Semonin, when the Philadelphia Museum faced bankruptcy in 1848 it was sold to a Darmstadt, Germany, museum where it remains on exhibit today, with correctly oriented tusks. Following its Baltimore exhibition between 1814 and 1845, Rembrandt's mounted Mastodon specimen was eventually acquired by the American Museum of Natural History.

As stated by Arthur Perry Latham, in 1855, "In these days, Geology became the rage. It was talked on every steamboat and canal pocket, and at every public watering hole."[74] Intriguingly, even before the Peales had erected skeletons of the mastodon, discovery of its grinders on the Hudson River valley farm of the Rev. Robert Annan prompted a visit in December 1780 from General George Washington, who collected Bone Lick fossils![75] Discovery of the New York skeletons attracted publicity. Consider, for example, the outcome of Rembrandt Peale's search for the *Incognitum*'s elusive, yet most symbolic lower jaw, as workmen probed into the morass with

> long-pointed rods and cross handles; after some practice, we were able to distinguish by the feel whatever substances we touched harder than the soil; and by this means, in a very unexpected direction, though not more than twenty feet from the first bones were discovered, struck upon a large collection of bones, which were dug to and taken up with every possible care. They proved to be a humerus, or large bone of the right leg, with the radius and ulna of the left, the right scapula, the atlas, several toe-bones, and, the great object of our pursuit, a complete UNDER JAW!
>
> After such a variety of labour [*sic*] and length of fruitless expectation, this success was extremely grateful to all parties, and the unconscious woods echoed with repeated huzzas, which could not have been more animated if every tree had participated in the joy. "Gracious God, what a jaw! How many animals have been crushed between it!" was the exclamation of all: a fresh supply of grog went round, and the hearty fellows, covered with mud, continued the search with encreafing vigour [*sic*]. The upper part of the head was found twelve feet distant.[76]
>
> Grog and fossil bones—it doesn't get any better than this!

Even as late as 1852, the discovery remained unsurpassed as to the "intense interest" it had stimulated in the Newburgh, New York, region. As paleontologist Ellis Yochelson declared in 1991, "In displaying to the public the second and third mounted fossil specimens in the history of vertebrate paleontology, C. W. Peale created a tremendous impact. From the published accounts, the height of the dinosaur craze of today is far milder than the 'mammoth' fever which gripped Philadelphia. The crowds stood in line for hours and paid the enormous sum of fifty cents to view this creature."[77] Mammoth fever swept Philadelphia, and natural history-supporting Thomas Jefferson suffered the well-deserved moniker of "Mammoth President." The mammoth, "The largest of terrestrial beings ... the ninth wonder of the world," was indeed a marvelous pre–Barnum spectacle. Advertising billing the new "Antique Wonder of North America," dramatically highlighted the animal's ferocity and size. For our American Monster was "huge as the Precipice, cruel as the bloody panther, swift as the descending Eagle and terrible as the Angel of Night."[78]

In February 1802, before the advent of his not so financially successful trip to England, Rembrandt Peale gave the second skeleton a marvelous send-off, showcasing a banquet in the belly of the carnivorous mastodon itself, a symbolic act of (Americanized) carnivory! Thirteen individuals sat down to dinner under the thorax of the beast, (one of them inventor and artist John I. Hawkins father of famed British artist Benjamin Waterhouse Hawkins (1809–1889), who staged a similar yet more famous grand stunt of his own on the Crystal Palace grounds at Sydenham, England, over half a century later).

Joseph Smit's 1893 restoration of the mastodon appeared very elephantine; a century following Cuvier's early detailed analyses, mystery surrounding the *Incognitum* had been relieved (from H. N. Hutchinson's *Extinct Monsters*, 1893 ed.).

By 1833, three decades before publication of Jules Verne's *Journey to the Center of the Earth* (1864) — with its scenes featuring mastodonts living at the Earth's central sea — the mighty mastodon had even entered the annals of fantastic literature as a creature of mystery inhabiting a prehistoric polar setting in Swiss draftsman Rodolphe Toeppfer's (1799–1846) novel, *Voyages et Aventures du Doctor Festus.*[79] Many life restorations of the mastodon would follow, although by the late 1860s, interest and intrigue surrounding this creature — then known to be herbivorous and elephantine in appearance — was on the wane, becoming supplanted by a burgeoning dinosaurian craze.

Elephantine mastodons were restored by artist Edouard Riou (1833–1900) for 1860s editions of Figuier's *Earth Before the Deluge.* By 1868, when Benjamin Waterhouse Hawkins sketched restored fossil vertebrates projected to appear in a "Palaeozoic Museum" display he was designing for New York's Central Park, the extinct pachyderm's herbivorous nature was implicit in his restoration. By then and increasingly with each passing decade, America would celebrate its esteemed savage past through the massive, toothy jaws of genuine prehistoric, reptilian super-carnivores.

Nevertheless, with mastodon's popular appeal ostensibly out there for public consumption, by the early 1840s, the stage was set for America's mightiest Leviathan-Behemoth yet conceived.

FOUR

Behemoth and Leviathan — The Beauty of His Beasts

The Missourium, as I have described it, was a creature of enormous magnitude, ferocity and strength, as well as fleetness in swimming; and by reason of his great weight and strength, could attack the largest animals with impunity.
— Albert Koch, *"Description of the Missourium, or Missouri Leviathan,"* 1841

By the 1840s, the mastodon's alleged nasty disposition became established on the strength of imaginative recreations conceived by one attuned to Nature's most improbably colossal denizens of land and sea. The genus Missourium, or the Missouri Leviathan — a larger-than-life mastodon-like animal, which had at least three assigned species names (M. theristrocaulodon, Leviathan Missourii, and M. kochii), was found and assembled by 19th century archaeologist, amateur paleontologist and museum proprietor Albert C. Koch, who also later gained undesired notoriety for his 114-foot-long fossil skeletal display of a sea serpent (i.e., Hydrarchus sillimani). While Hydrarchus (or "water king") infiltrated late 20th century popular accounts of the history of paleontology, Missourium remained a shadowy figure. With such startling, larger-than-life fabrications in tow, Koch anticipated the immensely popular cinematic dinosaurian icons of a century later. For his make-believe Missourium and Hydrarchus were intellectual antecedents of artificial prehistoria like King Kong and Godzilla, which we begin to explore in chapters nine and ten. In essence, Koch was the original "special effects" master.

Albert Koch has been referred to as a perpetrator of frauds, a forger, a charlatan showman in the mold of P. T. Barnum, and a scoundrel. And yet because his 1841 booklet *Description of the Missourium*[1] is written in a factual tone and seems entirely well intended, to what extent can these allegations be true?

Not much is known about this "man of great enterprise" today. Koch, a German who evidently had little scientific training, came to the United States in 1835. He settled in St. Louis and became a peddler of old bones, forming a fossil dealership. Such commercial enterprise was scorned by scholars who instead valued fossils and artifacts for scientific reasons. Koch traveled the American south digging bones, especially in Missouri and Alabama. He toured the country, England and Europe with museum displays; then, after initial public interest in these displays waned, he sold the exhibits to museums. By 1857, Koch became chairman of the Academy of Science of St. Louis's committee on comparative anatomy. Most popular accounts claim he died in 1866.[2]

Besides an unsigned newspaper article announcing his discovery of human artifacts found in conjunction with Ice Age mammal bones[3] and self-published booklets describing Missourium and Hydrarchus, Koch also wrote about his experiences.[4] By 1844, Mr. Albert Koch evidently began referring to himself as Dr. Koch. During his latter years he claimed to be a professor of philosophy. For a time his unsold collections were maintained at Chicago's Wood Museum where they were destroyed in the great fire of 1871. According to Carl Zimmer,[5] Koch died of a "lingering torpor of the liver, and later, when workmen disinterred his body for reasons unrecorded, they were shocked to find that it had petrified." The old fossil had actually become one!

But how did this amateur paleontologist become excommunicated by the scientific community? It started with Missourium, a poorly reconstructed creature for which Koch proposed as the genuine biblical creature — "Leviathan," incarnate. In his *Description of the Missourium*, Koch also referred to Indian traditions. Native Americans who formerly lived in this region had encountered skeletons of Ice Age mammals, creating lore explaining their occurrence long before Koch's arrival. They spoke of terrible battles waged between huge monsters along the Missouri and Mississippi rivers. Afterward, the natives burned the beasts' bloodied bodies, sacrificing them to the Great Spirit. This became an annual ceremonial rite. Then, in 1839, long after settlers occupied the fertile, sacred land of Kimmswick, a mastodon tooth turned up, leading to Koch's initial examination of the area in March 1840.

The bones of Missourium, excavated in 1840, were very real, not fabricated. However, Koch erred in compositing the bones of several Mastodons together into one enormous skeleton having an excessive number of vertebrae and ribs. This 32-foot-long, 15-foot-high monster was first displayed in St. Louis. Thereafter it was shipped to Louisville and then Philadelphia. Koch's Missourium was probably the first paleontology exhibit to travel through America.

Koch explained in his 1841 account that "the bones were found ... near the shores of the river La Pomme de Terre, a tributary of the Osage river, in Benton county, in the state of Missouri." Koch deduced from stratigraphy that Missouri Leviathan lived long enough ago for 6 or 7 types of strata to be subsequently deposited, with plant remains recording shifting environments. Therefore, Missouri Leviathan was perceived as a tropical animal which lived when the La Pomme de Terre was deeper (at least 46 feet) and much wider (about ¾ of a mile wide). It seemed to Koch that the mighty beast died during late summer or early fall.

Realizing that size does matter, Koch advertized the gigantic stature of his horned (i.e., tusked) beast. Missourium was about one foot taller than the tallest mammoth now known, and, owing to the extra vertebrae, about 10 feet longer than the largest mastodon. Koch listed several morphological features distinctive to Missourium, including a protuberance on the lower part of the jaw, and, most striking, its extra ribs, totaling 48 in number, 10 more than the American mastodon. The Missouri Leviathan supposedly did not have an elephant-like trunk and was aquatic; "it would be utterly impossible for him to exist in a timbered country ... and possessed ... like the hypopotamus [*sic*], the faculty of walking on the bottom of rivers, and ... [rising] occasionally to take air." It ate softer substances, having maxillary bones that were proportionally smaller than the mastodon. Curiously however, Koch stated that "His food consisted as much of vegetables as flesh, although he undoubtedly consumed a great abundance of the latter." Like the giant ground sloth *Megatherium*, Missourium was believed to have been protected by bony armor. Furthermore, the "singular position of the tusks has been very wisely adapted by the Creator for the protection of the body from the many injuries to which it would be exposed while swimming or walking under the water." Koch thought the toes on all 4 feet were webbed. To my knowledge, no life restorations of the Missourium were done.

Broadside illustrations contrasted the huge skeleton standing beside a diminutive African elephant. During its 1843 stay in Dublin, Ireland, Koch claimed,

> This unparalleled Gigantic remains, when its huge frame was clad with its peculiar fibrous integuments, and when moved by its appropriate muscles, was Monarch over all Animal Creation; the mammoth, and even the mighty Iguanodon may easily have crept between his legs, and now is universally acknowledged by all the European and American men of Science to be the greatest Phenomenon ever discovered in natural history. On viewing this vast relic, which after lying prostrate in the bosom of the Earth for Thousands of Years, now standing erect in all its grandeur, the beholder will be lost in wonder and astonishment, at its immensity and perfect preservation.[6]

Who wouldn't pay to see that!

Back when fossilized vertebrates were more of a novelty, it was common for naturalists to interpret the old bones in relation to the Bible. So, much like paleontologist Jerry Paul MacDonald did over a century later in his book, *Behold the Behemoth: The Quest That Solved the Mystery of the Dinosaurs of*

Job (PaleoGenesis Press, 1999), Koch rationalized remains attributed to Missourium with a biblical creature — Job's "Leviathan."[7] After dismissing extant whales and crocodiles as being the iconic Leviathan described in verses of the 41st chapter of the Book of Job, Koch proposed that Leviathan instead was none other than the Missourium! Relying on rather twisted interpretations of biblical description Koch concluded, "I am satisfied that this [Leviathan] could have been no other animal, either fossil or living, heretofore discovered, than the Missourium."

Criticism began dogging the Missourium soon after it became proudly displayed in Philadelphia. Here, Dr. Richard Harlan (1796–1843), discoverer of the North American aquatic Eocene whale *Basilosaurus*, became greatly inspired by Koch's imaginative handiwork. During his October 15, 1841, presentation to the American Philosophical Society, Harlan commented on the Missourium, claiming, "There is now exhibiting at the Masonic Hall in Philadelphia, one of the most extensive and remarkable collections of fossil bones of extinct mammals which have hitherto been brought to light in this country." Harlan diplomatically suggested that Koch had erred in attaching the tusks in upside down fashion. Harlan evidently noted a few additional mistakes as well. However, undoubtedly further researches "would enable [Koch] to rectify these errors."[8]

Had Koch merely deluded himself, or, instead, was he deliberately trying to fool others? Or, was Koch's presumption that a new significant fossil animal had been discovered just a futile exercise in wishful thinking?

By late 1841, Koch's Missourium occupied a place of honor in London's Egyptian Hall in Piccadilly, where a great throng beheld the marvel. Missourium seemed a natural wonder. Now, however, it became scrutinized by Richard Owen (1804–1892), who was on the verge of defining his new term "Dinosauria." Owen was less forgiving in his assessment than Harlan. At a meeting of the Geological Society of London on February 23, 1842, Owen denounced Koch's Missourium, suggesting it was an improperly mounted mastodon. Koch defended his handiwork in an address to the Geological Society on April 6, 1842. Regardless of the behind-the-scenes squabbling, Missourium continued to attract crowds through the summer of 1843, when Koch decided to exhibit elsewhere.

After a show in Ireland, Koch returned to his native Germany with Missourium in tow. By May 1844, Koch pulled up stakes to return to the United States. On the return journey, he passed through London where he struck a deal with the British Museum. Even though Museum officials knew Missourium was incorrectly reconstructed, they realized there was at least one very complete composite mastodon present within the assemblage. So they purchased it for a $2,000 down payment, plus an additional $1,000, paid annually until his death. Thus, Koch earned a handsome royalty of $23,000 accumulated between 1844 and 1866.[9] British Museum curators stripped

Missourium of its excess skeletal pieces. The mastodon remains on display in the Fossil Mammal Gallery, one of the world's finest composited specimens.

One wonders today whether Koch actually had sufficient Mastodon material on hand to realize whether he had more than one specimen in his graveyard. For instance, the presence of three leg bones (say, all right femurs), and three (say, two left and one right) upper armbones in the same deposit would have indicated there were at least three mastodons present. Therefore, all those ribs and vertebrae shouldn't have been composited along a single vertebral column because they couldn't have belonged to the same individual animal. In his 1841 *Description*, Koch never accounted for the number of bones discovered. Instead, his first sentence revealingly focuses on Missourium's great size! ("This gigantic skeleton measures 32 feet in length and 15 in height.") So it could be that Koch simply may have become blinded to the beauty of his beast.

But it's not as easy to forgive Koch for a more deliberate, twice-committed forgery — the "Hydrarchus sillimani of 1845," which was really an overly composited skeleton of a fossil whale Harlan had already described as *Basilosaurus cetoides* (the "whale-like king lizard"). Perhaps Koch fully realized he was building an incongruously long (114 feet, no less) vertebral column from bones he had dug in Alabama deposits. Although he tried to get away with it, this imaginary construct would prove his downfall. During its whirlwind overseas tour of America and Europe between 1845 to 1847, outrageous Hydrarchus would be blasted by luminaries such as Richard Owen, Charles Lyell and Gideon Mantell (1790–1852). Furthermore, Yale University mineralogist Benjamin Silliman (1779–1864) was insulted that Koch had named a sea serpent in his honor and so Koch renamed it Hydrarchus harlani. (Harlan, who had recently died in 1843, couldn't object.)

The original Hydrarchus was purchased by King Frederick Wilhelm IV of Prussia for the Royal Museum of Berlin, although according to Herbert Wendt,[10] "the Prussian king was already suffering from softening of the brain, which probably made the deal easier." But soon, in 1847, Koch was back in Alabama building another 96-foot-long Hydrarchus. This second specimen met its fiery end on Chicago in 1871. Intriguingly, Koch's magnificent, semigenuine beasts— Mastodons and Hydrarchus— proved influential with American originators of paleo-fiction!

Fossil whale publicity became manifested in Herman Melville's (1819–1891) *Moby-Dick* (1851), one of America's greatest novels.[11] Science historian Dennis R. Dean noted that Melville even referred to the whale as a "salt-sea Mastodon."[12] Consideration of Melville's paleontological musings will circuitously lead us back to deserved consequences of Koch's infamy.

Melville, who had served aboard a whaling vessel, and, as manifest through his classic novel, may most certainly be regarded as a contemporary expert on the natural history of whales, delved into the immortality of whale

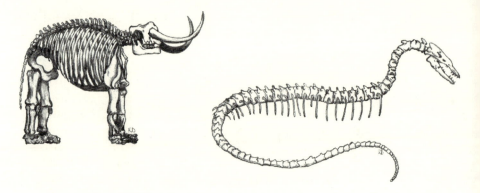

Albert Koch's prehistoric monsters were fictional because they were constructed from excessive numbers of fossil bones (e.g., vertebrae and ribs). At left, we see Koch's Missourium, while at right is his gigantic sea serpent, Hydrarchus (illustration, 2008, by Kristen L. Dennis).

races in two chapters. After proclaiming in chapter 104, titled "The Fossil Whale," that it is nigh time to magnify the whale by discussing "him in an archaeological, fossiliferous and antediluvian" context, through protagonist Ishmael, Melville asserts his "credentials as a geologist," mentioning former experiences as a stone-mason, ditch digger, digger of canals and wells, wine vaults, cellars and "cisterns of all sorts," implying that such exertions granted opportunity to encounter geological deposits in the earth. Next comes discussion of the Tertiary Period, which, as stated, connects all the fossil whales then known to science with post-antediluvian forms.

Melville tackles fossil whales then known to science, honoring Richard Owen, the British anatomist who realized that Harlan's 1832 description of *Basilosaurus* was incorrect. Realizing the fossil's mammalian (as opposed to allegedly saurian-reptilian) skeletal features, Owen renamed *Basilosaurus* as "Zeuglodon" instead (but Harlan's assigned name carried priority according to the rules of nomenclature). In 1839 Owen stated that the new animal was "one of the most extraordinary of the Mammalia which the revolutions of the globe have blotted out of the number of existing beings." Curiously, what Owen did with Zeuglodon is rather the opposite of what Baron Georges Cuvier — the only other paleontologist mentioned by Melville in "The Fossil Whale"— performed in the case of the "Maastricht animal" (discovered in 1770), thought by Petrus Camper (1722–1789) to be a fossil whale but which was subsequently pronounced by Cuvier to be a marine reptile, *Mosasaurus*.

In his short digression on fossil whales Melville seems reasonably informed about Zeuglodon's occurrences as well as a few other fossil whale discoveries made prior to Owen's assessment, and their scientific significance. Intuitively, he stated that the confusion created by *Basilosaurus*' skeletal anatomy nicely illustrated how, as in the case of modern species— particularly

Sperm Whales—whale skeletons (both extant as well as those of "departed species") offer "but little clue to the shape of his fully invested body." Relying on a distinct extrapolation of Charles Lyell's glacial theory, in "The Fossil Whale," Ishmael's thoughts glide "by a flood" back to an Ice Age time when (as he thought) the terrestrial regions of the globe would be uninhabitable, and "when the whole world was the whale's." Rather shockingly, Melville (through Ishmael) also speculated that humped whale species would most likely outlast man, suggesting that, unlike the case with the declining populations of humped American buffalo, whales would assuredly endure man's hunting practices. Furthermore, even though some paleontologists had theorized life had biologically degenerated since former geological ages (when species seemed in certain ways somehow more perfected) this wasn't so in the case of whales, as Melville claimed in chapter 105, titled "Does the Whale's Magnitude Diminish?—Will He Perish?" This is because, generally, modern adult whales only exceed in magnitude "pre-adamite" skeletal forms exhumed from Tertiary deposits.

In his book *The Raven and the Whale*, literary historian Perry Miller (1905–1963)[13] noted thematic relevancies between Mathews' odd publication and Melville's *Moby-Dick* (1851). First there is a revealing passage in Mathews' novella where warrior Bokulla spies a very whale-like Mastodon from afar in the waves, "sporting with the ocean." And in another of Melville's novels, *Mardi* (1849)—of which we'll discuss paleontological implications shortly—King Media makes associations between behemoths of land and sea, stating that "in antediluvian times, the Spermaceti whale was much hunted by sportsmen, that being accounted better pastime, than pursuing the Behemoths on shore." So perhaps its not too unreasonable to presume that Melville may have indeed had Cornelius Mathews' *Behemoth: A Legend of the Mound-Builders* in mind when writing paleontological passages in two of his novels a decade later.

Certainly by the late 1830s there were already a number of fictionalized paleontological writings of various flavors in print. According to Miller, most influential indeed to young Melville was Cornelius Mathews' (1817–1889) short novel, *Behemoth: A Legend of the Mound-Builders* (1839).[14] Mathews' school of thought and scholarly contribution was to facilitate transformation of American writing so as to distinctively and appropriately reflect its *American* identity. *Behemoth* exemplifies his intentions. At the time, the American creative writing style was regarded as inferior to the British. But in *Behemoth* the proud, idyllic, sublime majesty of America's western frontier landscape shines forth. Paleontology lingered at the heart of the young American literary movement.

Essentially, Mathews' "main design was to make those gigantic relics, which are found scattered throughout this country, subservient to the purposes of imagination." The author, "dared to evoke this Mighty Creature from the earth and striven to endow it with life and motion. Simultaneous and co-eval with this the great race [i.e., the mound-builders] that preceded the red men as the possessors of our continent, have been called into being." In other words, Mathews cautiously attempted something which relatively few had done before — he vividly related the story of an encounter between mastodons and (anatomically modern) paleo-men. Mathews was aware "of the great difficulty and magnitude of his undertaking." Accordingly, his tale turned out to be more allegorical and romantic as opposed to one purely founded on fact. Beyond mere author recognition, he hoped to have "accomplished some slight service for the literature of his country" by honoring our country's past archaeological "monuments." While in his imaginative novella, Matthew extolled the sublime majesty of America's restless frontier, *Behemoth* also reflects fears contemporary men may have harbored concerned savage nature and frightening primitives to be then encountered in the old, wild West, where hideous monsters from prehistory — metaphorical for what was then considered anthropologically primitive — dwell.

But what a whopper of a tale (or should I have said a whale of a tale?) is *Behemoth*! There is little effort to describe the Mastodon's physical appearance. Its size seems to fluctuate with circumstances as encountered by a great civilization of paleo-men, the mound-builders, led by chieftain Bokulla. From short descriptions we do know it eats vegetation and that it has a trunk and two tusks, but at times it seems half as tall as a mountain — or maybe just Godzilla-sized! It ferociously knocks over giant oak trees, buildings and city walls with reckless abandon, casting doom and gloom among the populace in the wake of its mighty footsteps. So many weapons wielded by an army of soldiers are brought to bear upon the Behemoth, the malevolent mastodon. Bokulla's mound-builders are under oath to destroy the monster, much like the *Pequod*'s crew were in the matter of the white whale's destruction. Yet the Mastodon seems invulnerable to everything, scattering the legions assembled by Bokulla. Although published prior to Koch's 1841 "Description," Behemoth behaves as if it were Missourium incarnate! Over a century before Toho released its influential film, *Gojira* (1954), Behemoth may have been the first *daikaiju* of fantastic literature. (See chapter nine for an in-depth definition of *daikaiju*.)

Partly the reason for why such liberties could be taken with a prehistoric vertebrate was because, as so keenly exemplified by Koch's monstrosity, Missourium, by 1839, questions still lingered as to the mastodon's skeletal anatomy. And there was no concrete evidence for the illustrious citadels and vast cities built by paleo-indians, of which certain tribes actually built funereal mounds. Mathews was simply suspending disbelief, vaunting the conjectured great and sublime majesty and awe of a relatively unknown heroic age out of the American past, that is, as known or shamelessly extrapolated from scanty evidence.

Most unusual measures are needed to dispatch the mighty beast, presumably the last of its kind. Eventually, the paleo-men wall the mastodon into a mountain valley meadow by blocking off the narrow pass leading into the valley with gigantic blocks of stone, where it eventually starves to death. Bokulla even proposes to adopt the monster as "our national idol." *Behemoth* concludes with a bit of theologically themed prose, not uncommon for its time. For on the fortieth day of confinement within the stony prison, "Behemoth died and left his huge bones extended on the plain like the wreck of some mighty ship stranded there by a Deluge, to moulder century after century, to be scattered through a continent by a later revolution, and, finally, to become the wonder of the Present Time."

Intriguingly, *Moby-Dick* wasn't Melville's only (nor his first) foray into things paleontological. In *Mardi*,[15] a rather humorous tale concerning an unusual sailing adventure to Polynesia, Melville served up a literary feast, Chapter 132, titled "Babbalanja Regales the Company with Some Sandwiches." When the party lands at the Isle of Fossils, which from a distance off "looks black as a whale's hump in blue water," they spy fossilized deposits, bizarrely and comically described, triggering discussion about paleontological history and geological processes. Here they see fossil objects, or a "stately banquet of the dead." "Like antique tablets ... ciphers ... bas-reliefs of beetles, turtles, ant-eaters, armadillos, guanos, serpents, tongueless crocodiles: — a long procession, frosted and crystallized in stone, and silvered by the moon." They also identify a (dinosaurian) three-toed footprint preserved in slate alongside ripple marks. Then geologically minded Babbalanja is urged to "expound these rocks," to explain "the origin of all the isles? how Mardi came to be?"

First outlined is a volcanic (plutonic) theory of geological origin. He claims "These were tombs burst open by volcanic throes." Continuing, the crucible that became Mardi "was charged with vapors, nebulous, boiling over fires volcanic. Age by age, the fluid thickened; dropping, at long intervals, heavy sediment to the bottom; which layer on layer concreted, and at length, in crusts, rose toward the surface. Then the volcano burst; rent the whole mass; upthrew the ancient rocks; which now in divers[e] mountain tops tell tales of what existed ere Mardi was completely fashioned. Hence many fossils on the hills, whose kith and kin still lurk beneath the vales." Here, Melville's interesting phrase, "kith and ken still lurk beneath the vales" is revealing for its pre–Darwinian publication, although Melville was probably never an evolutionist in that vein.

Relying refreshingly on food metaphor throughout the rest of the chapter, next Babbalanja contrasts the Plutonic possibility with a "celebrated sandwich system"—or a sedimentary Neptunian theory, offering a synopsis of Life's history. "Nature's first condition was a soup, wherein the agglomerating solids formed granitic dumplings which, wearing down, deposited the primal stratum made up of a series, sandwiching strange shapes of mollusks, and zoophytes; then snails, and periwinkles: — marmalade to sip, and nuts to crack, ere the substantials came." Those more substantial courses (i.e., "sandwiched" layer by ensuing layer — now take us from the primordial time up through the Pliocene. So, following, we read of the Old Red Sandstone "sandwich, clapped on the underlying layer ... imbedding the first fish," with Devonian *Cephalaspis* and *Pterichthys* offered as examples. Then there's the New Red Sandstone (Permo-Triassic) sandwich, "spread over with old patriarchs of crocodiles and alligators — hard carving these, — and prodigious lizards ... swimming in saffron saucers."

And if this "rare gormandizing" proves insufficient, there's further fare to whet your appetites. For the "Ool" or "Oily sandwich" comes replete with "fat old joints, and hams, and rounds, and barons of sea-beeves and walruses." Of course there were neither walruses (nor whales) living in Oolitic Mesozoic (Upper Jurassic) time, but how mouth-watering to partake of the "fillets and briskets, rumps ... haunches: shoulder to shoulder, loin 'gainst sirloin, ribs" dating from this period. Melville may perhaps be excused for another error in his Cretaceous menu of the "Chalk, or Coral" Period, in which he describes rich side-courses referred to "eocene, miocene, and pliocene." The eocene was noted for its "very savory" wild game; the miocene "second side-course" included "marine mammalia, — seals, grampuses, and whales, served up with sea-weed on their flanks, hearts. And flippers friccasied." Then the pliocene was rich in "whole-roasted elephants, rhinoceroses, and hippopotamuses, stuffed with boiled ostriches.... Also barbacued [*sic*] mastodons and megatheriums, gallantly served up with fir-trees in their mouths, and tails cock-billed."

Much of Babbalanja's amusing explanation is stylistically similar to Mark Twain's (1835–1910), as Melville was surely striving for humor in these passages from *Mardi*. Yet more significantly, Elizabeth S. Foster[16] felt such writings indicated that Melville meticulously researched geology texts prior to sprinkling in the geological snippets we find so entertaining today in *Moby-Dick* and *Mardi* (as well as in a few of his other books). But to what degree was Melville of the progressive geologist mind set?

Foster notes a revealing section of *Mardi* where Babbalanja states "But my ancestors were kangaroos, not monkeys.... Among the deepest discovered land fossils, the relics of kangaroos are discernible, but not relics of men. Hence there were no giants in those days; but on the contrary, kangaroos, and those kangaroos formed the first edition of mankind, since revised and corrected." In part, these are references to a then, contemporary debate concerning the *Chirotherium*, a New Red Sandstone creature described solely from its fossil tracks, resembling human hands with an extended thumb. The now invisible track maker was regarded in turn as a marsupial (e.g., large opossum or kangaroo), a cave bear, a great ape, a huge toad and later, in 1841 by Owen, as either of two forms of armored amphibian —*Labyrinthodon*—

walking with a crosswise gait. (This paleontological mystery remains unresolved.)

In this "kangaroo passage," Foster suggests that while Melville may have been "poking fun at the theory of the descent of man from earlier and lower forms, he was at any rate *considering* the idea of evolution."[17] (Reconsider the pointed meaning of his phrase "hence there were *no* giants in those days," for example.) We also catch a glimmer of his (for its time) rather precocious thinking in *Moby-Dick's* "The Fossil Whale," where Melville speculates as to similarity in forms between cetacean "intercepted links." However, before jumping to a seemingly all too easy conclusion here, two decades following publication of Charles Darwin's (1809–1882) *Origin of Species*, Melville dismissed Darwin as mere poetry. Melville carefully reflected upon paleontology's intriguing yet thorny ideology; personal speculations and inner conflicts concerning Nature, Genesis and Geology spilled over into his writings.

While it's not certain whether Koch ever read *Mardi* or *Behemoth*, intriguingly, the last portion of Koch's 1841 *Description of the Missourium* was devoted to "Evidences of Human Existence Contemporary with Fossil Animals." Both in October 1838, and also during his exhumation of Missourium fossils, Koch had noticed artifacts and evidence testifying that humans formerly coexisted with prehistoric (e.g., antediluvian) animals such as the mastodon.

In fact, Koch did "deem it my duty to lay before the world what facts I have been able to gather on this interesting subject, which will be strong evidence in favor of my belief, that there was a human race existing contemporary with those animals." On the bank of the Burbois River in Gasconade County, Missouri, a farmer discovered objects which Koch later identified as a stone knife and an Indian axe, as well as "several bones belonging to an animal of an unusually large size." When Koch disinterred the animal remains, he noted they had been burned. Koch even found a 6 to 12 inch thick ash layer situated 9 feet below the surface in which "Indian implements of war ... stone arrow heads, tomahawks, etc., etc." were mingled with 150 rocks, the latter thought to have been thrown as missiles by paleo-indian hunters. Completing the picture for his readers, Koch stated, "I found the fore and hind foot standing in a perpendicular position; and likewise the layer of ashes, so deep in the mud and water that the fire had no effect on them."

Rejecting a possibility that the bones had been consumed in a volcano-caused fire, the only logical conclusion was that humans had roasted the animal for food. In other words, the animal had apparently become mired in mud, killed by Indians and then cooked on the spot. (How intriguing that the fossil's burned condition matches Indian tradition.) Although according to Robert Silverberg, Koch speculated the mired animal had been a *Mastodon* (per Koch's Jan. 12, 1839, article in *The Presbyterian*), the kill specimen is not identified as such in his 1841 booklet. However, Silverberg further suggested that the unfortunate mired and eaten animal was instead a variety of ground sloth.[18]

Koch also found an "arrow-head of rose-colored flint, resembling those used by the American Indians, but of larger size," associated with Missourium's remains. This arrow was found lying under a femur bone, but in the same formation. In this stratum, yet 5 to 6 feet distant from skeletal remains, Koch noted 4 more arrowheads, three of which were co-deposited in the same formation. Koch documented that he had unearthed the arrowheads found with Missourium himself, and that they were "indisputably the work of human hands."

Unfortunately, Koch's very significant discovery, reported in *The Presbyterian*, went virtually unnoticed by the scientific community until the Spring of that year when Silliman republished the article in *The American Journal of Science and Arts*. Silliman inquired who the author of this "interesting and important" report may have been (because he didn't know Koch had written it and *The Presbyterian*'s article had no byline). Koch may have never learned that anyone found his work so illuminating. By the time Koch pronounced upon the coexistence of ancient man and extinct mammals (i.e., in his 1841 *Description of the Missourium*), scientists had already begun questioning the validity of the alleged genus "Missourium."

Stunts like Koch pulled with Mastodon and *Basilosaurus* bones were exemplary of what American paleontologists Joseph Leidy and Othniel Marsh (1831–1899) cautioned would burden American science with a dismal reputation. To real scientists intent on doing honest, credible work that could be verified by topnotch European scientists, quacks like Koch were unappreciated, only worthy of scorn. What profiteer Koch suffered is a harsh lesson, to be remembered in any age. Because he tried to dupe others (possibly) with Missourium and certainly with a second extraordinary fossil exhibit, Hydrarchus, Koch was branded a charlatan; later, serious-minded scientists were reluctant to credit Koch for a valid scientific discovery — proving that man coexisted with the likes of mastodon and ground sloth.

FIVE

Apish Antediluvia, or Darwin's Great Caveman Psych-Out!

Like all modern fauna, *Homo sapiens* (comprising all modern races) had a biological origin, an evolutionary ancestry and a geological past. Man most certainly had a common prehistoric ancestor who lived alongside other, now extinct mammalian fauna. While, to at least 20 percent of Americans,[1] these statements would ring true, far fewer would have comprehended such notions, deemed utterly preposterous over a century ago when the idea of *prehistoric* men derived from apes seemed sacrilegious and therefore most *monstrous*.

Why should cavemen be feared? After all, unlike many dinosaur species they weren't daunting by virtue of enormity. And while 19th century artists and certain paleoanthropologists have infused cavemen with ferocity, they were most likely less bellicose than ourselves. The mysterious extinction of at least one famous lineage (the Neanderthal, aka Neandertal, race) has been heavily contested over the years. Does Neandertal still dwell within us, and if so, rather ironically, should their noble genetic traits be most valued in modern society?

So, if cavemen weren't huge, fierce, or (in a genetic sense) even entirely extinct, what's the fuss?

While in comprehending the caveman phenomenon, *Star Trek*'s logical Mr. Spock might have stated, "Fascinating," instead, humans are emotional, extrapolating close knit, often self-serving theories about human origins from incomplete evidence. For, in essence, the caveman concept implies the abhorrent — undignified human bestiality and a disquieting association with apelike ancestors. Cavemen have represented Darwin's repugnant evolutionary ideals and even suggest civilization's tenuousness. Even though cavemen weren't huge like dinosaurs, they're depicted as hairy, naked, filthy and living in trees. Most alarmingly — they were Us! (Not that there's anything wrong with that.)

Prehistoric humans as prehistoric monsters? Because of those heartfelt human speculations — yes! Why? Consider.

Fear of the caveman lurking within, as ideas about them unfolded particularly during the late 19th century, stemmed from psychological elements. For the first time, unsuspecting civilized man confronted a deep, dark physical past — shockingly our own, so different from the trappings of prim and proper, lily-white Victorian society. The instinctive reaction to staring into antiquity's mystic mirror is to reject the unkempt, brutish face reflected therein; to deny that which is instilled within our souls, entrained within the genetic fiber of our being. And yet, fascinated, we are strangely drawn — like the moth to a flickering flame — to witnessing what was regarded as our worst, most monstrous nature through a variety of means, first in portraiture, later in museum reconstructions, then in contemporary fantastic literature and especially in film. Penetrating the veil, peering at such revealing restorations of hideous or uncouth ancestors many have felt, well, singed! "Get your damned hands off me and unleash my soul, you filthy apes," many have wanted to shout out at our accursed ancestry.

And so without further ado, arguably much scarier than any dinosaur, enter the Caveman!

The term *caveman* is nebulous due to inexactness, much as the reality of prehistoric man and hominids is fraught with scientific complexities. Just as one might use the term *dinosaur* to mean any of the hundreds of genera known to science, *caveman* is a relatively nonspecific term. But in a pop-cultural context the word *caveman* is not without pejorative meaning.

Today, *caveman* conjures an image of a hirsute, barefoot, club-wielding, primitive-looking man, wearing shaggy bear- or leopard-skin clothing, who lives in a cave shelter — the standard Hollywood version. He's already tamed fire, invented stone tools and heroically mastered the art of mammoth (or in cartoons and sci-fi films — dinosaur!) killing. And it's icy cold outside the cave, because, after all, it's the Ice Age. (Too bad our caveman hasn't invented shoes to keep his feet warm.) Yet any modern text on paleoanthropology will clarify that over the past 3.5 to 4 million years there have been many species of ancient humans evolving and surviving through the ages, linked genetically to ourselves. They didn't all live in caves, wield clubs or kill mammoths. And they never hunted (Mesozoic) dinosaurs or rescued Raquel Welch from gigantic pterodactyls. Yet life for them was precarious.

Over a century ago, we find that the public and many scientists lacked such knowledge, or sometimes rejected the implications of fossil evidence they possessed. Because it was all so emotionally jarring, so psychologically alarming, comprehension could only be taken a few steps at a time, and the first, most important stride of all was simply to prove man's geological (prehistoric) antiquity. Was man contemporaneous with prehistoric, extinct mammalian fauna? While the correct answer may seem so obvious today, even by the middle 19th century after numerous occurrences of ancient human bones and fossils had come to light, this remained heretical territory.

Actually, the concept of truly ancient men harkens back to the time when early scientists still pondered the possibilities of dragons and giants; the *idea* of fossilized men is as old as the scholarly study of fossils, winnowed in tandem with the emerging geological sciences. Not that these specimens were initially viewed as prehistoric monsters; the idea of a prehistoric past had yet to germinate. Rather than monstrous, like the remains of giants and dragons known to medieval naturalists, relics attributed to fossilized humans were considered genuine curiosities of natural history. Scholars probed for their true meaning, although especially in the early going misinterpretations resulted.

For instance, by 1726, Swiss naturalist Johann Jacob Scheuschzer (1673–1733) thought he had identified the incomplete remains of a man preserved on a stony slab who died in the biblical Great Deluge. Eight decades later Baron Georges Cuvier (1869–1832) proved the alleged Flood witness remains were instead bones of a Miocene salamander. When flint hand-axes were discovered in eastern England, John Frere reported in 1797 to the Society of Antiquaries that these dated from a "very remote period indeed; even beyond that of the present world." French savants professed that while the axes predated the biblical Flood's "sudden revolution," human history was still necessarily of short duration. Scientists also considered the bones of Guadeloupe Man, found embedded in a block of concretionary limestone on the Caribbean mainland of Guadeloupe in 1805. The partial skeleton was found with pottery fragments, stone arrow-heads and wooden ornaments. Did this individual possibly live before the "general cataclysm" which exterminated the great, extinct mammals? Another witness of cosmic wrath? In concurrence with Cuvier's theories, British scientists (correctly) concluded the remains were recent, dating from a time following the great mammalian extinctions, because the human bones weren't fossilized.[2]

As more specimens came increasingly to light, by the 1860s, scientists ran afoul of forgeries and natural perplexities, all of which conflated paleontologists' best intentioned efforts to shed light on fossil man, a subject which became exceedingly ticklish thanks to Charles Darwin (1809–1882).

It was easier then despite the degree of one's training to become misled or beguiled. On one hand, there were tempting specimens like the ancient human remains found along the Mississippi River near Natchez in 1845, intermingled with fossils of *Megalonyx* and mastodon. Although some pondered whether this represented evidence of a man who had lived with prehistoric fauna, Charles Lyell instead suggested that the human remains originated from an Indian grave. The human bones had eroded and fallen into a talus pile, co-deposited and re-buried alongside remnants of the pairing of far more ancient prehistoric beasts. Lyell was reluctant to accept Natchez Man as a denizen of the Pleistocene, although he stated, "Should future researches, therefore, confirm the opinion that the Natchez man co-existed with the *mastodon*, it would not enhance the value of the geological evidence in favour of Man's antiquity."[3] Recent radiocarbon dating has shown that the Natchez human bones are 5,580 years old, or post–Pleistocene in age.

Alternatively, when fossil huckster Albert Koch really did find evidence in 1838 of genuine mammoth-killing humans, colleagues were reluctant to believe. This was understandable because Koch was becoming notorious for tampering with fossil whale and elephant bones, creating fictitious prehistoric monster-sized skeletons for profit.[4] Most insidious but least convincing to the modern casual observer perhaps, was the reporting in 1869 of the discovery of a gigantic stone human, which became known as the Cardiff Giant. This curiosity was a human shaped figure carved by pranksters from a 12-foot-long block of gypsum rock, that was salted surreptitiously into a New York farm field.[5] Fraudulent fossils have certainly plagued the history of paleoanthropology.[6] For example, one such affair involving the alleged Piltdown Man, was so cleverly perpetrated during the 1900s that trickery wasn't deciphered for nearly half a century.

Realization that remnants of humans found in caves or grave sites weren't always historically recent or of an anatomically modern variety eventually took hold among the British and European scientific community by around 1860. In particular, naturalists' attention was drawn to the strangely shaped cranial domes and brow ridges of certain fossil men — perhaps the most famous of all — later named Neandertal (sometimes spelled "Neanderthal"). Science's introduction to Neandertal dawned in 1829 when physician and anatomist Philippe-Charles Schmerling (1791–1836) discovered their remains in Belgium's Engis Cave. Lyell, who eventually accepted the remains as proof of man's former coexistence with antediluvian fauna came decades later, at the time of his conversion stated apologetically of Schmerling's discovery that "I can only plead that a discovery which seems to contradict the general tenor of previous investigations is naturally received with much hesitation."[7] From 1848 onward, Neandertal's unsettling remains came increasingly to light, while their acceptance as early men as well as their evolutionary status came increasingly under fire.

However, rather than Neandertal's unimposing anatomy, the *ideas* generated by its former existence caused deep consternation, that is, when taken in context with Charles Darwin's and Alfred Russell Wallace's (1823–1913) theories concerning man's evolutionary descent from ape-like ancestors. After reading Darwin's stimulating book, *The Origin of Species* (1859), which clarified how organisms prone to survival through a process naturally selecting

fittest individuals of any species while exterminating lesser
varieties would over vast geological stages *evolve*, readers
wondered whether the outrageous theory applied to our
own species as well. Did man
evolve too? Brute-like Neandertal certainly didn't match
the contemporary public's perceptions of our prehistory.
Man may have at one time
been primitive, or survived in
a relatively uncivilized state,
but he'd always been, well,
manlike—surely not shaped
anatomically like the monstrous apes of Africa! Right?

Well, in his great book
Darwin waffled, concluding,
"Much light will be thrown on
the origin of man and his history."[8] You see, the time just
wasn't right to set the story
straight then, but from 1860
onward mankind faced his
progenitors coming to light—
monsters out of our own dim

A previously unpublished, original illustration
of savage, brutish Neandertal Man sketched by
Charles R. Knight (collection of Allen G. Debus,
photographed by the author).

ancestry. An ideological war spawned from public rejection of Neandertal
and evolution's related ape monsters.

Even following Lyell's aforementioned conversion on the matter of Neandertal's coexistence with prehistoric mammalian fauna, it was difficult for
many scientists to accept the true meaning of the fossil evidence rapidly mate-
rializing before their very eyes. In fact, Neandertal seemed more primitive
and monstrous for its apelike characteristics—seemingly an anthropoidal
man or troglodyte—following Marcellin Boule's (1861–1942) highly influential analyses (1911–1913) of the "Old Man" specimen. This was an adult skeleton found at La Chapelle-aux-Saints in 1908 (discovered in the same year as
the alleged Piltdown Man, clearly a setback year in the annals of paleoanthropology!); it later was judged as abnormal, crippled from arthritis. Boule
presided over the bones with scholarly intentions, yet he was decidedly wrong
in establishing this particular aged specimen as the type representing the
Neandertal species.

As Trinkaus and Shipman later opined, Boule "took the Old Man ... as the 'type' of Neandertals—not in a formal sense, but psychologically—and painted a detailed picture of Neandertal anatomy that became the received truth. And this truth showed Neandertals as terribly primitive and apish, in no way a possible ancestor of the glorious Cro-Magnons who followed them so quickly in time. To Boule, Neandertals could only be an extinct and remote relative of modern humans."[9] Boule's Neandertal was stooped over, bow-legged and bent-kneed. Neandertal's hips wouldn't permit upright posture, their spines were apelike and they suffered from severe mental deficiencies relative to Cro-Magnon. In short, Boule had created a monster. It wasn't until 1957, however, only a handful of years following Piltdown Man's debunking, when paleontologists proved Boule's Old Man skeleton was arthritically deformed and therefore that many of Boule's conclusions concerning the Neandertal race were erroneous.[10]

The history of paleoanthropology, archaeology or the evolutionary sciences isn't intended to be emphasized herein. Rather, pop-cultural fashioning of the caveman into a heralded prehistoric monster is the focus. Accordingly, consideration shall next be devoted to the progressive imagery of fossil man in pop-culture, ultimately leading into filmic renditions.

While common 20th century perceptions of prehistoric men projected from Boule's impressions of the Neandertal skeleton found in 1908 at La Chapelle-aux-Saints, we find that the practice of image-making of men set in reconstructed landscapes deemed prehistoric had been in vogue for centuries, long before the American Museum's Charles R. Knight and Dr. J. H. McGregor began, respectively, painting and sculpting their visages from skull fragments. In this regard, Stephanie Moser regards ancient portrayals of the mythological hero Hercules as highly significant because "Herakles embodies many of the qualities of primitive humans."[11] However, while the ancient Greeks pondered man's primeval origins and explored prehistory in mythological writings, more relevant here is the historically modern imagery and perceptions of fossil men viewed as monsters.[12]

Curiously, symbolic iconography which we immediately recognize in Hercules' attire and accouterments, as well as respond to today in conventionalized restorations of prehistoric humans, namely, clubs and other crudely fashioned tools, hairiness, clothing made from animal skin, cave accommodations underscoring the "wildman's existence outside humanity,"[13] evidence of culture and technology-building use of fire, and often a bare-breasted "Madonna-like" mother holding a prehistoric baby, were developed centuries ago. We do so respond rather in Pavlovian fashion to these innately understood symbols partly because, "iconic images ... retain their popularity and continue to be reproduced even when new data is found."[14] But also, "pictures can give us the 'pleasure of understanding the conditioning to which our imagination is subject.'"[15] Furthermore, these icons continued to be "used because they were quickly understood and it appears that the construction of a notion of the primitive or the prehistoric was essentially visual."[16]

In early views, after scholars defused belief in monstrous races of creatures such as the half-man, half-donkey,[17] prehistoric man became, in essence, equated to wildmen, humans living outside, or on the brink of civilization — such as in the American West, Africa or the Pacific islands, exotic lands where artists gleaned further scientific data honing their sense of prehistoric perspective. And so, especially after the 1500s, as artists refined their views of the stereotypical warrior of antiquity, the apparel and appearances of, say, Celtic warriors of the British Isles were, for example, often restored with tattoo markings, noted characteristics of Native Americans. These images represented an early form of visual archaeological restoration, speculatively blending many lines of evidence such as relics and classical source documents with field observations of foreign, primitive or even more barbaric cultures and people. And not unlike perceptions of mythical Hercules, an "icon of the savage but heroic warrior was a useful tool for projecting back into the more distant eras of human existence."[18]

At times, inspirations for prehistoric man imagery became downright weird. One Dutch writer, John Picardt, among the first to associate prehistoric humans with landscape ruins and monuments (i.e., of Drenthe in the Netherlands), published a book titled *Short description of several forgotten and hidden antiquities* (1660). Perhaps inspired by Athanasius Kircher's writings on giants, Picardt implied that modern humans were descended from real giants who dominated olden times.[19] A century later, artists went one step beyond.

Next, by 1777 we find artist John Strutt summarily copying and combining all the old popular restorations of ancient Britons and Germanic warriors as perceived by several historical authorities into single didactic *groupings* replete with background landscapes projecting prehistorical flavor. The authenticity of Strutt's engravings is paradoxical however, because individuals restored in each picture could never have all co-existed. To better understand this, it would be as if a modern artist combined five or six drawings of *Tyrannosaurus*, as that dinosaur had been scientifically regarded at various historical periods from 1905 through 2005, except all merged into a *single* drawing. Despite the seeming authenticity of such a diagram, perhaps not a single one of those individual, mutually exclusive representations would be accurate!

Certainly, repetition of icons (i.e., clubs, animal skin clothing, caves, etc.) in early representations of man illustrated in a primeval state helped affirm through visual cues what constituted a sense of the prehistoric. Just think about those Geico commercials televised during the summer of 2007, for example, where an unkempt caveman making his way through an airport spies a poster of himself wielding a club, wearing animal hide clothes adjacent to the printed words, "So easy a caveman could do it." Drawing information from primitive societies, three centuries ago (and before) there was also a similarly offhanded, insulting (politically incorrect) manner to the early representations of mankind — judged akin to savages of far away countries.

Evolution implied that the earliest humans on our lineage must have differed anatomically from the earliest anatomically modern humans (the earliest traces of which were discovered in 1868, i.e., Cro-Magnon Man, which first appears in the fossil record 40,000 years ago). Furthermore, distressingly, Victorian evolutionists insisted that man had a simian ancestry, comprising waves of ape-like men, or man-like apes living long before those Celtic warriors artists had been striving to accurately restore. It wasn't just that man's antiquity dated back to the time of the extinct mammalian fauna, a fact which had become generally accepted by the mid–1860s.[20] Now certain paleontologists suggested that this early human breed was as brutish and savage as the monstrous fauna they shared their primeval world with. And that humans were genetically allied to apes. And, what's more, there may have been different *species* of humans alive then and before, born of suspect human qualities. Who would stop the madness!

As popular science texts and other publications prospered during the mid–19th century, artists turned their attentions to man's dubious ancestral roots, for isn't seeing nearly believing? As Moser stated, toward the goal of conveying evolutionary theories of man's origins to the public, "By presenting the theory of human descent from apes in terms of long-established visual traditions of understanding, illustrators performed a vital task. Essentially what they did was make the unbelievable believable."[21] A weighty challenge was to overturn scripturally inspired artistic renditions of Man's divine origins, replacing these with imagery of hairy, half-witted ape-like monster men eking out bare means of survival in godless prehistory.

While French writer Louis Figuier wrote several editions of a lavishly illustrated popular text, *Earth Before the Deluge*, concerning geological history and featuring the world's paleontological marvels, he reserved opinions on the subject of man's antiquity, initially envisioned as Edenic. However, by 1870 he acquiesced on the subject of man's savage, primitive past. Now, as conveyed through Emile Bayard's sequence of restorations published in Figuier's *Primitive Man*, readers learned that man had staged through a series of technological ages of stone, bronze and iron.[22] Also, the earliest men were evidently caucasoid and lived much like the creatures they slew — as animals themselves. And, significantly, they waged a symbolic war with extinct beasts. As stated by Nicolaas Rupke, "it was as though the geological reconstructions were journeys of discovery into a past as savage and unruly as the contemporary colonies. The picture of savagery enhanced the Victorians' images of people as civilizing rulers."[23]

In his (post–Darwinian) popular paleontological writings Figuier refrained from presenting text or imagery featuring man anatomically differing from his present form. Yet while Figuier demurred on evolutionary principles, in 1861, only two years prior to publication of his most famous book, *Earth Before the Deluge*, fellow French writer, botanist and geologist Pierre Boitard's (1789–1859) *Antediluvian Studies — Paris Before Men* was published. Here, Boitard incorporated his own gruesome illustration, titled "Fossil Man," of a hideous cave-dwelling simian family, reflecting objectionable evolutionist views. Although Boitard rejected the basis of the hypothetical image, "The impact of the picture clearly lay in its shock value, being an extremely powerful statement on the primitive."[24] As translated by Rudwick, Boitard viewed his stone axe-wielding creation not as a noble savage (as Figuier would have preferred) but as a primeval monster, "the most singular and horrible animals."[25] Following Boitard and Figuier, beginning in 1873, artists increasingly fascinated readers with glimpses of the "ferocious-looking, gorilla-like," yet genuinely prehistoric Neandertal, deemed "savage" and "always ready for attack or defence [*sic*]."[26] Over the years, Neandertal's shape-shifting image would become, first — by 1909, more primitive, aggressive and animalistic, then, especially after 1957, more congruent with the modern human form.

Between 1887 to 1891, fossil newcomer *Pithecanthropus alalus* heightened consternation over man's troubling evolutionary origins.[27] And by 1908 with entirely fictitious Piltdown Man's ("Eoanthropus") addition to the pantheon of famous prehistoria, the suite of monster fossil men haunting modern man in popular culture through later decades of the 20th century was essentially complete. Latent, brooding fear of less civilized, savage races capturing and ravishing our women in the present also became a more prominent theme during this period as reflected in fantastic imagery of fossil men and prehistoric apes, but perhaps culminating with RKO's classic 1933 film, *King Kong.*[28]

Pithecanthropus, or the "speechless ape-man," conceived in 1876 from Ernst Haeckel's (1834–1919) theoretical notions as to how a human missing link predecessor should appear,[29] precipitated an illustration published in 1887 clearly borrowing from the African race's features — darker skin and fuller lips. Gabriel von Max's 1894 restoration resulted from Dutch anatomist Eugene Dubois' (1858–1940) discovery of Hackel's supposed link, publically introduced as Java Man. Here, a dismal, naked and hairy ape-like family seeks shelter in a cave; a female holds a suckling infant.[30] Max's eye-opening painting wasn't so terrifying to behold but the expressions of his cave people mirrored modern man's depressing outlook on our hidden origins.

Trinkaus and Shipman have stated that in the case of Neandertal (although certainly a statement generally applicable to other races of prehistoric humans as well), "Infuriatingly, the fossils do not speak for themselves. It is the examining scientists who bring them to life, often endowing them with their own best or worst characteristics. Each generation projects onto Neandertals its own fears, culture, and sometimes even personal history. They are a mute repository for our own nature, though we flatter ourselves that we are uncovering theirs rather than displaying ours."[31] Although Neandertal — by the 1920s the most popular caveman type of all, thanks to Boule and an American Museum of Natural History contingent[32] — was framed in contemporary art as a veritable human being,[33] in popular consciousness Neandertal existed and was stereotyped as the monstrous brute painted by Frantisek Kupka in 1909. This was resultant of lasting impressions of Boule's erroneous scientific reconstruction as well as mounting public concerns over Darwin's monkey theory.[34]

Following the horrors of World War I, artists were more inclined to imbue Neandertal with stoic traits, a hardy yet still rather terrifying creature persevering through a soulless, prolonged (and perhaps metaphoric) Age of Ice. Paul Darde's 1931 statue displayed at France's Prehistoric National Museum in Les Eyzies-de-Tayac, for instance, "emphasizes the sheer physical power and bulk of Neandertals.... It is a symbol of endurance.... [Neandertal] looks off into the distance into the future, with only the faintest glimmering of hope. The statue is an image created by a world that had survived — but only just — a great and terrible war."[35] Neandertal, perhaps like ourselves— ultimately, would not survive its war with nature. Yet, perhaps more than any other race of prehistoric humans, Neandertals have come to represent what modern native peoples formerly symbolized "the 'other' or alien counterpart to the modern selfhood."[36]

But if prehistoric man appeared brutish or even monstrous in many restorations, it was because there was an ulterior motive in assigning primitive qualities to cavemen. Moser notes that as "with the classical images of foreign peoples, nineteenth-century artists used pictorial conventions to convey physical distance and also to imply historical separation. Thus, the imagery reinforced the belief that white European culture was superior and that people with darker skin were not only less civilized but that they represented an earlier stage of existence."[37] Publically, Darwinian evolution was an anathema because whether or not such ape-men (or man-ape) species were directly on our ancestral tree, based on paleontological evidence we were genetically linked rather closely to them all and therefore, analogously, even more so to the modern races of mankind judged (as if cultured societies had any right to say) "inferior."

Beginning especially with Charles R. Knight's dramatic, tradition-setting caveman paintings (circa 1915–1920) for the American Museum's Hall of the Age of Man, prepared under advisement of Henry F. Osborn (1857–1935), pop-cultural views of prehistoric man shifted more toward anatomically correct perspectives based on refined analysis of fossil evidence. Yet it would be years before the Piltdown fiasco would be untangled and until other races of early men (e.g., *Australopithecus*) would win undisputed acceptance on the human family tree. At this juncture, "visual reconstruction had become the primary means of communicating the topic of human evolution to the public."[38] Knight fathered a new generation of scientific illustrators of fossil humans, followed ably by distinguished artists Zdenek Burian, Maurice Wilson, Jay Matternes and John Gurche.

Besides painted museum restorations, often reproduced in popular books where they were consumed in mass, by the early 20th century artists of varied persuasion and skill tackled alternate means for communicating how cavemen lived. Both in literary form as well as projected through film, these alternate paleoartists of fantastic literature and also the movie industry often resorted to portraying fossil men as, well, primeval monsters.

Not wasting golden opportunities to cast man in light of the new discoveries and ideas, skilled writers of fantastic fiction published tales of rugged prehistoric men and ape-men. For as paleoanthropologist Pat Shipman noted, even as late as the early 1900s, the very idea of human evolution was "still a dangerously controversial subject." Here was fodder precisely needed for the most powerful and imaginative, mind-dazzling stories.[39] The best examples of imaginative paleoanthropological literature function "on a deeper, psychosocial level, as a means of understanding our evolutionary ancestry."[40]

In 1981, Marc Angenot and Nadia Khouri[41] referred to the "ape-man tales" genre collectively as "prehistoric fiction." They relegated one such subcategory to the (paleoanthropological) narrative formula known as the "lost world tale." Typically, lost world dinosaur fiction where ape-men appear is often most favored by readers. Most, but not all, of the associated narrative formulae (involving atavism and simian societies of prehistory or even modernity) originated in consequence to the Darwinian dispute over "polemics on the simian genesis of man." Thematically, Angenot's and Khouri's defined "prehistoric fiction" would seem evolutionary to readers, although sometimes laced with racial undertones.

Let's survey some of the more significant examples. Besides mirroring general disgust for implications of Darwin's theory, such fantastic literature also reflected man's fearful fascination with gorillas. A live specimen had been delivered to the London Zoo in 1858 and the mountain gorilla, the largest gorilla species, was discovered in 1901. Adding to the furor, two British paleontologists—anti-evolutionist Richard Owen (1804–1892), and ardent evolutionist Thomas H. Huxley (1825–1895), aka Darwin's bulldog—debated the nature of gorilla anatomy in its relation to man's during the 1860s, emphasizing brain morphology. And so, following Boitard's aforementioned *Paris Before Men*, one of the earliest to dabble in speculative fiction concerning prehistoric man was Jules Verne. In his popular *Journey to the Center of the Earth* (1864, 1867 eds.), Verne entertained readers with suggestions of terrifying prehistoric (or antediluvian) humans and apes, including one phantasmagoric, club-wielding anthropoidal giant driving a herd of Mastodonts. And in a chapter not written by Verne, added to early 1870s British editions of *Journey*, another creature — the Ape Gigans, or the antediluvian gorilla — joined the pantheon of subterranean prehistoric monsters.[42]

The vibrant yet sometimes strident ape-man evolutionary theme reverberated through popular period writings. Robert Louis Stevenson composed *The Strange Case of Dr. Jekyll and Mr. Hyde* in 1886, now a frightening classic concerning the "brute that slept within me" (meaning the monster plaguing Henry Jekyll's soul, in essence part of Jekyll).[43] That Hyde played "apelike tricks" has been seen allegorically as a "mythical reinterpretation of [Charles Darwin's] *Descent of Man*."[44] By 1887, H. Rider Haggard wove a "hideous little monkey frame" into his haunting projection of the immortal Egyptian lady Ayesha, in *She*. Later, in his 1901 novel *The Village in the Treetops*, Verne postulated an extant population of hostile, missing link *Pithecanthropus*-like apes known as Waggdis, living in Africa's mysterious recesses. And by 1906, in his *Before Adam*, Jack London envisioned an atavistic mental "dissociation of personality" process transferring a modern man predisposed to atavism through a falling-in-space dream back into the mid–Pleistocene. He is incarnated into the body of a male ape-man, exhibiting characteristics not unlike contemporary understanding of *Pithecanthropus*. Even *The Wizard of Oz*'s creator himself, L. Frank Baum, couldn't refrain from charging his imaginative 1900 children's tale with simian monsters — winged monkeys (which were especially terrifying to behold in the 1939 film version).

More elaborately than Verne, in his novels and short stories Herbert George Wells (1866–1946) displayed uncanny fascination for the implications of Darwinian evolution. Whereas the providentially minded Verne rejected evolution, heretical Wells — who had during the 1880s been a student of Huxley — embraced evolutionary ideals. Besides masterful, evolutionary themes such as *The Time Machine* (1895) and *The War of the Worlds* (1898), Wells also crafted *The Island of Dr. Moreau* (1896) and a short tale, "The Grisly Folk" (1921), both noted for (paleo-)anthropological perspectives.[45] While *Time Machine* and *War of the Worlds* extended evolutionary principles into the far future as well as distant space, anthropoidal monsters featured in the satirical *Dr. Moreau* and "Grisly Folk" instilled contemporary readers with a sense of primitive man's inhumanity, and outright awfulness.

Sadistic mad scientist Moreau never fully explains to a shipwrecked British gentleman named Prendick the true goings on of his despicable island habitat. Turns out that Moreau surgically and biochemically transforms animals into quasi-human monsters. This form of pseudo-evolution or transformation of lower animals, one of them an ape man, into more advanced forms (i.e., man) is termed heterogony. Ancient Greek and Roman philosophers and poets[46] believed such species transformation would at times be possible. But in a post–Darwinian world the idea could only be, at most, science fictional. At the heart of the reaction to Darwin's objectionable idea was the core question — "Is there a god?" Here, we find Moreau playing a Frankensteinian god while at the same time revealing how cultured mankind, not unlike human-like pigs resulting from his painful experimentation, can be taught to read and write, recite commandments and ponder a deity — the mad scientist himself. Are *we* then merely self-taught, intelligent pigs cast without right of free will into a random, godless existence? Stevenson and Wells opened our eyes! Curse what knowledge of the ape-man had wrought!

In effect then, Wells has critiqued human society. Moreau has taken the most uncultured souls of all — raw animal stock — and through brutal, tyrannical means striven to create rudiments of civilization from a filthy, cannibalistic horde. Because Wells ultimately believed that the natural evolutionary course carried benevolent powers, the grand vivisection experiment backfires. When island god Moreau dies, the quasi-human animals revert to natural, primitive, savage tendencies. Rescued Prendick is haunted by the experience, forevermore believing "that I too, was not a reasonable creature, but only an animal tormented with some strange disorder in its brain, that sent it to wander alone, like a sheep stricken."[47] The rest of the globe is then but a projected Moreau's island, although cleverly disguised through cultural trappings and societal creature comforts. And, most disturbingly, beast-men lurking throughout, walking among us, are little more man-like than were the ape-men of prehistory.

Wells' "The Grisly Folk" conveys the brutality of men out of prehistory, where Neandertal spars with Cro-Magnon for very existence. According to Leon Stover, Wells "inaugurated the fashion in science fiction of portraying Cro-Magnon as a conquering enemy at victorious war over Neandertal man."[48] In "Grisly Folk," borrowing from Marcellin Boule's "Old Man" misconstruction, Neandertals are characterized as bear-like, canine and grotesque, downright grisly folk. Cannibalism is a recurrent theme in Wells' tales, and so we learn that the hulking predatory Neandertals devour early modern men's children. On the fringe of the early modern humans' final triumph over Neandertal, Wells suggests that a basis of the ogre legend (i.e., a gigantic "man-eating monster") could be dim geological recollections of the former primitive race "justifiably" exterminated by our ancestors.[49] Was Neandertal the "archetypal ogre" as Wells suggests, merely a prehistoric form of Morlock? Futuristic, cave-dwelling Morlocks coupled with the submissive Eloi race share a fascinating relationship approximating that of Wells' Neandertal and Cro-Magnon, although in *Time Machine* the technological Morlocks are the ogres.

Charles DePaolo suggests that Wells conflated the Neandertal race with another alleged variety of fossil man which by then had perplexed paleoanthropologists and evolutionists for over a decade — Piltdown Man. Piltdown Man was a quixotic species that may have deceitfully inspired another race of subhumans of fantastic literature inhabiting a lost fictional plateau in South America, as conceived by Sir Arthur Conan Doyle (1859–1930). After all, isn't it curious that Piltdown Man's remains were found, ironically, only 30 miles from Charles Darwin's former home, and even closer to Conan Doyle's?

By the 1910s, most prehistoric or dinosaur fiction involved men out of prehistory interacting with other prehistoria or sometimes anachronistically with modern humans. The chief example from this period is Conan Doyle's *The Lost World* (1912), in which a London paleontological expedition to an uncharted region of South America encounters living examples of dinosaurs, pterodactyls, Cenozoic fowl and mammals extinct everywhere else, and even prehistoric men. For our purposes, while in the throes of certain doom and with stiff upper lips, two feuding scientists (Challenger and Summerlee) debate whether the frightening anthropoids who are about to murder them are *Dryopithecus* or *Pithecanthropus*. Ultimately, with Professor Challenger leading as commander, these ancient, violent peoples are exterminated by a more modern, yet cave-dwelling race of Indians. In this final bloody battle, the males of Ape Town are slaughtered; remaining ape-like females and children are led away in bondage.

In 1983, Conan Doyle was implicated as a perpetrator of the infamous Piltdown Man hoax, which according to one of the gullible paleontologists— E. Ray Lankester (1847–1929)—dated from the Pliocene or possibly Miocene epoch. Under evolution's dark specter, Piltdown Man's attributes satisfyingly projected a more modernistic, larger brained human species having evolved in Britain, predating more primitive Neandertal. So Piltdown's (as opposed to Neandertal's or the more primitive *Pithecanthropus'*) descendants would have led to the exalted, dominating *civilized* races of mankind. According to John E. Walsh, however, "The supposed intimate link between Piltdown and Doyle's novel *The Lost World* proves to be far overstated."[50] Significantly, *Lost World's* relatively un-evolved ape-men don't resemble contemporary reconstructions of the more advanced Sussex Dawn Man (e.g., Piltdown).[51]

While Conan Doyle's lost plateau represents the classic prehistoric environment where time—meaning evolution—has stood still for eons (or at least since the Jurassic), Edgar Rice Burroughs (1875–1950) chose to write considerably more dinosaur fiction invoking vibrant evolutionary processes generating (and preserving until modernity) primitive species of mankind. In, arguably, Burroughs' best novel, *The Land That Time Forgot* (1918), followed by a pair of sequels published that same year including *The People That Time Forgot*, ghastly proto-men are conjured from prehistoric times. Whereas Burroughs' acclaimed "Pellucidar" inner–Earth series of novels relies more on Conan Doyle's time-stands-still gimmick, in *Land That Time Forgot*, all living creatures indigenous to the South Pacific lost island Caspak are incessantly evolving, mirroring Ernst Haeckel's since falsified Biogenetic Law. Darwinian Haeckel espoused that "ontogeny recapitulates phylogeny." Except on fictional Caspak, organismal *growth* may play out the entire course of geology's "ladder" of evolution within individual lifetimes!

Fictive fossil apes and men resulting from Burroughs' Caspakian science fictional gimmick — an alternate evolutionary conceptual model termed "phylosynthesis" by DePaolo— progress from reptilian and lower mammalian grades through nine hominid sub-species stages, including the "Bo-lu," likened to Neandertal, pithecanthropine "Sto-lu," "Band-lu" which are Cro-Magnon, and "Galu" or modern humans.[52] Rather predictably, the lower grade humans whose characteristics were founded on contemporary 1910s paleo-archaeological knowledge, prove most menacing to the party of European humans who have become exiled on Caspak. And, conversely, while Burroughs' Caspak island phylosynthesis generates continually higher forms of organic life from "lower" forms, author and mathematician John Taine (1883–1960) envisioned a *retrograde* evolutionary process for his 1930 novel, *The Iron Star*.[53] Here, tribes of modern humans regress through increasingly more primitive and hostile evolutionary ape-like stages after being exposed to a new extraterrestrial element, asterium, found in a meteorite deposit.

In *Tarzan the Terrible* (1921) Burroughs dabbled further into evolutionary themes, as exemplified by the inhabitants of a mysterious African land known as Pal-ul-don. This novel featured lost tribes of *Pithecanthropus*, as well as even more ancestral anthropoids— hairy black men known as Waz-don, beastlike men called Tor-o-don, and odd lizard men known as Horibs.

Ever since Boitard's *Paris Before Men*, prehistoric or fossil cryptozoological apes have become a staple of fantastic literature. Numerous generations and variants of bestial anthropoids or ape-men have appeared — symbolizing evolution's harsh realities— in published novels and all manner of science fiction stories. Notably, for example, in Russian geologist Vladimir Obruchev's (1863–1956) 1924 novel *Plutonia*, concerning an inner–Earth world in which past geological periods may be witnessed, modern explorers escape a marauding pack of Neandertals. Delos W. Lovelace's (1894–1967) *King Kong* (1932), a novelization of the RKO movie script, featured the most colossal prehistoric ape of all! Then, during World War II, supposed Wellsian primitive ogre-like traits of our evolutionary ancestors were projected onto Allied forces enemies— the Japanese — in pulp stories such as Robert Moore Williams' "The Lost Warship" (1943), where U.S. battleship *Idaho* is cast backward to the dinosaur age through a time-slip. And in John York Cabot's "Blitzkreig in the Past" (1942), a U.S. tank transported into the past battles dinosaurs and cavemen ruled by a female "Hitler."[54]

During the 1960s space-race, man's evolution from ape ancestors took on extraterrestrial flavor under the talents of Pierre Boule (*Planet of the Apes*, 1963), and Arthur C. Clarke (*2001: A Space Odyssey*, 1968).[55] In particular, Clarke's classic tale projected vignettes of an australopithecine tribe thwarting extinction and evolving into ourselves only after mastering the untamed art of conspecific aggression. Rather cynically, as the anthropoids' killing instincts are refined by alien technology (i.e., the Monolith) through use of weaponry, in essence, man is born.

When will the madness end? Not anytime soon, apparently. Lin Carter thrilled late 1970s and early 1980s readers with five novels inspired by Burroughs' Pellucidar subterranean realm concerning primitive tribes (and dinosaurs) existing in an inner–Earth realm named Zanthodon. Since the 1990s, fascinated readers have been treated to acclaimed ape novels such as Petru Popescu's *Almost Adam* (1996), John Darnton's *Neanderthal* (1996), Stephen Baxter's *Evolution* (2003), Joe DeVito's and Brad Strickland's *Kong: King of Skull Island* (2004), and Jonathan Green's *Pax Brittania: Unnatural History* (2007).[56]

By the 1910s cavemen of fantastic fiction fueled Hollywood fare and other (more scientifically oriented) documentaries. Thanks to popular press accounts, perhaps the greatest period of escalation in popular intrigue over the astonishing reality of cavemen (circa 1900 to 1925) coincided with the film industry's rise. Most fictional, filmic examples seem cliché, casting cave people, so often witnessed in conflict with dinosaurs, as somehow monstrous and inferior to moderns and therefore justifiably extinct. In fact, conclusions offered by Warren D. Allmon in the case of *dinosaur* art and restorations would certainly apply to 20th century representations of cavemen in film. Allmon stated that "Dinosaur art ... increasingly pictured dinosaurs as primitive because that is what the evolutionary views espoused by most turn-of-the-century vertebrate paleontologists appeared to require ... old forms had to be primitive, and dinosaurs at this time therefore came to be overwhelmingly viewed as primitive, ponderous, and poorly adapted, a view that persisted until the 1970s."[57] Simply exchange Allmon's word *dinosaur* for *cavemen* or *fossil men* and you'll understand Hollywood's winning formula.

Due to special effects limitations (and also perhaps because of general audience distaste toward obvious Darwinian implications), cavemen in films have been usually featured not as ape-like but instead as anatomically modern in appearance — the way Figuier thought of them during the 1860s. Decades ago, more creative costuming sometimes resulted in a Neandertal cast but only rarely have prehistoric humans antecedent to Cro-Magnon and Neandertal been portrayed in science fiction and horror films (my emphasis here). According to Michael Klossner, author of *Prehistoric Humans in Film and Television* (2006), between 1905 to 2004, 581 titled dramas, comedies and documentaries were produced featuring various kinds of cavemen. Typically, directors offer viewers lessons about ourselves— modern mankind, although the primary underlying purpose for making such movies is simply to make money. And so, preying on innate fears of our dark uncivilized ancestry, and in spite of several good faith attempts to portray prehistoric men with scientific accuracy in documentaries, a sure-fire way for luring audiences into theaters remains to portray cavemen as instinctually beneath ourselves—behaving as savage brutes or even as prehistoric monsters menacing attractive, more modernistic looking women (cave-damsels in dino-distress, and modern beauties ravaged by primitives). Yes, now although projected into another medium our favored, fantastic themes of yesteryear pervade! Fear of Darwin's cavemen and our dark visions of man's uncivilized origins truly are vitally, *psychologically* ingrained.

Reduced to ragged primitiveness on an island of lost souls, menaced by savage, prehistoric (huge, scary and nearly extinct) ape Kong years before Raquel Welch's and Carole Landis's analogous roles, Fay Wray's "scream queen" character Ann in *King Kong* (1933) is where the successful Hollywood "formula" originated. Yet, while to most readers the term *caveman movie* may conjure delectable images of scantily clad Raquel Welch as she appeared in the blockbuster *One Million Years B.C.* (1966) or the equally enticing Carol Landis starring in *One Million B.C.* (1940), films such as *The Lost World* (1925) and *The Neanderthal Man* (1953) epitomize the essence of centuries-honored, prototypical, post–Darwinian fears mirrored into modernity.

In *The Lost World* silent movie, rather than having interacting tribes of apemen and indians there is only a solitary, stealthy apeman played in a nondescript hairy ape suit by actor Bull Montana. It is not clear what species Montana is playing although it appears to be a cave-dwelling form. From its grimacing, glowering werewolf-ish countenance and a notable lack of tools and cave art, however, it most likely is not Neandertal nor Cro-Magnon (despite its evident upright posture). So the apeman could be *Pithecanthropus*, or possibly "Eoanthropus."[58] Whatever he's supposed to be, *Lost World's* apeman would seem a veritable manifestation of evil—*Darwinian* evil—apeman personified as an unchanging, primeval Mr. Hyde, ruffian ruler of the plateau's prehistoric denizens.

While *Lost World's* stop-motion dinosaurs are usually the only prehistoria viewers enthusiastically recall from this film, the apeman is one of its most significant, symbolic characters. The foreboding, frightening apeman acts menacingly toward the scientific party, including Professors Challenger and Summerlee who represent modern paleontological science and, of course, as in the novel—have the audacity to espouse Evolution. The long-toothed, canine-fanged apeman consorts with a chimpanzee pal in their cave lair. Thus a ladder of evolution is evident, with apeman serving as link (no longer missing) between simian (i.e., chimp) and modern humans.

The territorial apeman leaps about agilely in a monkey-like manner, spies on the invaders deceptively from treetops, rolls a boulder on the party from cliffs above and (foreshadowing a sequence later filmed for *King Kong*) hauls a rope ladder back up, upon which members of the scientific expedition are attempting to escape the plateau. The apeman is shot *twice* by sportsman John Roxton—first in the arm and later, fatally, in the chest. Although

Lost World's (1925) apeman never makes a clear play for the expedition's token female, Paula White played by Bessie Love, he is the only prehistoric creature killed by any member of Professor Challenger's party, suggesting it is the most awful prehistoric creature of them all and therefore most worthy of extinction. Incidentally, Roxton's role (both in Conan Doyle's novel and the subsequent film) as the prototypical hunter who shoots wild game out of the past (i.e., the apeman), would be followed by scores of other metaphorical hunting parties journeying into prehistory, notably as in Ray Bradbury's 1952 time travel classic "A Sound of Thunder," and Richard Marsten's *Danger Dinosaurs!* (1953).[59]

In *The Neanderthal Man*, a Dr. Jekyll–like mad scientist named Groves self-injects serum transforming him into a Neandertaline monster. In previous experiments, he administers the drug to his pet house cat, converted into a sabertooth tiger, and even his mute housekeeper who turns into an apelike woman. Groves theorizes that modern "man is not himself. He's part of every ancestor he's ever had.... Man has lost nothing of his emotions from the dawn of history." And his serum can restore those states of mind. Despite the large brain of his reverted Neandertal persona, Groves' animalistic instincts come to the fore. He's plagued by a "hungry urge to kill." Groves-as-Neandertal is later killed by his equally primitive sabertooth cat. The lurid movie poster for this film shows the converted ape-scientist seeking the unwanted attentions of a buxom young lady with unbridled passion while her seductively dressed companions turn away with horrified expressions. As Trinkaus and Shipman later stated, "Coincident with the flowering of the new evolutionary synthesis, this movie's theme echoes both the exhilarating excitement of scientific progress as well as its vaguely disturbing and threatening aspects." Furthermore, this poster displays "fear of the rampant sexuality that was equated with physical primitiveness."[60] In the similarly themed Jekyll and Hyde rip-off, *Monster on the Campus* (1958), another mad scientist converts into a Neandertal-like humanoid after inadvertently exposing himself to the experimentally tainted blood of a primitive Coelacanth fish.

Perhaps the most hideous anthropoid of all, hinting at mankind's most ineffable, recessed evolutionary origins, was the Gill Man, featured in *The Creature from the Black Lagoon* (1954) as well as two sequels. Paleo-scientists have prominent roles in *Creature*, in which a living fossil anthropoidal Gill Man is discovered in the Amazon. (Indeed, perhaps the Black Lagoon lies in the vicinity of Conan Doyle's fabled land of Curupuri!) Moreover, while scientists suggest a Devonian (400-million-year-old) origin for the Gill Man, they also speculate that its fossils date from 150 million years ago (which would be the Jurassic Period instead). Ichthyologist Dr. Reed (played by Richard Carlson) hopes, idealistically, that by studying the Gill Man, not only would such knowledge contribute to our understanding of how life evolved on Earth, but also offer clues as to how man may eventually adapt to other worlds in outer space having different surficial conditions. To suspend disbelief, zoological comparisons are made to the Kumonga lungfish, which also lives in the Amazon, physically unchanged since the Devonian — evidently like the Gill Man.

According to monster movie lore, the idea for the Gill Man sprang from Director William Alland's recollection of "an obscure South American legend about a mysterious creature that was reputed to be extant in the swamps and jungles of that continent."[61] But this is Hollywood, remember, where it is difficult to denote anything with objectivity. And there *were* other contemporary, pop-cultural influences at large. Only sixteen years earlier, news headlines had made quite a splash with sensational discovery of a veritable evolutionary missing link between marine and terrestrial vertebrates—(or fish and humans)— the living Coelacanth, thought to have gone extinct during the Mesozoic Era.[62] This natural history discovery proved every bit as significant for its time, and duly heralded, as were W. Douglas Burden's mid–1920s observations of the Komodo Dragons. Also, Howard P. Lovecraft's (1890–1937) horrific *The Shadow over Innsmouth* (1936), concerned a New England coastal colony; people who in adulthood steadily transform into piscine, anthropoidal varieties from the "Palaeogean." And Karel Capek's (1890–1938) *War with the Newts* (1937) satirically chronicled the marine, anthropoidal Newts' decisive military triumph over the human race.[63]

So aquatic anthropoidal amphibian monsters had been anticipated by others; analogously, through exploring the Gill Man's prehistoric persona, its propensity for unveiling otherwise hidden (crypto-)aspects of humanity heightened popularity of Universal's latest monster. In *Creature*, the tragic Gill Man character stalks a human female played by pretty Julia Adams, and even develops a fascination for her beauty — much like Kong did with Fay Wray two decades earlier. As Glut suggests, the Gill Man is aroused by her invasion of its waters. John Baxter has stated of their underwater ritual that: "Gliding beneath her, twisting lasciviously in a stylized representation of sexual intercourse, the creature, his movements brutally masculine and powerful, contemplates his ritual bride, though his passion does not reach its peak until the girl performs some underwater ballet movements, explicitly erotic poses that excite the Gill Man to reach out and clutch at her murmuring legs."[64]

Glut also notes that, "Though the Gill Man should be indifferent to a female of an alien species, he is most definitely attracted on the physical level to Kay. The implication is that there is a human being awaiting birth underneath the green scales and fins, even though he might not surface until the next 500 million years."[65] In other words, Gill Man is a terrifying manifestation of mankind's most remote evolutionary past, analogous, yet far more prehistoric than any species of Caveman known to science. Instilled genetically, mankind has a lurid gill man past, leading in non–Darwinian ladder-like

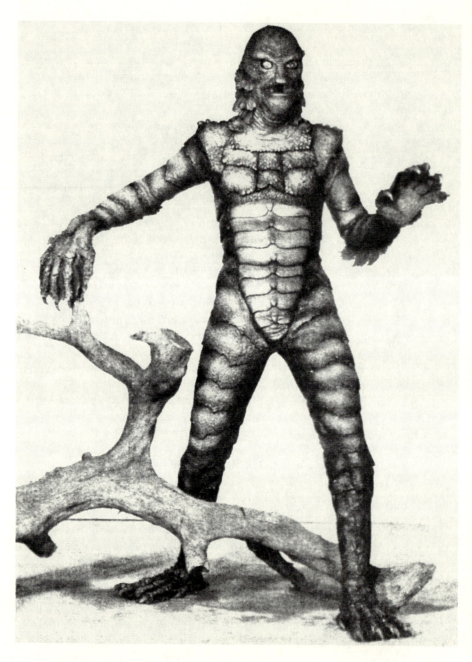

The Gill Man — "living fossil" from the Devonian Period — thrilled 1950s audiences in a trilogy of *Creature from the Black Lagoon* Universal features. Because of its apparent anthropological ties to modern man, Gill Man may be considered the most ancient "caveman" example. By the 1980s, public interest in gillmen had faded, while fascination over fictional, intelligent "dinosauroids" escalated. (See note 65, Debus, 2009.)

fashion directly to ourselves. In fact, remaining in an amphibian living fossil state is evil. For, as explained by Captain Lucas (Nestor Paiva) in *Revenge of the Creature* (1955), "The Beast exists because it is stronger than — than the thing you call evolution. In it is some force of life, a demon, driving it through millions of centuries. It does not surrender so easily to weaklings like you or me."

This idea is reinforced in the series' concluding, Dr. Moreau–like, entry, *The Creature Walks Among Us* (1956). Here, scientists conclude that the Gill Man has human-like, underlying tissues that may be accentuated. So, as one mad scientist (played by Jeff Morrow) suggests, bypassing hundreds of millions of years of intervening (contingent) evolutionary steps as recorded in the fossil record, simply, "with some careful surgery, the Gill Man might be evolved to the next step on the scale of life."[66] Yet even then, Gill Man (now sans gills), like all monstrous cavemen, retains an objectionable infatuation for the allures of modern females! Gill Man is the most primordial caveman.

The lesson? Beware the cruel caveman lurking within, for he is evil!

By the 1980s, gill-men of fantasy and fiction couldn't stem the rising tide of intrigue over more intelligent dinosaurs (so suffused with fascinating evolutionary implications). Meanwhile, in fiction, "dinosauroid" lizard people (resulting from an *erasure* of the Late Cretaceous catastrophe 65 million years ago) emerged. The idea of crafty, yet sometimes monstrous and alien dinosaur-men became standard, eclipsing former pop cultural fascination with amphibious-men. (See note 65, Debus 2009.)

The caveman refuses to die or go extinct, its popularity heightening in many shapes and forms through the decades since then through new publicized discoveries, books, magazine articles, toys and collectibles, museum displays and documentaries. By the 1960s, through sensational discoveries and more accurate reinterpretation of old evidence, paleoanthropologists added more specimens and species to man's prehistoric family tree. Prehistoric man was becoming firmly embedded within modern human culture and consciousness, even entertaining us via episodes of Hanna-Barbera's hilarious cartoon, *The Flintstones*.

And if you doubt this—consider! Before we knew one famous *Tyrannosaurus* specimen in name — "Sue" — we were already familiar with a 3.5-million-year-old ancestor on a first name basis—"Lucy" the australopithecine — represented by bones discovered in 1974 in Ethiopa.[67] Unlike *Tyrannosaurus*, Lucy was no monster, although if we could restore Lucy to life, many would still fear her visage and meaning. And this is thanks to Darwin's dangerous ideas.

SIX

Triumph of
the Leapin' Lizards

In its infancy, paleontology thrived in the hearts of men and fossilists popularly as well as scientifically, on the foundation of its disinterred fossil vertebrates—principally prehistoric mammals such as the giant ground sloths, mastodons and mammoths. Prehistoric mammals remained more popular than dinosaurs and allied fossil sauria for decades following Richard Owen's first published naming of the Dinosauria in 1842. Throughout this early period there was no perceived popularity contest, per se, staged between dinosaurs and fossil mammals; people weren't anxiously waiting for dinosaurs to enter the arena of public opinion so they could switch allegiance from the furry to the scaly (or feathery) side. Terms like *dinophiles* and *dinosaurabilia* did not exist.[1] Fossil mammals held their own in the emerging popularity sweepstakes through the late 1920s. Why? For one, you didn't have to fake a dinosaur (like Koch had with Missourium) to make it seem big. They, or many kinds, were so, naturally. Principally, the reason rests in the evolving nature of paleoart and paleoimagery, as it unfolded during the late 19th and early 20th centuries. Then, prehistoria most likely to enter public consciousness in various media as "monstrous" became, increasingly, the dinosaurs—which after the1930s triumphed, visually, over fossil mammals.

Why then, did fossil mammals begin to lose their edge in popular culture to those upstart dinosaurs? I've explored this topic before,[2] but here shall examine the most definitive reasons, entrenched in paleoimagery's mass appeal!

As published notices of numerous prehistoric organisms increased throughout the 19th century, thanks especially to the efforts of paleoartists, the identities of several key prehistoria captivated public audiences. Thanks to men such as such as Edouard Riou, Benjamin Waterhouse Hawkins, Joseph Smit, Henry A. Ward, Edward Cope, and Charles R. Knight, visual imagery of prehistoric life became increasingly, hauntingly familiar. Curiously, while magnificent fossil saurians were known from a time pre-dating Cuvier's and Wistar's elucidation of Great-Claw, prehistoric mammalian fauna (including mammoths and mastodonts) seized public attention for decades, even following official naming of the Dinosauria. In particular, the dinosaurs— so popular today — long played second fiddle to prehistoric mammals in popular culture; and herein, we'll explore underlying reasons, founded upon contemporary painted, sculptural, published and animated restorations of these beasts. Then we'll map the hierarchy of scientific discoveries through the 19th century, and view prehistoric creatures resulting from those discoveries fueling popular notions of prehistoria as being hellacious monstrosities ruling savage, prehistoric worlds.

Elsewhere[3] I suggested that by the end of the 1930s, "the prehistoric struggle was over. Dinosaurs had outstripped the prehistoric mammals. While, generally, interest in prehistoric life downshifted during World War II (both in academic institutions and in the popular vein), dinosaurs came roaring back with a vengeance following the war." Increasingly, "dinosaurs possessing 'iconic' value such as the "Brontosaurus" and *T. rex* were becoming increasingly familiar, while mastodons and mammoths weren't publically promoted to the same degree.... So, why not the mammals? Could it be that they are somehow less inherently interesting, *visually*?" While many prehistoric mammals, to whom we're more closely related in an evolutionary sense, may be imagined with relative ease, clothed in fur, "there are so many interesting unanswered questions concerning proper restoration of the dinosaurs. Maybe it is this ultimate mystery of the dinosaurs [i.e., we don't even know exactly what they looked like] ... [and] ... that accuracy of dinosaur restorations is often contested by scientists [which] makes them controversial and therefore more inherently interesting."[4] And, hence, arguably, dinosaurian monsters of popular culture stemming from contemporary imaginings of how they looked have varied considerably partly because underlying science evolved.

I've also explored this idea in *Paleoimagery* (2002), but here my thesis is underscored by individuals attuned both long ago, and recently. When commenting on his statues, or geological restorations, made between 1853 and 1854 for the Crystal Palace display, Benjamin Waterhouse Hawkins hoped to

render the appearance and names of the ancient inhabitants of our globe as familiar as household names.... The inevitably fragmentary state of such (fossilized) specimens of course left much to the imagination, even to those who looked at them with some little knowledge of comparative anatomy ... that amount of knowledge is not found among the average acquirements of the public at large.[5]

The impact of Hawkins' inaugural, sculptural display of life-sized prehistoric animals (which incorporated both prehistoric mammals as well as sauria, including three dinosaur genera) cannot readily be imagined today in our dinosaur-saturated world. While at Sydenham today the huge (anatomically erroneous) dinosaur statues most captivates visitors, the prehistoric animal which inspired this display was not a fossil saurian, but Great-Claw, *Megatherium*.[6] More recently, professor of art history and paleo-aficionado Jane P. Davidson astutely stated in her 2008 book, *A History of Paleontology Illustration*, "There is no paleontology without illustration.... The illustrations were there before there was a professional science of paleontology, and thus we can say that there has never been paleontology without its illustration ... there is a genre of art that we may call paleontology illustration."[7]

Paleontologist and Historian of Science Martin J. S. Rudwick adds that such scenes from deep time

> signal the invention of a novel pictorial genre that has remained ever since a powerful medium of visual rhetoric, conveying a sense of the reality of geohistory not only to savants and serious amateurs but also in the long run to the general public too.... Part of the attraction of such scenes was that they depicted exotic plants and unfamiliar animals, many of them gigantic in size and some of them also monstrous in appearance. This accentuated the alien "otherness" of the deep past: it really had been in sharp contrast to the present world, a foreign country where nature did things differently.[8]

How did fossil mammalian "monsters" of yesteryear lose celebrated status to the anointed dino-monsters in the pop-cultural war?

Elucidating Fossil Sauria

For decades prior to naming of Dinosauria, fossil sauria were known to naturalists, yet as we have seen, the emerging geological and allied paleontological sciences pivoted on understanding fossil mammals instead. We've already seen which fossil mammals the sauria were chiefly up against on into the mid–1840s, but what were their principal dinosaurian challengers?

In 1676, British chemist Robert Plot's (1640–1696) account from *The Natural History of Oxfordshire*, incorporated a strange Plate showing "formed stones." One of these detailed a curiously shaped stone which Plot thought might have been a petrified equine bone, or possibly a bone from an elephant led by the Romans into Britain, or perhaps even an oxen. This long lost specimen is now regarded as the earliest dinosaur bone — quite possibly belonging to a megalosaur — published by a naturalist.[9] Then, a century before British scientists rather accurately described aquatic, long necked plesiosaurs, William Stukeley (1687–1765) published "An Account of the Impression of the Almost Entire Sceleton [*sic*] of a Large Animal in a Very Hard Stone" (1719).[10] While many observers thought the skeleton was human, Stukeley noted its reptilian nature, stating it "...seems to be a Crocodile or Porpoise," which drowned in the biblical Flood. The bones were figured in his publication, which Jane Davidson suggested is the first published plesiosaur fossil. From such humble beginnings.... Far more dramatic, however, were circumstances surrounding discovery of the "Maastricht animal," later described as the *Mosasaurus*, the first fossil saurian for which pictorial record of its excavation resulted. In 1766, within St. Peter's Mountain, quarrymen laboring at a chalk limestone quarry outside the Dutch town of Maastricht uncovered the 4-foot-long, dagger-toothed jaws of an animal that anatomist Adriaan Camper (1759–1820) declared in 1800 was a large marine lizard. However, as Desmond stated in 1975, "since lizards of this size do not inhabit the world today, others thought it more probable that the jaws had belonged not to an antediluvian toothed whale, but a prehistoric crocodile."[11] The sensational fossil trophy changed jealous hands several times within the Netherlands before being seized from its glass-enclosed shrine by France's conquering army in 1794. British anatomist William Daniel Conybeare (1787–1857) later named this mysterious creature, *Mosasaurus*. We'll return to this important antediluvian monster shortly.

The alleged megalosaur and "Maastricht animal" were augmented by an "equally celebrated" discovery of another strange Bavarian Solnhofen fossil originally described by Cosimo Collini (1727–1806) in 1784. This new, batlike creature evidently possessed elongated bony arms thought to have supported wings. It wasn't a bird nor a marine animal, but a winged dragon of sorts later known as a wing-fingered pterodactyl. Cuvier, who distinguished "wing-finger's" reptilian features in 1809, declared there had been an age of reptiles preceding the dawn of mammals. For many years, pterodactyls, or pterosaurs, were "not represented *like*, but *as*" dragons.[12] Nearly 15 decades later, inspired by such specimens, a distorted fantastic progeny would transform into Toho Productions' Tokyo-destroying winged terror-saur monster, Rodan. So, unfamiliar reptilian creatures known from the Secondary Rocks (i.e., Mesozoic Era and, initially, latter periods of the Paleozoic Era combined) now established from terrestrial, sea and sky habitats, were already known decades prior to Great-Claw's elucidation. Of course, because the largest fossil mammals, known from the Superficial deposits, were not as geologically old, there were more and better preserved specimens available for contemplation.[13]

By the 1810s, two additional creatures joined the growing pantheon of prehistoric monsters. One was long-necked, flippered marine saurian *Plesiosaurus*, gaining notoriety at the talents of principal investigators, William Conybeare and Henry De la Beche (1796–1855). Of this creature, Cuvier informed Conybeare that "one shouldn't anticipate anything more monstrous to emerge from the Lias quarrries."[14] This remarkable animal was neither an ancient crocodile, nor an ichthyosaur. It seemed "strangely intermediate" between lizards and "fish-lizards."

For his 1824 paper, Conybeare sketched the already described fish-lizard's skeleton adjacent to a skeletal reconstruction of *Plesiosaurus dolichodeirus*. Imaginatively, Conybeare remarked (as had Jefferson, analogously, a quarter of a century before in the case of his *Megalonyx* battling American *Incognitum*), that "We now also learn for the first time, that the head of this animal was remarkably small, forming less than the thirteenth part of the total length of the skeleton; while in the *Ichthyosaurus* its proportion is one fourth. This proportional smallness of the head, and therefore of the teeth, must have rendered it a very unequal combatant against the latter animal."[15] In public view and for many decades, thanks to contemporary premier paleoartists and writers, the two marine saurian antagonists would remain eternally locked in an imaginary mortal combat.

Fish-lizard *Ichthyosaurus*, or Proteo-saurus as originally named informally, had been known for over a decade by the time of plesiosaur's indoctrination.[16] Although *Mosasaurus*' discovery predated that of *Plesiosaurus*' scientific description, because of the latter's geological association with *Ichthyosaurus*, artistic renditions of the more intriguing possibilities for marine saurian combat soon followed. Vivid life restorations and portrayals of *Plesiosaurus* and *Ichthyosaurus* battling one another proved especially popular, especially through the 1870s, and beyond.

Perhaps the first of these that has survived from Great Britain's heroic age of geological investigations was drawn by De la Beche in 1830. Titled "An Earlier Dorset" (otherwise known as *Duria antiquior*), it offered an aquarium perspective, marine setting with palm fronds waving on the shore, inhabited by invertebrates, fish, crocodiles, pterodactyls, turtles, and, most prominently, a central scene showing an ichthyosaur biting bloodily into a plesiosaur's neck.[17] Rudwick refers to this as the "first true scene from deep time."[18] Reproduced as a lithograph, De la Beche's composition proved inspirational, for as Rudwick noted, through the next several decades artists either copied the scene, or based their own more original vignettes—although adding more fauna as more fossils came to light — upon "An Earlier Dorset." Scenes of battling marine sauria continually appeared in scattered scientific references, and especially in popular books such as artist John Martin's (1789–1854) lurid, Hadean frontispiece prepared for Thomas Hawkins' *Book of the Great Sea-Dragons* (1840). Here, in a scene titled "The sea-dragons as they lived," the "phrenitic" appeal of battling marine sauria perhaps reached a crescendo. Of this scene, Thomas Hawkins (1810–1889) wrote,

> Looking back retrospective far over the wintry Ocean, into Pre-adamic Shades, we encounter execrable and dreary things in the abounding Chaos. Through briny clouds incumbent impetuous Monsters gleam phrenitic, livid or green, or swarthy snakes, quadrupedal and deadly. Wide over the desolate Seas warring Dragons innumerable and hideous, enacting Perdition.... Martin has ... attained, with all his stupendous Powers, the utter hideousness.... Giant of

Wrath and Battle, behold! The Great Sea-Dragon, the Emperor of Past Worlds, maleficent, terrible, direct, sublime.[19]

Paleontologist Christopher McGowan has stated that

Such sanguine imagery of the ancient globe was based on a distorted view of the fossilists' nineteenth-century world. Today, thanks mainly to television, we know that animals in their natural habitat are not locked in eternal battles to the death. We know, for example, that combat between meat-eaters, or between any other groups of animals, rarely becomes an overt conflict of tooth and claw. Rather, it is a subtle competition for resources and for living space. People in Regency England rarely had the opportunity to witness interactions between animals in the wild. Some individuals had seen lions and tigers in menageries, and could see how fierce they were. It was therefore natural for them to suppose that such animals spent most of their time in open conflict in the wild. It is not surprising, then, that the early fossilists were so obsessed with depicting combat among the denizens of the prehistoric world.[20]

Paleoartists have never abandoned this ever-popular theme, although how it was conveyed certainly evolved in tandem with emerging science.[21] For example, Riou's (1833–1900) scene showing a huge monster battle waged between *Ichthyosaurus* and *Plesiosaurus*, published in 1860s editions of Loius Figuier's (1819–1894) *Earth Before the Deluge*, influenced Jules Verne's (1828–1905) writing of memorable passages within his *Journey to the Center of the Earth* (1864). Riou also depicted this pairing of marine monsters for the 1867 edition of Verne's classic novel. Indeed, Zdenek Burian's (1905–1981) 20th century paintings of battling mosasaurs and plesiosaurs harken the mind back to De la Beche's fantastic paleoimagery![22]

However, it was Benjamin Waterhouse Hawkins whose artistic efforts placed marine sauria prominently into public consciousness for a time during the mid–19th century. We note here that relatively neglected *Mosasaurus* staged a minor comeback of sorts during this interlude.

The tale of Benjamin Waterhouse Hawkins' restorations of prehistoric life has been retold numerous times, appropriately because he is the 19th century's most significant popularizer and promoter of "in-the-flesh" prehistoria. More so than any artist of his time, Hawkins had the good fortune of having his fabulous work placed prominently on public view, where it has been experienced by many millions through the present day. His spectacular life-sized recreations done for the 1854 Crystal Palace Exhibition were particularly influential. Hawkins' models were designed as grandiose exhibits highlighting all that was known or that should be feted concerning vertebrate paleontology until that juncture. While dinosaurs, prehistoric mammals, and other prehistoria were all placed on display, most haunting and impressive to visitors were fossil sauria. So, in a sense, for the first time, fossil sauria emerged over the most famous turn-of-the-last century prehistoric monsters. Hawkins established scientifically authorized perspectives of each

fossil animal; his imagery of fossil sauria remained virtually unchallenged through the early Darwinian period (circa 1860–1890), as relatively complete dinosaur specimens began turning up, first in America and shortly afterward in Belgium.

Both Hawkins' Crystal Palace fossil menagerie, and sculptures intended over a decade later for New York's Central Park — the ill-fated "Palaeozoic Museum" — were conceived as life-through-time displays.[23] Such three-dimensional exhibits, museum mural series, or printed layouts in popular books emphasize the pageant of life through geological time, presentations which necessarily incorporate prehistoric mammals to represent the planet's more recent geological epochs.

For his Crystal Palace displays, Hawkins did not showcase the older generation of prehistoric monsters—fossil mammals. Instead (under advisement of Richard Owen, and to a certain extent physician Gideon Mantell) Hawkins placed the first three charter member genera of dinosaurs prominently on view (i.e., *Iguanodon, Megalosaurus,* and *Hylaeosaurus*). Hawkins also sculpted numerous creatures representing the Secondary rocks, although, collectively, reptiles for the "Secondary Island" (dinosaurs, crocodilians, pterodactyls and marine saurians) caused the greatest sensation. Other creatures inhabited Tertiary bounds, mammalians described years before by Cuvier (i.e., *Paleotherium* and *Anoplotherium*). Here, Owen sought to honor his illustrious, intellectual forebear. And the great Irish Elk ruled over most recent geological domain. However, there were no cavemen; a Great-Claw relative of Jefferson's *Megalonyx*— South American *Megatherium*— was relegated in status as a lone tree-hugger left apart from the main grouping. Now not quite so monstrous, *Megatherium* guards a children's zoo. Despite *Megatherium*'s presence, its ancient nemesis— American Mastodon — was absent. Owen and Hawkins dramatically featured relatively well known prehistoric monsters formerly populating the British Isles, then most recently known to science. So, at Crystal Palace, dinosaurs scored an early victory over other prehistoria. Yet they hadn't won the war.

Mosasaurus was included among the marine sauria at Sydenham, a rather curious addition because the genus wasn't found in Britain. The genus also wasn't very well known to science. Accordingly, the sculpture itself is incomplete, reflecting only its 4½-foot-long head and ornately scaled back. As Steve McCarthy and Mick Gilbert stated, to diminish guesswork, Hawkins "ingeniously used the water to mask the incomplete nature of his model, allowing the public's imagination to fill the gaps."[24] In his later 1860s through 1870s

Opposite: Top —*Mosasaurus*; middle —*Ichthyosaurus* and *Plesiosaurus* inspired by Edouard Riou's similar restoration for Figuier's *Earth Before the Deluge*; bottom — "Laelaps" inspired by Knight's 1897 restoration (restorations by F. Long, circa 1905; originally printed as collectible cards by the Kakao Company of Hamburg-Wandsbek, Germany, collection of Allen G. Debus, photographed by the author).

prints, miniature replicas of several of his Crystal Palace creations, and paintings, Hawkins featured the more familiar marine sauria (plesiosaurs and
ichthyosaurs), while *Mosasaurus* faded into relative obscurity.

That is, it faded into obscurity until Edward D. Cope, who had by the
late 1860s described several varieties of marine sauria from eastern North
America, published a marvelous drawing showing "Fossil Reptiles of New
Jersey." In this scene from deep time, a "Laelaps" (now, correctly, *Dryptosaurus*) standing along the shoreline is menaced by a snaky looking plesiosaur
known as *Elasmosaurus*, while fork-tongued *Mosasaurus* taunts human viewers. The 45-foot-long, "reptilian whale" *Elasmosaurus* upon which the sketch
may have been based was a specimen that Benjamin Waterhouse Hawkins
helped clean preparatory to his research investigations for populating the
Palaeozoic Museum. In an early publication, Cope erroneously placed the
animal's skull at the tail end, instead of at the end of the neck bones—as indicated in this figure. In his accompanying *American Naturalist* article, Cope
described such denizens of (as he would have referred to it), the fifth day of
the "Mosaic record of Creation," highlighting the relatively common and, by
then, well known form and probable demeanor of *Mosasaurus*.

> The gigantic Mosasaurus, the longest of known reptiles, had few rivals in the
> oceans. These ... were the sea-serpents of that age, and their snaky forms and
> gaping jaws rest on better evidence than he of Nahant can yet produce.... The
> Mosasaurus was a long slender reptile, with a pair of powerful paddles in
> front, a moderately long neck and flat pointed head. The very long tail was flat
> and deep, like that of a great eel, forming a powerful propeller.... A loose flexi-
> ble pouch-like throat would then receive the entire prey, swallowed between
> the branches of the jaw.[25]

As a caption for his artist's illustration (whose work was no doubt based
on Cope's own surviving preliminary sketches), Cope stated, "*Mosasaurus*
watches at a distance with much curiosity and little good will."

Hawkins intended to recreate both dramatically posed, life-sized sculptural versions of *Elasmosaurus* and, evidently, a "nondescript" mosasaur for
his Palaeozoic Museum, but with demise of that herculean undertaking, no
such models resulted.[26] Roughly, from this time forward, although certainly
not in every case, scenes featuring restorations of marine sauria became suffused with more naturalistic undertones, while genera of marine sauria
became increasingly subservient to the more impressive *terrestrial* sauria that
were being dug up in the American west.

Despite slow, incremental gains made between the 1820s and 1842, even
by the time of Hawkins' Crystal Palace statuesque masterpieces, dinosaurs
were generally not yet on equal footing with the more familiar fossil mammals, and wouldn't be for another half-century. While Verne, for example,
incorporated fighting, 100-foot long marine sauria, pterodactyls and fossil
mammals within his *Journey to the Center of the Earth*, no dinosaurs resided

Benjamin Waterhouse Hawkins' small statue representation of his 1854 Crystal Palace *Iguanodon*; this model is displayed at the Burpee Museum of Natural History in Rockford, IL (photograph by the author).

within the planet or at its "Central Sea" (or more correctly the "Lidenbrock Sea"). And despite the intriguing nature of Victorian dinosaurs, discoveries of horned fossil mammals made in the American west during the 1870s caused a secondary wave of sensation; resulting furor rivaled press granted to America's first dinosaur discoveries.

But, first, how did the earliest dinosaurs known to science fare in their pop-cultural battle for survival and prominency versus reigning mammalian varieties such as Great-Claw and the Mastodon?

Enter the Dinosauria

Although we now know that by 1812 Cuvier had described fossil "crocodiles" from Honfleur which, in hindsight, are dinosaurian, the first fairly well known specimens of fossil animals later categorized as "dinosaurs" were announced to scientific societies during the early 1820s. Much intrigue surrounded discovery and identification of the first two such named genera, *Megalosaurus* and *Iguanodon*. The story of how these great beasts were elucidated has often been recounted[27]; for our purposes it will be more instructive to outline the evolving course of paleoimagery resulting from such scholarly considerations.

Cope's kangaroo-limbed "Laelaps" at center; snake-tongued, serpentine *Mosasaurus* at right; *Elasmosaurus* with head placed at tail end at left. In the background at left, standing, is kangaroo-lizard *Hadrosaurus* munching on vegetation (from *American Naturalist*, 1869).

Unlike today when the latest dinosaur discoveries are excitedly broadcast on local televised news or through press releases, reports of *Megalosaurus* and *Iguanodon* at first were largely restricted from public circles. Eventually word spread from fossilists who learned of them. Collectors searched for fossils which were then sold to better trained naturalists who could more effectively preside over their remains to learned peers. Eventually, through the medium of illustrated "scenes from deep time," imagery of these animals became increasingly available in popular books. For the most part, the general public then didn't comprehend prehistory with facility like we do today; fossil reptiles known from such scanty evidence were difficult to decipher, especially without pictures suitable for laymen's comprehension. And for the literate, books were not as plentiful or as available as is the case today. And so, understandably, until the mid–1850s when Hawkins and Owen flaunted early perceptions of dinosaurs for those who contemplated prehistory, the champion prehistoric monsters of the 1820s and 1830s remained large mammalian forms.

Dinosaurs, although not referred to by that name until (1841 or) 1842, made their first appearance in living (as opposed to skeletal) form in one of the earliest depictions of a restored paleo-ecological setting, including the indigenous flora and fauna. Fittingly, since it was first of the clan to be properly described, *Megalosaurus bucklandi* became the first dinosaur introduced into a "scene from deep time." Patterned distinctly after De la Beche's *Durior antiquior*, this reconstruction was rendered by Nicolas Christian Hohe (1798–1868). It appeared in the third installment of August Goldfuss' (1782–1848) *Fossils of Germany* (1831). This illustration, titled "Jura Formation," blended contemporary paleontological understanding of the Solnhofen limestone with information on British Liassic strata.[28] A quadrupedal *Megalosaurus*, rather subdued in background scenery and hardly so "monstrous" looking, lurks toward a lake teeming with invertebrate life and aquatic reptiles while pterosaurs soar above. Looking more like a slender crocodile than the dinosaur which the current dinophilic generation would find familiar,[29] Hohe's *Megalosaurus* hardly heralds the unparalleled interest in dinosaur art that would come a century and a half later. Between 1833 and 1834, other artists couldn't help themselves from reintroducing Hohe's creeping megalosaur into their own depictions of extinct fauna printed, for example, in the widely circulated and affordable *Penny Magazine*. While *Megalosaurus* was regarded as a great carnivorous lizard, like the large crocodile-shaped animal figured by Hohe, it was generally judged smaller than its saurian rival, the far more mysterious *Iguanodon*.[30]

Sometime around 1832, Gideon Mantell (1790–1852) reconstructed *Iguanodon*'s probable skeletal anatomy in articulated form as a horned, *quadrupedal* creature. However, Mantell himself relied on a fascinating restoration, dated 1833, for his well received public lectures, a watercolor painting by George Scharf titled "Reptiles Restored," left unpublished by Dennis Dean until 1999.[31] This recreation of the Weald-Tilgate Forest flood valley region shows the earliest known life restoration of a 100-foot-long *Iguanodon* evidently based on Mantell's skeletal reconstruction looking much like an enlarged rhinoceros iguana. It is joined by plesiosaurs, birds, turtles and an ichthyosaur. At right, the latest addition to the soon to be named "dinosaurian" ranks, spiky-armored *Hylaeosaurus*, creeps toward the meandering river.

In 1838, more elaborately styled depictions of *Iguanodon* influenced by Scharf's 1833 painting appeared as frontispieces in popular treatments by George Fleming Richardson (1796–1848), a former curator of the Mantellian Museum, and then Mantell's own *Wonders of Geology*. Richardson's publication, *Sketches in Prose and Verse*, was graced with an illustration by George Nibbs, engraved by George Scharf titled, "The Ancient Weald of Sussex."[32] Here, a huge smiling *Iguanodon* appears front and center, surrounded by usual suspects belonging to the Secondary Period—crocodilians, ichthyosaurs, pterodactyls, ammonites and plesiosaurs. This creature resembles Hawkins' later 1854 sculptures of this genus, although significantly, as in Scharf's restoration, the animal's limbs appear more lacertilian than elephantine—as Owen thought they were correctly and theoretically positioned.

Meanwhile, John Martin completed another scene from deep time, "The Country of the Iguanodon," for Mantell's book, offering a triad of huge iguanodonts grappling in dimly lit, nightmarish din. Then, in the year in which Owen named the Dinosauria, Richardson's latest popular book, *Geology for Beginners* (1842), featured Martin's stylistic depiction of warring reptilian monsters (marine sauria and probably both an iguanodont and a megalosaur), this a new frontispiece titled "The Age of Reptiles."

Ralph O'Connor recently rediscovered a wall-chart titled "The Comparative Sizes of Extinct Animals," prepared for William Buckland's (1784–1856) lectures dating from "not long after 1835," showing restorations of prehistoric animals then known to science figured adjacent to several of their reconstructed skeletons.[33] In 1833, Mantell guesstimated that *Megalosaurus* grew to an astonishing 50 feet in length, while his horned *Iguanodon* grew from 70 feet to 200 feet long for his monster! *Iguanodon*'s presumed immensity, reflecting Mantell's speculations, dominates the Buckland wall-chart, as the figured 100-foot long horned iguana-like monster appears longer than the underlying mammoth, deinothere, and mastodont, all positioned in the same row.[34] Although in the case of the latter Mantell stated his belief that "its magnitude is here under-rated; for like Frankenstein, I was struck with astonishment at the enormous monster which my investigations had, as it were, called into existence ... we can scarcely err in assuming, that the living *Iguanodon* bore considerable resemblance in form to the Iguana of the present day."[35] By 1842, Owen had scaled down his calculation of *Megalosaurus'* great length to a relatively modest 30 feet, while *Iguanodon* had been whittled down to a mere 28 feet.[36] Through 1848, however, Mantell asserted that *Iguanodon* had the bulk of an elephant, occupying a station in its own "Age of Reptiles" analogous to that held by modern pachyderms.[37]

Of course, as we now know, the Iguana wasn't a good paradigm for *Iguanodon*. After better specimens were collected it became possible to see just how significantly Mantell, Owen and Hawkins had erred in their reconstructions. For, by the early 1880s, meticulous study of *Iguanodon* then known on the basis of many complete specimens, coupled with knowledge of two dinosaurs from the American east coast, *Hadrosaurus* and "Laelaps" (both known from relatively incomplete remains), helped scientists formulate clearer visions— leading to erection of a *bipedal* prehistoric dinosaurian monster.[38]

Evidence for the possibility of bipedal dinosaurs *had been* available, although men studying these trace fossils can't be blamed for their erroneous interpretations. Dinosaur tracks had been found in Massachusetts dating from the Late Triassic Red Sandstone. The Rev. Edward Hitchcock (1793–1864) devoted years of scholarly research to the study of fossil trackways, and in 1836 published a descriptive paper, "Ornithichnology: Description of the Footmarks of Birds." Yes, because so many of the tracks were three-toed like those of modern birds, Hitchcock and others believed the tracks represented shoreline movements of ancient walking birds. Except some of these birds were evidently quite massive, perhaps "almost twice as heavy and high as the ostrich."[39] Consequently, Hitchcock created the first fantastic bird-like monster (which really, unwittingly, was a bipedal, theropod dinosaur) entering the then scanty annals of prehistoric monster literature through his 1836 poem, "The Sandstone Bird," published in *The Knickerbocker*.[40]

By 1856, Philadelphia paleontologist Joseph Leidy (1823–1891) had described a handful of dinosaurs known from scanty remains from the Nebraska Territory. Two of these were assigned to genera destined, a century and a half later, to create a sensation in page-turning science fiction novels and blockbuster horror films; teeth belonging to tyrannosaurid *Deinodon*, thought then to be a megalosaur, and "raptor" genus *Troodon*. Few would learn of these discoveries outside of scientific circles.[41] But Leidy's next 1858 announcement would gain lasting impact, when he introduced America's first relatively complete, although skull-less, dinosaur—*Hadrosaurus*.

Hadrosaurus was of the type later known as "duck-billed," a herbivorous variety related to the *Iguanodon*, although most commonly represented today by such genera as *Maiasaura* or *Edmontosaurus*. Most startlingly, Leidy noted the disproportion between *Hadrosaurus*' front and hind limbs, which differed so significantly that he "was at first inclined to believe they belonged to different animals."[42] Certainly they didn't resemble the limbs as restored in any of the Crystal Palace dinosaurs restored only four years earlier. Furthermore, "the disproportion is even greater than in the *Iguanodon*, and as indicated by comparison with the remains of an individual of the latter...." Leidy was certainly on the right track, much closer to the truth than Owen had been with his rhinocerine *Iguanodon*, as would be corroborated two decades later by Louis Dollo (1857–1931) and L. F. DePauw. Leidy hypothesized that presumably amphibious *Hadrosaurus* stood "kangaroo-like," tripodally resting on its tail while browsing. This suggestion of a kangaroo lizard would have lasting implications, for by 1866, Cope would extrapolate with his impressions of the carnivorous 18-foot-long "Laelaps," which was "furnished with great powers of running and leaping."[43]

Not only did Laelaps seem kangaroo-like, owing to its relatively short front limbs as compared to the more massive hind limbs, but imaginatively Cope strayed further in assigning to it a leaping *kangaroo* style of locomotion. To him, just as carnivorous *Megalosaurus* pursued herbivorous *Iguanodon* in the European Weald and Oolite, in the eastern regions of Cretaceous North American, "Laelaps" proved the "most formidable type of rapacious terrestrial vertebrata" capable of dispatching its fateful "enemy"—web-toed *Hadrosaurus*—with its sharpened foot claws and saber-shaped teeth. The precocious, artistic, and fertile-minded Cope who enjoyed sketching the animals he described in his writings[44] imagined a probable prehistoric scene, focusing on "Laelaps," whose massive tail, points to a semi-erect position like that of the Kangaroos, while the lightness and strength of the great femur and tibia are altogether appropriate to great powers of leaping.... If he were warm-blooded, as Prof. Owen supposes the Dinosauria to have been, he undoubtedly had more expression than his modern reptilian prototypes possess. He no doubt had the usual activity and vivacity which distinguishes the warm-blooded from the cold-blooded vertebrates.

We can, then, with some basis of probability imagine our monster ["Laelaps"] carrying his eighteen feet of length on a leap, at least thirty feet through the air, with hind feet ready to strike his prey with fatal grasp, and his enormous weight to press it to the earth....

It will readily occur to the paleontologist, that the existence of creatures of the form of *Laelaps, Iguanodon,* and *Hadrosaurus,* would amply account for the well known foot-tracks of the Triassic Red Sandstone of the Connecticut Valley.[45]

Accordingly, in the aforementioned "Fossil Reptiles of New Jersey" illustration commissioned by Cope for his 1868/1870 *American Naturalist* article, one sees a pairing of kangaroo-postured dinosaurs—the first life restoration of Leidy's *Hadrosaurus,* and Cope's "Laelaps" standing on the opposite shore, perhaps poised to spring savagely upon *Elasmosaurus.* Both in *Iguanodon's* and presumably *Hadrosaurus'* cases, quadrupedal and kangaroo-lizard paradigms were soon to be essentially disproved — or seriously placed in doubt — by Dollo and De Pauw. As we'll see, thanks to the melded artistry of Cope and influential animal artist Charles R. Knight, the incongruous hopping theropod dinosaur view persisted into the 1930s.[46]

By 1868, Benjamin Waterhouse Hawkins mounted a skeletal reconstruction of *Hadrosaurus* in Philadelphia. Several plaster replicas were made for other institutions. Hawkins had intended to create life-sized sculptural restorations for a "Palaeozoic Museum" that would have been exhibited in New York City's Central Park. This display would have been entirely analogous to his Crystal Palace animal recreations completed a decade and a half earlier, although this time North America's prehistoric life would have been emphasized. *Hadrosaurus* would have been one of the central attractions. Hawkins learned more of the anatomy of his new fossil charges through onerous activities such as preparing Cope's *Elasmosaurus* skeleton, and mounting (heavily retouched) skeletons of the *Hadrosaurus* and erecting an armature for an in-progress "Laelaps" skeletal mount.[47] In giant ground sloth fashion, Hawkins' Philadelphia *Hadrosaurus* skeleton clutched a fabricated tree for added armature support.

Although the Palaeozoic Museum never reached fruition, as Hawkins' handiwork was demolished on May 3, 1871, several remarkable examples of paleoart (photos and illustrations) have survived indicating the vision and grandeur Hawkins contemplated for his museum's design. Commenting on

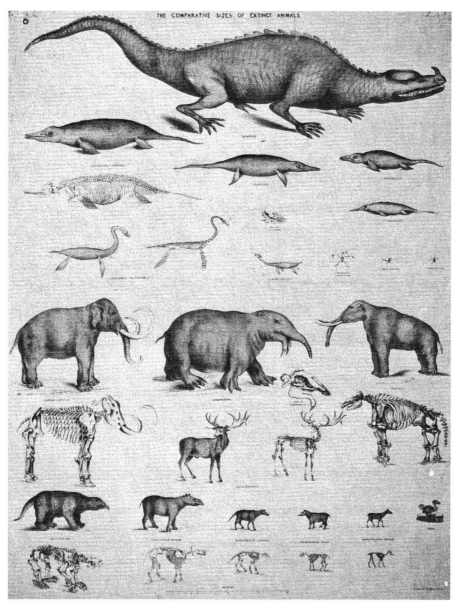

Buckland's chart of prehistoric vertebrates as known to mid–1830s science. At top is an immense *Iguanodon*, restored per Gideon Mantell's views, while a variety of other prehistoric creatures (mostly mammals and the already extinct Dodo bird) lie underneath. While images of prehistoric mammals outscore fossil sauria in this poster, note that the imposing dinosaur owns the top rung, outstripping even the marine reptiles in scale. Mastodon, near middle row at right, shown both as life restoration and skeletal reconstruction, seems diminutive relative to *Iguanodon Deinotherium* at center of the image. © Oxford University Museum of Natural History, Buckland Papers (Drawings).

the symbolic importance of one particular 1868 illustration, a piece of design work for the Commissioners of Central Park showing the planned interior, Paul Semonin stated in his 2000 book, *American Monster*, that "Hawkins' drawing of his tableau depicts the eclipsing of the mastodon by the dinosaurs. The American monster, with its trunk uplifted, appears forlornly in the background, upstaged by the immense, kangaroo-like figure of *Hadrosaurus*, turning its hind side to the former rulers of the antediluvian world.... Drawn to scale, the lithograph shows the intimidating size of the new saurian monsters, which tower over the spectators viewing the prehistoric scene from behind a railing."[48]

Perhaps in a symbolic sense (and only in the Mastodon's case), Semonin may have grasped something here. Or maybe this illustration merely represented how contemporary circumstances appeared in Hawkins' mind. As we shall see, fossil mammalia hadn't generally been eclipsed by dinosaurs yet — and wouldn't be until the early 1930s. Contemporary 1860s readers could be highly entertained by Jules Verne's *Journey* novel, for instance, with its fighting marine lizards and haunting mastodons, with nary a mention of dinosaurs even though through his reading of Figuier's *Earth Before the Deluge* (1863), Verne certainly knew of them. (English translations of Verne's novel appeared by 1871.) Few individuals actually saw the 1868 Commissioners' piece of artwork then (or for that matter any of the other similar sketched designs Hawkins kept in his scrapbook for this and other similar museum displays). So, arguably, a disproportional, (slightly) forced perspective favoring of dinosaurs in the 1868 illustration didn't necessarily reflect consensus popular opinion. In fact, prehistoric mammals remained a powerful, imaginative force to be reckoned with for decades. (Besides, the mastodon doesn't really look all that diminutive relative to the dinosaurs in the cited picture anyway.)

Less debatable than the presumption that Hawkins might have made a visual, editorial statement concerning the dinosaurs' triumph over the Mastodon, however, was imagetext projecting more accurately posed, bipedal dinosaurs. Parting with Owen's Victorian ideals, Hawkins had erected the world's first bipedal dinosaur ("Laelaps," then under construction, would have been the second). Then, between 1875 to 1877, Hawkins completed a series of 17 paintings vivifying the geological succession of life. While *six* of these paintings feature fossil mammalian fauna from the Tertiary and Pleistocene, only two paintings ("Jurassic Life in Europe" and "Cretaceous Life of New Jersey") feature dinosaurs. There, in both depictions, amidst the usual assortment of winged dragons and fighting marine reptiles, are early renditions of fierce looking ruling reptiles. In "Jurassic Life in Europe," a restoration of *Megalosaurus,* propped on all fours rather slavishly in deference to Owen snarls over an unlucky victim — a quadrupedal *Iguanodon, sans* its formerly characteristic nasal horn. In Hawkins' "Cretaceous Life of New Jersey," however, a vicious pack of decidedly *bipedal* "Laelaps" dispatch prey from a herd of (bipedal) hadrosaurs.

Clearly, the concept of bipedality in dinosaurs was in a state of flux by the mid–1870s. While by 1877, according to the most period's most prominent paleoartist, at least two dinosaur genera had assumed upright posture and locomotion, *Iguanodon* and *Megalosaurus* in popular view, were still relegated to "on all fours" status. Yet, by 1886, a restoration appearing in Camille Flammarion's (1842–1925) *Le Monde avant la creation de l'homme* would have seemed entirely radical. Here we see *Megalosaurus* and *Iguanodon* battling in upright, erect position, quite a departure from how Riou positioned this, already by then, traditional pairing of prehistoric monsters nearly a quarter of a century earlier! What had happened between 1877 and 1886?[49]

In 1878, a marvelous deposit of *Iguanodons* had been discovered feet over 300 meters underground in a Belgian coal mine near the town of Bernissart. Thirty-one individuals were excavated from the subterranean setting (a circumstance which Verne must have appreciated!). By 1883, the indefatigable Louis Dollo meticulously demonstrated that this genus was a bipedal animal, and furthermore that its alleged horn was really a spiky thumb that may have been a defensive adaptation, concluding with conviction that "*Iguanodon* would, therefore, belong to the bipedal category due to the structural difference between its fore- and hind limbs.... Therefore, this Dinosaur must have walked upright (been bipedal)."[50] Furthermore, Dollo asserted that "not only did *Iguanodon* stand upright, but, further, that they *walked* and *did not jump....* Our Dinosaur must have simply allowed its caudal appendage [i.e., its tail] *to drag* along the ground, and the impression thus formed was certainly very weak because it has not been preserved."[51]

Dollo's words, and circulation of his 1883 report with De Pauw's definitive published skeletal reconstruction certainly set dinosaurs on a new footing, while unwittingly paving the way for 20th century fandom's most beloved upright, tail-dragging prehistoric movie monsters. Dollo's bipedal "reptilian-giraffe" dinosaur model differed considerably from Owen's theoretical elephantine perceptions. As Allmon stated in 2006, Owen, inventor of the Dinosauria, suffered as he "was never again a significant influence in dinosaur science."[52] From this time forward, restorations of the most monstrous seeming (carnivorous) dinosaurs became characteristically bipedal — as exemplified in early artistic restorations and reconstructions of the early 1890s— such as Joseph Smit's *Megalosaurus*, and O. C. Marsh's (1831–1899) *Ceratosaurus*.[53] Regardless, the world wouldn't easily let go of the idea of leaping "kangaroo-dinosaurs."

As in the case of sparingly few complete specimens, especially armored forms such as spiky *Hylaeosaurus*, paleontologists realized that not all dinosaurs were bipedal. Given that a few stegosaurs were known before 1854, one may wonder why Hawkins and Owen chose not to add a plated variety of dinosaur to the "Secondary" monster island at Sydenham? And here is the story of the world's third (or arguably fourth) most popular dino-monster

Stegosaurus—so often matched with its traditional horned *Ceratosaurus* adversary.

While *Stegosaurus* remains one of our most familiar dinosaur genera today, at the outset there was considerable confusion as to its accurate appearance, especially in the early years following its 1877 scientific description. Such flux and controversy in opinion was captured in period paleoart and paleoimagery, some of which will be recounted here. Years later, thanks to the magical skills of Marcel Delgado (1901–1976) and Willis O'Brien (1886–1962), who in turn were inspired by Charles R. Knight's artistry, *Stegosaurus* became a "star" dino-monster in two films signaling the prehistoric mammals' waning bid for pop-cultural supremacy. But we'll come to that. For now, few may realize that stegosaurs were known from South Africa and England decades prior to *Stegosaurus* and it is curious that none were ever restored in paintings or in sculpture by the period's master paleoartist—Hawkins.

A Crystal Palace Stegosaur?

Stegosaurus is *so* familiar and so unmistakably a dinosaur, yes, *the* quintessential plated dinosaur in fact, that many casual dino-philes may not realize *other* plated dinosaur genera have been described, and that remains attributed to at least 3 genera of stegosaurs (i.e., "Omosaurus" as equaling *Dacentrurus*, *Paranthodon* and *Craterosaurus*) were discovered before discovery of *Stegosaurus*. The usual perception of the stegosaurian clan is that they were exclusive to North America. But European *Dacentrurus* and *Craterosaurus*, as well as African *Paranthodon* stegosaurs entered the picture early, even before *Stegosaurus* had been described and well before the now familiar picture of staggered double-row plate geometry with defensive tail spikes had been conceived.

What did contemporary scientists think about these unusual remains? Why didn't European scientists first devise our iconic image of a plated dinosaur decades before Othniel C. Marsh described *Stegosaurus*, or before Edward D. Cope described "Hypsirophus" (1878, really a composite of *Allosaurus* and *Stegosaurus* material)? There was even opportunity for Benjamin Waterhouse Hawkins to have sculpted a Victorian stegosaur alongside his triad of British dinosaurs—*Hylaeosaurus*, *Megalosaurus* and *Iguanodon* for the Crystal Palace's prehistoric island landscape during the early 1850s. But fate wouldn't allow this opportunity, allowing stegosaurs to become popularly regarded as a variety of dinosaurs indigenous to North America, although that perception is utterly false.[54]

During the mid–1970s flowering of the dinosaur renaissance, Adrian J. Desmond wrote of a fossil reptile described by Sir Richard Owen—"Omosaurus" (or "humerus saurian," a name first applied by Owen in 1874 to these remains), which he (Owen) had originally suspected was another variety of marine whale-lizard, in today's terms a "sauropod" known as a Cetiosaur, although now understood to have been a long-necked sauropod. Because you too may find the tale "humerus," it shall be recounted here.

From the five fused hip bones, found in 1874 in a Kimmeridgian age (Late Jurassic) clay pit at the Swindon Brick and Tile Company, Wiltshire, Owen correctly reasoned that his new alleged cetiosaur was another genus of dinosaur. Desmond cited Owen's contemporary description of the animal's well preserved tail, that "would probably be exercised, as in the largest living Saurians in delivering deadly strokes on land, as well as in cleaving a rapid course through the watery element."[55] Owen never realized how precocious was his assessment of the beast's "attack organ." For "Omosaurus" actually was not another genus of cetiosaur. Instead, "Omosaurus" was a plated dinosaur, akin to North American *Stegosaurus*.

Owen's 1875 published description of the partial remains of "Omosaurus" *armatus* became the first scientific description of a stegosaur understood to be a dinosaur instead of some other kind of fossil reptile. In 1890, British paleontologist Richard Lydekker (1849–1915) reassigned "Omosaurus" *armatus* to the genus, *Stegosaurus*, although this interpretation was reversed three years later by H. G. Seeley (1839–1909). The name "Omosaurus" was later found to be already in use and changed to *Dacentrurus* ("very prickly tail") in 1902.[56]

Meanwhile, the first fossils from South Africa later recognized as dinosaurian in nature were of a stegosaur. In 1845, nine years before the Crystal Palace exhibition of 1854 opened, two British naturalists Dr. William Guybon Atherstone (1813 or 1814–1898) and Andrew Geddes Bain (1797–1864) in the company of three children, identified fossilized remains of a stegosaur at a Lower Cretaceous locality along the Bushmans River near Port Elizabeth on the Eastern Cape. The actual fossils were apparently found by Miss Jeanie Bain.

Although Atherstone and Bain couldn't envision a plated dinosaur from the incomplete remains, as related in Dr. Billy de Klerk's article, "The first dinosaur fossil discovered in South Africa: The Stegosaur *Paranthodon africanus*,"[57] Atherstone suspected the new animal was dinosaurian, possibly even related to *Iguanodon*. Hence, the given name—"Cape *Iguanodon*." Hoping to learn more of their discovery, represented by an upper jawbone, two additional skull fragments, yet altogether several "bones bigger than those of an ox" (including a leg and hip bones), in 1849 and 1853, Bain shipped fossils including those of the "Cape *Iguanodon*" to Richard Owen at the British Museum. The leg and hip bones were evidently lost prior to Owen's scientific description of the creature.

Owen published a description in 1876, renaming the "Cape *Iguanodon*" as "Anthodon" serrarius (or flower tooth), and in 1882, American paleontologist Marsh included "Anthodon" in his erected dinosaurian family — Stegosauridae. However, as Lydekker noted in 1890, Owen mistakenly mixed the skull of a pareiasaur with "Cape *Iguanodon*" material. Realizing that pareiasaurs didn't survive beyond the Permian period, Lydekker surmised the Lower Cretaceous dinosaur fossils came from the Bushman's River locality. (Bain had collected the pareiasaur skull in the Karroo formation.) Today we know that Bain's pareiasaur was a hundred million years older than the "Cape Iguanodon."[58]

Now stegosaurian *Paranthodon* (as "Anthodon" was renamed in 1929), had Owen ever gotten around to describing it shortly after Bain shipped him its fossils, would have made a marvelous addition to the menagerie still exhibited on the prehistoric island at Sydenham. So why wasn't one attempted?[59] The circumstances may seem even a bit more condemning after learning that yet another stegosaur, this time a *British* discovery, was made about the time Owen became aware of South Africa's *Paranthodon*.

Fourteen decades following Hawkins' Crystal Palace successes, in 1993, artist and writer George Olshevsky (with Tracy Ford) determined that two other genera known from British Upper Wealden (Early Cretaceous) specimens were probably stegosaurid in nature. These dinosaur genera, the poorly known *Regnosaurus* and *Craterosaurus*, in fact, according to Olshevsky, may have been congeneric (i.e., *Regnosaurus* equaling *Craterosaurus*). *Regnosaurus northamptoni*, known from a single portion of right lower jaw discovered in 1838,

A restoration of battling *Iguanodon* (right) and *Megalosaurus* (left), for the first time drawn, together, with upright posture. This figure was published in Flammarion's *Le monde avant la creation de l'homme. Origines de la terre. Origines de la vie. Origines de l'humanité* (Paris: C. Marpon and E. Flammarion, 1886).

was identified by Gideon Mantell in 1841 (and later redescribed in 1848) as an iguanodont. But Owen regarded *Regnosaurus* as synonymous to *Hylaeosaurus*. In 1969, the specimen was reclassified as "hylaeosaur" and then subsequently, in 1971, as sauropod. Olshevsky referred to *Regnosaurus* as "a rare Early Cretaceous stegosaur ... by a wide margin the first stegosaur ever to be scientifically described."[60]

So why didn't Owen persuade Hawkins to sculpt a plated dinosaur for the Crystal Palace exhibition of 1854? Given that Owen didn't describe *Paranthodon* until 1876, was he simply preoccupied with other matters, (as he was years later at the time of armored *Scelidosaurus*' 1858 discovery)? Is the reason for why *Paranthodon* (known about as incompletely as were *Hylaeosaurus* and *Megalosaurus* at the time) became neglected by Owen for so many years simply because he had too many conflicting obligations and professional responsibilities? Or was Hawkins himself too strapped for time to complete yet another sculptural undertaking of a dinosaur that in his mind may have appeared quite similar to his *Hylaeosaurus*? Was the Crystal Palace funding insufficient for construction of more than the original trio of life-sized dinosauria?

Regnosaurus was not recognized for its stegosaurian affinities until 1993, so its existence couldn't have reinforced impressions of *Paranthodon* (for which no plates or spines were then associated) as a unique breed of plated dinosauria known to Victorian science.[61] Therefore, Owen couldn't conjure a meaningful or distinctive image of a dinosaur that was plated as in the North American *Stegosaurus* that would be described a quarter of a century later, with which to inspire Hawkins' artistic visions. In personal communication dated 2002, Dr. Ken Carpenter said, "Regarding *Paranthodon* at Crystal Palace, considering that all that is known of it is the snout, I doubt that Hawkins could have made a reconstruction, certainly not as a spiny animal." And in fact, during the years 1842 to 1858, Owen did regard *Regnosaurus* as a junior synonym for *Hylaeosaurus*.

A century and a half later, such questions are difficult to answer.

Earliest Stegosaur Restoration?

Reading Marsh's brief and rather sketchy published 1877 *Stegosaurus* description today reinforces how little was then known of this new animal. Marsh even wrote to Owen, noting that "Stegosaurus was one of the strangest of animals." Most of its early remains were discovered in the famous Morrison formation of Upper Jurassic age. The same beds from which bones of "Brontosaurus," *Diplodocus*, *Allosaurus* and *Ceratosaurus* were disinterred during the later 19th century. More *Stegosaurus* remains were discovered between 1879 to 1887, especially from fossiliferous Quarry 13 at Como Bluff,

Wyoming. In 1878, Marsh's great scientific rival, Cope, discovered specimens which he named "Hypsirophus," later reassigned by Marsh to *Stegosaurus.*

While the first relatively accurate image of the plated dinosaur prepared under Marsh's direction appeared as Plate 9 in vol. 42 of *American Journal of Science* (1891),[62] few are aware of or have seen another image published in *Scientific American* only seven years after Marsh announced the new genus. This description was published on November 29, 1884, page 343, in an article featuring news about *Iguanodon* and "Brontosaurus." A short caption to the illustration in question (Fig. 3 in the two page article) indicates, "American landscape of the Jurassic Epoch with reptiles and plants of the period." The artist was A. Tobin, whose illustration was engraved by "Vermorckin." A peculiar bipedal animal at foreground seizes attention, as this appears to be an upright sauropodous creature, outfitted with many rows of spines. What hath nature wrought?

Further explanation was provided on page 344 of the *Scientific American* issue.

If, through the admirable discoveries that have been made in recent years, we endeavor to bring to life again the fauna of the Upper Jurassic period in the United States, we shall find one that is no less rich and strange than that of the Old World. Here we have, amid araucaria and cycads, the gigantic stegosaurus, with a body clothed with bony plates and spines, that formed a powerful armor for it, and with forelegs much shorter than the hind ones; the compsonotus [*sic*], with forepaws equally as well developed as the hind ones; and the strange flying reptiles, the pterodactyles (Fig. 3).

The next paragraph in the article goes on to describe the "Brontosaurus," referring to Marsh's 1883 skeletal reconstruction, labeled as Fig. 4 in the article. So, evidently, the upright, spined sauropod in Fig. 3 was indeed intended to represent *Stegosaurus*!

When Marsh described *Stegosaurus armatus* in *American Journal of Science*,[63] his 1877 article was not illustrated. But he speculated as to *Stegosaurus'* life appearance, claiming "The limb bones indicate an aquatic life.... The body was long, and protected by large bony dermal plates, somewhat like those of *Atlantochelys* (*Protostega*). These plates appear to have been in part supported by the elongated neural spines of the vertebrae." He continued, "The present species was probably thirty feet long, and moved mainly by swimming." Marsh acknowledged that *Stegosaurus'* bones were found in association with a sauropod, "near the locality of the gigantic *Atlantosaurus montanus*, and in essentially the same horizon." Immediately following his published description of the *Stegosaurus* in the *American Journal of Science* issue, in the same issue, Marsh proceeded with description of sauropod genus *Atlantosaurus*.

The curious animal depicted in the 1884 *Scientific American* issue as Fig.

Frank Bond's curious 1899 restoration of *Stegosaurus*, showing excessive numbers of armor plates covering its back and tail, with equally excessive numbers of tail spikes protruding, porcupine-like, from its dorsal region. Note how similar this restoration appears to Toho's *daikaiju* "Anguirus," introduced in 1955, albeit without the array of facial horns (from Charles W. Gilmore's "Osteology of the Armored Dinosauria in the United States National Museum, with special reference to the genus *Stegosaurus*," *Bulletin of the United States National Museum* 89, 1914).

3 is adorned with 5 rows of spines from the shoulder area to the hips, where the longest vertebral row gives way to 18 to 19 dermal plates. Marsh evidently was confused by *Stegosaurus*'s "teeth" several of which "are cylindrical, and were placed in rows ... are especially numerous, and may possibly turn out to be dermal spines." Although the 1884 *Stegosaurus* restoration has large plates over its caudal tail vertebrae which are noted as plates in *Scientific American* text, they appear as enlarged crocodilian scutes. Marsh stated that "one of the large dermal plates was over three feet (one meter) in length." From rough proportions, it is certainly possible that several of the most anteriorly positioned plates along the tail as seen in the 1884 restoration approached that size.

Still, there are far too many spines, all placed incorrectly in the 1884 restoration, none at the end of the tail. Possibly those extra "teeth" gave someone an idea that there were spines of varying sizes situated all over the body, like British artist Benjamin Waterhouse Hawkins restored the Crystal Palace quadrupedal *Hylaeosaurus* in 1853 under Richard Owen's guidance. Then too, close proximity of admixed sauropod bones may have resulted in the restored animal's elongated neck.

According to Marsh's chief collector, Arthur Lakes (1844–1917), when the block of *Stegosaurus armatus* was pried open, "We broke open the block in which it lay and exposed twelve long black enamelled spines.... There was a pair or two sets of these spines side by side six a piece with two small hour glass shaped bones close to them."[64] Because several fossils found in conjunction with the *Stegosaurus* remains turned out to be *Diplodocus* teeth and limb bones, is it possible that some of these spines may have been sauropod vertebral spines? In 1992, sculptor Stephen Czerkas suggested that conical spines were arranged in a row along the tail, neck and body of sauropods like *Diplodocus* and *Barosaurus*.[65] But in 1877, Marsh had apparently already noted these; he may have misinterpreted their nature. So Marsh may have had a mixed bag of sauropod teeth, spines and stegosaurid spikes, which he perceived collectively as stegosaur armor. (Incidentally, the 1884 restoration therefore also ranks as one of the oldest known showing a "sauropod" in bipedal pose.)

The *Scientific American Stegosaurus* restoration of 1884 seems to be a unique sauropod-stegosaur hybrid. Much of it was guesswork, and seems so curious today, especially when compared to all the paleoimagery to which we've become accustomed.

Within a few short years, by 1901, artists had completed more new restorations of *Stegosaurus* than of any other dinosaur. Since a representative sequence of such restorations has been presented elsewhere, the reader is referred to figures printed here with explanatory captions and the accompanying footnote for an outline of the pictorial/conceptual (imagetext) pathway taken toward *Stegosaurus*' proper anatomical restoration.[66]

By the 1880s, both dinosaurs and their fossil mammal cousins had been proven extinct. Some dinosaurs, like "Laelaps" especially, doubtlessly had fierce dispositions in life, while (discounting Koch's Missourium) representatives of both groupings (e.g., *Stegosaurus* and *Megatherium*) truly were massive beings. But in size no terrestrial mammal could match Dinosauria's greatest: brontosaurs, or more properly, the sauropods! By human standards, contemplating such monstrosities, their absurd giraffe-like necks and utter hugeness, must have seemed intimidating. By 1877, knowledge of these ruling Jurassic giants was germinating in the minds of foremost paleontologists such as Owen and Thomas Henry Huxley (1825–1895), who debated the provenance and natural history of the British sauropod *Cetiosaurus*.[67]

Next, we'll outline how the tale of how the late 19th and early 20th centuries' fourth (or arguably third) most popular dino-monsters, for a time collectively called "brontosaurs" and later sauropods, rose to prominence as dino-monsters.

Most Colossal!

While Owen and Dollo considerably trimmed Mantell's wilder speculations as to *Iguanodon*'s length and proportions, eventually dinosaurs approximating the 100-foot-long milestone were discovered. A century ago, however, it was North American "Brontosaurus" (now, for purists, known as *Apatosaurus*) that fired imaginations. Both Marsh and Cope discovered representatives belonging to this colossal clan. In 1877, Cope pictorially reconstructed *Camarasaurus*, while in 1883, Marsh published a skeletal reconstruction of "Brontosaurus," the earliest, clear true sauropod portrayal to be widely circulated by the press. One of the earliest life restorations of "Brontosaurus" soon followed, appearing in Hutchinson's *Extinct Monsters* (1893).[68] Here, artist Joseph Smit depicted one brontosaur partly submerged underwater while its companion stands upright on the shore. Thanks to the popular press and other promotional devices, this dino-monster generated so much furor and became so widely known that it was eventually cast in various "special effect" starring film roles, especially in the two science fictional movies facilitating and in a sense celebrating the dinosaurs' pop-cultural triumph over fossil mammalian challengers.

Sixty-foot long "Brontosaurus" was soon eclipsed in length by its evolutionary cousin, 90-foot long *Diplodocus* which during the 1900s became the world's first cosmopolitan dinosaur. Relying upon several examples of paleoimagery and paleoart, paleontologists hotly debated how such massive creatures moved and the extent to which they may have been amphibious.[69] Outside scientific circles, however, through the late 1930s, popular consensus maintained that "Brontosaurus" preferred to wallow in swamps to lighten the corpulent load upon their weary limbs and tired feet. Largely, this was resultant of influential Charles R. Knight's many striking, turn of the century examples of his magnificent artistry — particularly one atmospheric, painted restoration completed in 1898 for New York's American Museum of Natural History. Here, a herd of brontosaurs feeds tranquilly in swampy muck haven; one individual stands on the distant shore standing sideways to the viewer, while a central figure cautiously regards the human intrusion.[70]

Coupled with Knight's 1897 illustration of Cope's *Amphicoelias* for *The Century Magazine*, as described in William Ballou's accompanying article, the image of aquatic, sauropods became ingrained in the hearts and imaginations of America's first generation of dino-philes. Ballou wrote,

> The amphibious dinosaur, *Amphicoelias altus* [Cope], was one of the most remarkable of the tall types. It lived in water, but never swam; it walked on the bottom, indifferent whether its head was above the surface.... The weight of the animal might have been three or four tons. It probably never came out of the water. Had it done so, the great weight of the structure might have

caused a collapse. The water was its safeguard, for its surrounding weight held its gigantic frame together.[71]

While Marsh presumed that "Brontosaurus" walked with its legs positioned quadrupedally like an elephant, Cope initially (before the massive front limbs had been interpreted) considered even more dramatic poses, such as the contrasting possibility that similar dinosaurs such as the *Amphicoelias* may have also stood upright tripodally, or even walked upon its hind limbs. Cope's visions seemed to have germinated in Knight's mind, transforming into a live titan, as in the case of a (circa) 1900 restoration showing a sauropod standing in tripodal position.[72]

Following further brontosaur discoveries made in the American west (i.e., *Diplodocus*), in a November 1898, the *New York Journal* seized the day. Below the riveting headline, "Most Colossal Animal Ever on Earth Just Found Out West," an artist drew a Brontosaurus standing upright on its hind limbs, gripping the building structure with its front limbs while gazing into a skyscraper window. Diminutive people stand near the dino-monster's big feet. The caption to this eye-catcher reads, "How the Brontosaurus giganteus would look if it were alive and should try to peep into the eleventh story of the New York Life Building."[73] Of course, for psychological appeal, the dino-monster was drawn much larger than in reality for effect (and to increase magazine sales). This picture foreshadowed the Hollywood phenomenon of the coming century. For example, a similar stop-motion animated scene of a brontosaur trampling London would be filmed by Willis O'Brien and Marcel Delgado for their 1925 silent film, *The Lost World*. From here it was a short (yet colossal) set of strides to Godzilla, Gorgo and fellow *daikaiju* (to be discussed in Chapter Eight).

Emblazoned on the same *New York Journal* page, readers would have spied a horned, toothy skull attributed to the "Brontosaurus in Wyoming," when in fact this caption was erroneous. Instead, the figured skull belonged to one of North America's earliest carnivorous dinosaurs to be established on the basis of relatively complete remains and whose skeleton would be confidently reconstructed by Marsh in 1892 — *Ceratosaurus*. By then it seemed clear that such theropods were indeed bipedal tail-draggers. *Ceratosaurus* seemed a worthy would-be foe for the Jurassic sauropod giants as well as the plated, defensively outfitted *Stegosaurus*. Soon restorations of this new dinosaur began to appear, particularly those by Joseph Smit, Knight, Charles Whitney Gilmore, Gerhard Heilmann, and sculptor Joseph Pallenberg, fueling the public's thirst for knowledge of these amazing prehistoria.[74] A *Ceratosaurus* prop was later cast in Director D. W. Griffith's 1913 silent film, *Brute Force*, the first movie in which cavemen combated dinosaurs. This horned creature, which snapped its jaws and twitched its tail with accompanying eye movements, therefore became the first dino-monster movie recreation.

An 1884 idealized landscape populated by Jurassic denizens, including a "sauropodous-looking" Stegosaur (at center).

Griffiths' *Ceratosaurus* was relatively static, although it rotated seesaw-like at the hip.

Also during this period, Marsh published early descriptions of a Cretaceous horned dinosaur named *Triceratops* whose skeleton was reconstructed in 1891. But "three-horn" and its adversary — an even more famous contemporary — will be the subjects of the next chapter. Instead, let us consider how fossil mammalian forms fared during this late 19th century period of, admittedly, escalating dinosaurian challenge and intrigue. On a pop-cultural level, were fossil mammals finally falling by the wayside? Not quite.

During the early 1820s, Buckland had already popularized fierce, prehistoric mammals, resultant of his cave explorations. Despite their proverbial fifteen minutes of fame, however, Buckland's fossil hyenas couldn't enduringly capture the hearts and imaginations of fossil enthusiasts, not when beasts such as Great-Claw, Mastodon, horned woolly rhinos, and the soon to be named *Deinotherium* lurked within Time's haunting recesses. Intrigue surrounding western North America's even more exotic-looking bestial mammalian Tertiary forms, described later in the 19th century, eventually eclipsed all of these.

A Chimeric Dino-monster

One of the most famous "dinosaurs" of all time wasn't a true dinosaur, but instead a 250-million-year-old creature which in an evolutionary sense was much closer to humans than dinosaurs![75] This animal received lasting fame through Knight, whose Cope-inspired painting of a magnificent fin-backed reptile graced Henry Fairfield Osborn's 1897 *Century Magazine* article.[76] This animal, known as the "Naosaurus," in form so recognizable to dinophiles today, would seem dinosaurian in aspect. Despite its suggestive appearance, indeed, such animals were more mammalian than reptilian. Furthermore, perplexedly, Cope's and Knight's classic "Naosaurus" is a chimera that never existed. Because they were extinct *relatives* of mammals, it is appropriate to examine fin-backed "Naosaurus" among other "mammals" then coming to light and becoming increasingly popularized during this period when popular interest in dinosaurs rapidly escalated.

As stated by paleontologist Martin Lockley, 10-foot long, fin-backed creatures of the Permian period, such as *Dimetrodon*, "with its unmistakable fin [or] sail [are] irresistible to authors and artists compiling books on prehistoric animals. As a result, *Dimetrodon* is a celebrity and makes many frequent cameo appearances alongside various other famous dinosaurs."[77] The underlying reason for *Dimetrodon*'s celebrity status is Knight's superb artistry, beginning with his 1897 restoration of Naosaurus. Understand, however, that Cope was the mastermind behind Naosaurus. Cope unwittingly erred in early scientific descriptions of the sail belonging to *Dimetrodon* (described in 1878), and, later, Naosaurus (described in 1886). While the cascade of life restorations resulting from Cope's inaugural visions, along with the associated, evolving stream of scientific interpretations surrounding the natural history of such creatures has been addressed elsewhere,[78] it is appropriate here to outline how Naosaurus became such a familiar fossil form, before fading from the mainstream by the mid–1910s.[79]

Cope had mistakenly attributed bony "crossbars" positioned along the sail of a herbivorous pelycosaur known as *Edaphosaurus* with its carnivorous contemporary, *Dimetrodon*. Cope first sketched his vision of Naosaurus, later encouraging Knight to sculpt the animal. Knight then proceeded to a painted restoration which was published in *Century Magazine*. This Naosaurus, a most impressive sight, was nestled within pages of the same popular publication along with several of his other now most famous painted restorations (e.g., "kangaroo-limbed" *Hadrosaurus,* leaping "Laelaps," and horned, thorny-looking *Agathaumas*). Grinning "Naosaurus" reclines at front and center, while *Dimetrodon* lurks in the background. Both long-tailed, sharp-toothed creatures are evidently carnivorous. Knight modified this painting (although still dated 1897), altering the toothy, yet edaphosaurine head shape of Naosaurus and divesting it of those sail "crossbars," thus conforming with the appearance of *Dimetrodon* as understood today. However, now his former "*Dimetrodon*" standing in the distance (presumably transformed into an "*Edaphosaurus*") has those crossbars visibly added to its sail.

Knight went on to complete several other different looking naosaur sculptural varieties, as well as more accurate *Dimetrodon* depictions. In 1907, Henry Fairfield Osborn oversaw the mounting of "Naosaurus'" skeletal frame at the American Museum of Natural History. This resultant, a short-tailed carnivore with its admixed edaphosaur sail bones and dimetrodont head, well characterizes the confusion then surrounding this formerly mysterious creature.

Then, in a popular 1908 article, Dr. E. C. Case reconsidered the case for Naosaurus, suggesting that Naosaurus was anatomically more aligned with herbivorous *Edaphosaurus.* By 1914, Case produced an accurate restoration of *Edaphosaurus cruciger.* This led to Knight's essentially correct visions of *Dimetrodon* and *Edaphosaurus* (which he still stubbornly referred to as Naosaurus) of the mid–1920s on through 1942.

Quite possibly, due to Knight's early influential artistry promoting Naosaurus, a "fin-back"—*Dimetrodon*—would make its most sensational public appearance in 20th Century–Fox's *Journey to the Center of the Earth* (1959), since seen by millions, starring James Mason and Pat Boone. This wasn't the first or only time that fin-backed creatures had been indoctrinated into the movie-making arena. Nonetheless, *Journey's* 1959 dimetrodonts, thanks to movie special effects "magic" scaled much larger than life into a terrifying herd of 70-foot-long beasts, were the best pseudo-saurs ever captured "live" on film. ("Fins" were glued convincingly to the spines of live rhinoceros iguanas; Gideon Mantell, discoverer of the Wealden "iguana tooth," would have marveled at the scenes!). For producing their compelling visual special effects, in 1960, L. B. Abbott and James B. Gordon earned Academy Awards.[80] Ironically, however, despite *Dimetrodon's* proto-mammalian phylogenetic status, most audiences erroneously interpreted *Journey's* convincing fin-backs as dinosaurs!

"Theres" Were There

In amassing scientific laurels while sating deep ambitions, during the 1870s and 1880s, Othniel Marsh and Cope sustained general interest in North

American fossil mammals of the Tertiary Period. Consequently, strange horned mammals known as uintatheres gained equal footing with rival dinosaurs then coming to light. Marsh's and Cope's fossil reconstructions garnered widespread attention through the popular press, news of their maniacal competition for priority in scientifically naming new fossil species leaked its way into public opinion via newspaper editorials. While their feud — America's first major paleontological controversy — eventually became a rich source of ridicule, Marsh and Cope passionately stoked the fires of paleontological investigation. The spirit of these times was merrily caricatured, such as in an 1890 *Punch* cartoon portraying "Ringmaster Marsh" putting elephant-sized Tertiary uintathere and titanothere charges through their paces in a skeletal circus.[81]

Uintatheres were tusked, knobby-horned mammals, first evolving in North America 15 million years following the dinosaurs' Late Cretaceous extinction. At ten feet long, they were the largest mammalian varieties alive during the Eocene. They were impressive looking beasts, and in elucidating their natural history, Marsh and Cope zealously strove to outdo one another. Both men were brilliant. Cope was particularly prolific, although in his scientific judgments he was a bit more mistake-prone. Marsh is usually assessed as a rather more cautious and patient fellow. Among their peers, Cope was regarded as more congenial than Marsh. In the case of fossil unitatheres, Cope struck early, yet misguidedly.[82] It was as if men such as Cope, Marsh, and later Cope's protégé Osborn were the real "mad scientists" of their times, dashing off outlandish evolutionary theories, invoking strange forces they could hardly comprehend.

Writing to his father in September 1872, Cope described the uinthathere collected in Wyoming's Uinta Mountains, *Eobasileus*. "This was a monstrous animal, and Elephantine in size and proportions. Its skull is three feet long, & the hips five feet across. The head of the femur of one is as large as the top of my hat. It stood shorter in the legs than an elephant, and was proportioned more as in the Rhinoceros, but was twice as large as the living of the latter.... In a word Eobasileus is the most extraordinary fossil mammal found in North America, & I have good material of illustrating it."[83] Cope really knew how to rouse interest in a specimen that was relatively unknown, even in the absence of a good life restoration.

Illustration was forthcoming — although not until after a peculiar cascade of errors garbled the naming of one of the names he assigned in August 1872 to a unitathere, "Loxolophodon." Meanwhile, Marsh was studying bones belonging to uintatheres as well. He industriously beat Cope to the punch with his naming of "Dinoceras mirabile" and a larger form which he named "Tinoceras." Actually, both men were merely erecting new (junior) synonyms for the valid genus Leidy had already properly named only weeks before on August 1, 1872 — *Uintatherium*. Despite their ardor concerning who had named which (invalid) genus first, soon Cope and Marsh began to feud over how such beasts should be correctly restored in life.

While Cope believed that uintatheres were elephantine in nature (i.e., related to the Proboscidea), Marsh felt they were more closely allied to the Perissodactyla (e.g., horses and rhinos). Cope insisted that uintatheres had elephant-like trunks, whereas Marsh could find no evidence for such organs among his many "Dinocerata" specimens. Cope also believed that uintathere canine teeth were analogous to the tusks of elephants, a conclusion which Marsh emphatically denied. However, Marsh erred in speculating (as had scientists so many decades before in considering the Mastodon's presumed diet) that presence of such enlarged canines implied carnivorousness. And while Marsh soundly accused Cope of not paying heed to Cuvier's rules of anatomical correlation, as relating to elephant teeth, Marsh defied Cuvier's doctrine with his claim of a hoofed ungulate (which according to Cuvier's rules should be herbivorous) that had become (unusually) adapted to an car- nivorous diet. Marsh was evidently comparing uintathere canines to those of saber-tooth tigers, which were carnivorous.[84]

A section of Charles R. Knight's 1898 restoration of the "Brontosaurus" (i.e., *Apatosaurus*) wallowing in swampland (from H. F. Osborn's *The Origin and Evolution of Life on the Theory of Action, Reaction and Interaction*. New York: Scribner's, 1918 ed.).

Cope's visions of the "Loxolophodon" translated into at least two restorations published by Edwin Sheppard in 1873, and in 1879 by Mary Gunning. Cope's own sketches of *Eobasileus*, dated January 12, 1873, as published in Osborn's *Cope: Master Naturalist*, 1931 show a creature of elephantine proportions.[85] These early life restorations prepared under Cope's direction, or derived from his expertise, possessed both trunks and elephant ears. In Gunning's restoration the cranial horns have transformed into deer-like antlers. A decade later, under Marsh's advisement, museum technician J. H. Emerton's reconstructed a skeletal "Dinoceras" made of papier-mâché. Several "Dinoceros" copies were delivered to a number of institutions, such as Chicago's Field-Columbian Museum and the British Museum of Natural History. While Marsh experimented with two-dimensional, illustrated skeletal reconstructions, only on rare occasions would he condone or acknowledge the value of life-sized, three-dimensional recreations.[86]

Utility of Fossil Mammals

In the early going, dinosaurs simply were overshadowed by weight of evidence of fossil mammals and birds in the *theoretical* arena, that is, toward comprehending Life's history, the nature of our planet's evolution and magnitude of natural processes occurring in the geological past. Certainly it seemed, marine reptiles gave paleontologists a clear window into the nature of the Secondary Era, and during the 1840s Owen even enlisted the Dinosauria in an *ad hoc* crusade against the Lamarckian evolutionists. But Buckland's once captivating hyenas, and captivating occurrences of fossil rhinos, cave bears, the Irish Elk and mammoths were far more carefully considered in light of the 19th century's most puzzling faunal geological transition of all — not the famous K-T transition so heavily popularized during the 1980s, but that separating historical periods from darker ages of most recent prehistory, later elucidated as the recent Ice Age interval.

During the 1820s, Buckland had described vertebrate fossils collected in cave deposits, particularly the Kirkdale Cave in Yorkshire. Conybeare's resulting cartoon engraving, showing a frightened Buckland entering the cave only to witness, not old bones, but living restorations of four rabid-looking hyenas was fortified by a whimsical poem, "The Hyena's Den at Kirkdale near Kirby Moorside in Yorkshire, discovered in A.D. 1821." While for many years this was thought to be the first completed, accurate restoration of fossil mammals posed as realistic, animated creatures in an early example of a "scene from deep time," Ralph O'Connor has recently discovered evidence in another "now lost" illustration, quite possibly used as a visual aid for Budkland's lectures, dated 1823, showing lively restored fossil hyenas predating Conybeare's.[87] In another amusing sketch cited and reproduced by O'Connor, extinct bears are "resurrected" from fossil bones and then put through their paces in circus style by a paleontologist necromancer at Bavaria's Gailenreuth Cave. News of the startling discovery of "antediluvian" cave ecosystems was circulated among the scientifically literate as well as other interested "selected gentlemen both in Britain and on the Continent," and other "serious amateur naturalists."[88]

Beyond the humor imbued in these illustrations, and accompanying doggerel, was the import cave remains posed for theoretical geology. For Buckland, the consummate diluvial geologist, interpreted such remains as having resulted from a deluge that swept the bones and encasing gravels therein. To him, this great (allegedly antediluvial) event, suddenly depositing relics of the biblical deluge, represented the most recent of Cuvier's geological "catastrophes" (or "revolutions of the globe"). In the early debates, however, other naturalists took issue with Buckland's conclusions, claiming that "the allegedly sudden mass extinction might in fact have been a gradual and piecemeal process without any drastic physical 'revolution' at all."[89] Nevertheless, it remained perplexing to some why such events were no longer observable in the present, while others questioned at what point ancient man had made his advent in Europe, either before or following the deposition of hyena bones.[90]

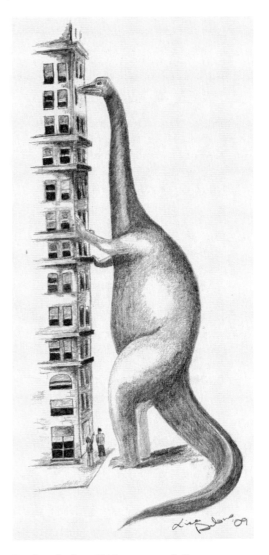

During the late 1890s, sauropod dinosaurs became recognized for their huge size. One artistic rendition appearing in a New York newspaper heralded the coming of the most colossal dinosaurian giants. In her illustration above, Lisa Debus has reproduced one of these early printed images that would have seemed bewildering to readers.

Later, during the 1860s post–Darwinian era, discovery of the Bavarian reptilian bird, *Archaeopteryx* tantalizingly provided glimpses as to how biological transformation, guided blindly by the force of natural selection, may have resulted in the development of modern birds. Furthermore, how were fossil birds such as *Archaeopteryx*, or for that matter Marsh's toothed birds of the dinosaur age, related to bird-like dinosaurs such as *Compsognathus* or *flightless* birds of the Cretaceous, Tertiary and modern times? Years after Mantell's death, Dollo delved into evolutionary implications posed by his "herd" of marvelously preserved *Iguanodon* specimens. But paleontologists such as Marsh and Huxley, so intrigued by prospect of Darwinian evolution, were more richly rewarded in their endeavors by the remains of (non-human) fossil mammals, rather than dinosaurs, because their well preserved remains were so plentiful in the American west. Accordingly, fossil mammals earned their own press headlines too; it was as if the adoring public wasn't predisposed or hadn't yet been trained by the media to afford dinosaurs the special prestige they have gained today. Through the decades, however, one finds that the most popular fossil mammals became those having physical characteristics bearing closest affinity by analogy to dinosaurs; huge size, razor teeth, horns, bony armor, or vicious disposition. (Less flaunted by museum curators and publishers, carnivorous, flightless fossil birds of the later Cenozoic, especially those known as "Terror Birds," or other forms such as the earlier genus *Diatryma*, seemed conceivably as frightening as were smaller varieties of theropod dinosaurs.[91])

For the evolutionists, Marsh opened significant windows into American prehistory, thus altering the course of fossil vertebrate evolutionary research during the latter 19th and early 20th centuries. Most famously, Marsh presented evidence for a linear evolutionary trend in fossil horses, leading to modern *Equus*. This evidence simply fascinated Huxley who was actively engaged in the great Victorian evolutionary debates against men such as Owen. Marsh had even prepared an influential diagram showing how equines lost their toes, gradually, from four in the ancestral "Orohippus" down to one as in all modern varieties. This trend correlated well with another, heightening of the tooth crown, reflecting a long term evolutionary change in diet. In 1876, Huxley lectured to a throng of socialites on the "Demonstrative Evidence of Evolution," for example, citing Marsh's fossil equine evidence. A headline, "Horses with Fingers and Toes Discovered in America...." soon appeared in the *Herald*. Claiming rather controversially, "to doubt evolution ... is to doubt science," Huxley declared Marsh's evidence as pivotal toward proving Charles Darwin's theories.[92] By the early 20th century, display cases exhibited at the American Museum under Osborn's direction would profoundly accentuate the idea of horse evolution, as well as that of other mammalian varieties—including man.[93] During the 1900s through the 1920s, fossil mammals were central to Osborn's theme and scheme.

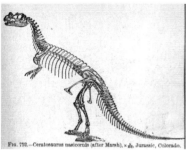

At left, Marsh's skeletal reconstruction of *Ceratosaurus* (from Joseph Le Conte's *Elements of Geology: A Text-Book for Colleges and for the General Reader*. New York: D. Appleton and Company, 1903). At right, the author's 1993 sculptural restoration showing modern views of *Ceratosaurus* engaging its "usual" rival adversary, *Stegosaurus*.

While horses aren't so monstrous, Marsh did elucidate another form of odd-toed ungulate mammals, later factoring significantly into evolutionary debates of the early 20th century. These Eocene animals were at one time known as "Titanotheres," otherwise known as "Brontotheres." Here, while Marsh got the ball rolling, it was Osborn who took things out on a limb. (Cope wasn't quite so active with titanotheres.) Brontothere remains, known principally from South Dakota's Badlands territory, proved so numerous that it was possible to demonstrate evolutionary relationships between the various genera — known as a "phylogeny." Although Marsh's labors resulted in a revealing skeletal reconstruction of the horned *Brontops robustus*, published in 1889, he died before the anticipated comprehensive monograph could be completed. That considerable cause was ardently taken up by Osborn, later published in 1929 as a massive two-volume monograph, titled *The Titanotheres of Ancient Wyoming, Dakota, and Nebraska*.[94]

Osborn recognized the importance of visual displays for the general public facilitating scientific ideas that otherwise must have seemed at best arcane, or rather incomprehensible. For numerous restorations and other illustrations were sprinkled throughout its 951 pages, including artwork by Knight, M. Flensch, Lindsay Morris Sterling, and Erwin S. Christman (1885–1921). While Osborn's verbosity therein remains difficult to decipher, their illustrations were remarkably easy on the eyes. Osborn launched a self-styled, idiosyncratic set of "orthogenetic" evolutionary principles in which non–Darwinian overspecialization, ultimately led to extinction. While Marsh may perhaps be praised for presenting evidence then fortifying Darwin's and Huxley's ideas, today, Osborn's theories based upon Cope's neo–Lamarckian ideals, may be judged awry. At the American Museum and in his publications, rather founded on his hero Cope's ideals, Osborn exemplified his illustrious ideas publically, dignifying Life's directional movement.

Charles R. Knight's first painted version of fin-backed reptiles, as published in *Century Magazine* (vol. 55, 1897). Chimeric "Naosaurus" stands at center.

During the 1900s, at the American Museum, Osborn commissioned artistic displays to educate the public about evolutionary processes. The monstrous looking, twin-horned brontotheres were central to Osborn's cause. While Knight's paintings vividly accentuated musty old bones visible within glass exhibit cases, Christman's sculptures proved most imposing. Analogous to Marsh's ladder or "progressive march"[95] of equine evolution, Christman prepared a series of skull portraits, showing restored heads of titanotheres, representing the earliest genus, *Eotitanops*, beneath two intermediate genera, then culminating in their zenith with fierce looking *Brontotherium*. Here, ostensibly, was evolution at work, producing a species of great size and with enormous horns, inevitably from more diminutive forms. According to Ronald Rainger, this display "became a unique landmark in the museum's Hall of the Age of Mammals ... [which] instilled visitors with fear and fascination."[96] The public came away with visions of fossil mammals dramatically reinforced by Knight's and Christman's life restorations, tied to evolutionary precepts, while the dinosaurs symbolized particularly Man's dominion over Nature. In later years, other artists such as Field Museum of Natural History sculptor Frederick A. Blaschke would sculpt an impressive brontothere family for museum display that in many respects seemed virtually alive to this writer.[97]

Especially during the 1890s and thereafter, prehistoric life became staple phantasmagoria in popular books, magazine features and other media accounts. Throughout this period (circa 1890 to 1935), prehistoric mammals were not neglected (as they are today), even while the dinosaurs eventually made significant gains by virtue of triumphant cinematic restorations (*The Lost World* and *King Kong*), as well as life-sized sculptural representations displayed at Chicago's "A Century of Progress" and elsewhere. Via a "naturally selective" process moderated by public consumers, certainly by the 1950s, dinosaurs emerged triumphant. Chief illuminators of the early period (e.g., Joseph Smit, Charles R. Knight, Bruce Horsfall, and later Zdenek Burian) offered artistic visions and personal impressions of how prehistoric life appeared through the ages. Their restorations set themes and style, conventionalizing our visions of familiar prehistoria for two generations in America, that is, until a Dinosaur Renaissance of the late 1960s through the 1990s reset the stage for reexamining traditional views of how many kinds of prehistoric animals lived and should therefore be accurately restored.

Now, in order to illustrate how prehistoric mammals were holding their own amid advances made in the paleontological sciences versus dinosaurs, several popular accounts will be examined (e.g., discounting contemporary textbooks and scientific journals), dating from this period. Next, we will sample how information about prehistoric monsters was reported to the public in the period from circa 1890 to 1935.

Fossil Mammals Versus Dinosaurs in Printed Matter

Although, today, dinosaurologists most revere the 1897 *Century Magazine* dinosaur articles illustrated by Knight, Osborn's 1896 article, "Prehistoric Quadrupeds of the Rockies," lavishly illustrated with nine of Knight's original painted restorations, scored an early victory (i.e., preempting dinosaurs) for fossil mammals in that publication.[98] However, as noted by David Rains Wallace, "Probably prompted by the *Century*'s lavish articles, 'dinomania' finally erupted in the sensationalist press."[99] Wallace then describes how dinosaurs stormed fully into mainstream America via the aforementioned December 11, 1898, *New York Journal* issue headlining the "Most Colossal Animal." Yet America's then stirring wave of mass hysteria over prehistoria may perhaps be dated to James Erwin Culver's "Some Extinct Giants" appearing in the April 1892 edition of *The Californian Illustrated Magazine*, as well as books by the Rev. Henry Neville Hutchinson, who prolifically popularized prehistoric life, particularly through editions of his *Extinct Monsters: A Popular Account of Some of the Larger Forms of Ancient Animal Life* (1892, 1893). While only 3,000 copies of (both editions of) Hutchinson's book were published, embellished by Mr. Smit's many pictures and life restorations, for its time *Extinct Monsters* generously captured the then escalating prehistoric monster phenomenon. Smit's magnificent Plates were equally apportioned to fossil mammals relative to dinosaurs and other winged and marine sauria. Although in other, more numerous figures sprinkled throughout the book's chapters, sauria outscored fossil mammals by a five to two ratio. Published in Britain, *Extinct Monsters* was also sold and distributed in America.

Culver, whose "Some Extinct Giants" article was illustrated by Carl Dahlgren, featured dinosaurs that were "being mounted to the astonishment and wonder of the unscientific public."[100] Culver continued, rhapsodizing, "The appearance of these early inhabitants and their size is almost beyond comprehension. It was a time of weird shapes—dragons, with all but the fiery breath; sea serpents one hundred feet in length; whale-like monsters that crawled along in shallow waters; uncouth reptiles with gigantic bodies and small heads ... others with enormous jaws lined with sharp fangs; birds with teeth and no wings; flying monsters with leathery wings twenty-two feet across." Such was the state of confusion at this time, however, that *Stegosaurus* was misconstrued as another of Cope's "kangaroo-lizards," while his horned *Agathaumas* was mistakenly reconstructed by Dahlgren as a uintathere. And could "Laelaps" really "cover nearly one hundred feet at a bound"? Indeed, exposure to dinosaur science then, learning of these relatively fresh discoveries, such bizarre animals and theoretical notions concerning all manner of prehistoria and the worlds in which they lived, must have seemed especially fascinating.

Ballou stirred imaginations with his another imaginative offering, this time on "The Serpentlike Sea Saurians" published in *Popular Science Monthly*.[101] Here, he related marine sauria, former inhabitants of America's great inland sea that disappeared when the "Mountains ranges and plains gradually arose, casting forth the waters and leaving the monsters to die and bleach in Tertiary suns."[102] Ballou emphasized mosasaurs, enhancing his article with an illustration by J. Carter Beard titled "The Great Cretaceous Ocean," depicting the feeding frenzy about to commence among hungry mosasaurs with gaping maws and enraged plesiosaurs, huge turtles and fish, while great-winged *Pteranodon* surveys frothing waves from above. The animals depicted in Beard's drawing were credited to paleontologist Samuel Wendell Williston's (1852–1918) scientific restorations.[103]

Hutchinson also wrote magazine articles, such as his "Prehistoric Monsters," appearing in the December 1900 issue of *Pearson's Magazine*.[104] Hutchinson's article was nicely illustrated by Lawson Wood, who to large degree, in turn, based his own restorations on Knight's 1897 *Century* restorations, and Smit's. Emphasizing dinosaurs, Hutchinson also addressed prehistoric mammalian monsters, *Megatherium* and *Mylodon*, the latter of which he suggested blurred distinctions between the prehistorical and modernity because

Mary Gunning's restoration of distinctively elephantine Uintatheres (adorned with reindeer antlers) published in William Gunning's *Life History of Our Planet* (New York: Worthington, Co., 1879). Note saber-toothed cat, *Machairodus*, at center.

it may still live in Patagonia. The Reverend Hutchinson demurred on evolutionary topics, invoking devout undertones facilitating readers' efforts to "gain wider views about Creation." He also referred reverently to that "oracle in Paris"—Cuvier's "law of correlation," which had essentially been disproved by evolutionist Huxley by that time. Lawson Wood's restorations of fossil sauria outnumbered his sole mammal depiction (*Megatherium*) by six to one.

R. I. Geare's 1910 *Out Door Life* article titled "Some Extinct Animals" viewed modern extinctions from a conservation perspective. Delving into extinctions of prehistory, he discussed the most popular fossil mammals (which he stated may have lived from 3 to 4.5 million years ago) followed by several dinosaurs (which may have lived from 6 to 15 million years ago). Each of the ten illustrations and figures appearing in Geare's article were of saurians, however.[105]

During the 1910s, several paleontology books were available for those who wanted to "bone up" on the in topic of paleontology, by then rapidly gaining critical mass following. British paleontologist E. Ray Lankester's *Extinct Animals* (1905)[106] seems inspired by Hutchinson's popularizations. Founded upon a popular lecture series, Lankester bid readers back from the recent Ice Age through the Time's corridors, to the primeval periods. *Extinct Animals* proved inspirational to Arthur Conan Doyle in his writing of a pivotal fantasy novel, *The Lost World* (1912). Not only did Lankester's persona lend itself to Conan Doyle's fictional character, the heroic Professor Challenger, but several creatures described in Lankester's book also make cameos in the novel. Fictional character Maple White's restoration of *Stegosaurus* (based on Lankester's own *Extinct Animals* restoration, which in turn is derived from Knight's by then 1897 *Century* entry), is presented as evidence sufficient to momentarily suspend disbelief, suggesting that prehistoric animals still live on the "lost" South American plateau. Harry Rountree illustrated the first serialization of Conan Doyle's classic, appearing in *The Strand Magazine* (1912). Rountree's pictures of modern humans confronted by monstrous living dinosaurs, sauria and other prehistoria, including anthropological primitives, served as early (paper) "special effect" visuals that we would so come to love in later years in stop-motion animated form. How fitting that Conan Doyle's influential thriller would become the first *major* Hollywood adaptation of dinosaur fiction.

Of special significance, Conan Doyle's *Lost World* introduced the first leaping dinosaur of science fictional history, evidently a megalosaur that jumps after its intended prey, journalist Ed Malone. In Conan Doyle's thrilling scene Malone narrates how, "A great dark shadow disengaged itself and hopped out into the clear moonlight. I say 'hopped' advisedly, for the beast moved like a kangaroo, springing along in an erect position upon its powerful hind-legs, while its front ones were held bent in front of it." Leidy, Cope and Knight remained at large!

Around the time that scientific investigations commenced at southern California's Rancho La Brea Pleistocene deposit, Hutchinson introduced *Extinct Monsters and Creatures of Other Days* (1911), expanding fossil mammals sections of his 1893 edition, while presenting more of Smit's restorations. Smit (and other early paleoartists) were also gaining recognition for restorations appearing in two of Henry Knipe's sumptuous volumes, *Evolution in the Past* (1912), and *Nebula to Man* (1905), both outlining and lavishly illustrating Life's magnificent historical panorama. Then Robert Bruce Horsfall's (1869–1948) fabulous paintings of fossil mammals adorned William Berryman Scott's *A History of Land Mammals in the Western Hemisphere* (1913). Scott's entry departed from convention, as he focused on fossil mammals, rather than Life's entire span through geological time.[107]

Paleontologist Barnum Brown (1873–1963) stoked dinosaur furor with his 1919 *National Geographic* article, "Hunting Big Game of Other Days," amply illustrated with eight of Knight's restorations, and numerous photographs taken in the field and museum halls.[108] This featured news of Brown's expeditions to Canada's Red Deer River, documenting his discovery and collection of Cretaceous reptiles.

By the mid–1920s, dinosaurs increasingly became main attractions in popular paleontological publications, even though authors intended to convey knowledge concerning the history of *all* life throughout geological time. In outlining the alien panorama ("otherness") of Life's history coupled with information concerning how scientists have explored the deep past, authors weren't intentionally framing the dinosaurs as the most significant prehistoria. One such example was B. Webster Smith's pocket-sized *The World in the Past: A Popular Account of What It Was Like and What It Contained* (1926, 1931).[109] The contents mirror usual concepts found in such books. Geological forces as then understood are outlined, followed by explanations for geomorphological features, leading to a presentation of Life's history from the "Dawn of Life" in the Precambrian culminating in the Pleistocene Ice Age and the Age of Man. Particularly interesting are the 266 Plates and figures, "73 of which are prepared in colour by W. J. Stokoe." Throughout the text, the usual dinosaurian suspects (e.g., *Ceratosaurus, Brontosaurus, Stegosaurus, Diplodocus*) are represented visually, although fossil mammals are emphasized as photographs of museum skeletons and line illustrations. In the medium of colored Plates, Stokoe's saurian restorations outnumber fossil mammals by only six to one.[110] Note that both editions of Webster's book appeared following release of First National's "The Lost World" (1925).

H. J. Shepstone's contribution, "Big Game of Other Days," appeared as a chapter in *Wild Life of Our World* (1934).[111] For illustration, Shepstone relied principally on photographs of both life sized and miniature sculptures of prehistoric animals. Photographs of several of Pallenberg's German dinosaur sculptures were printed adjacent to vividly posed likenesses of other prehistoria placed in miniature dioramas, attributed to "Camerascopes-Mondiale." Saurian figures and restorations in this chapter outnumbered fossil mammalia images by thirteen to six. Shepstone's is a typical entry of the time, wherein after outlining how incredibly old the Earth was, descriptions of several of the "queer" and "wonderful" creatures of the past were presented. Descriptions of such huge and bizarre animals were supplemented with particulars of how their immense bones are so painstakingly collected, prepared and finally mounted in museums. The author dramatically concluded, announcing discovery of a record-breaking, gigantic genus from Africa, reportedly 150 feet in length!

Of Knight's several popular paleontological books, his masterpiece clearly was the tour de' force *Before the Dawn of History* (1935),[112] offering marvelous insights—elaborate captions—concerning 43 of his great paintings and illustrations of the past. Here is a striking example of closely bonded "imagetext" (visual images) coupled with explanatory textual language. Only seven of the pictures in Knight's book showcased dinosaurs; another three dealt with marine reptiles. However, eighteen featured fossil mammals of one kind or another; another three dealt with prehistoric man exclusively. There was also a restoration of a "Permian Reptile Group—Texas," one of his finest, presenting "Naosaurus" and *Dimetrodon*.

In Knight's mind, evidently, fossil mammals certainly weren't on the wane.[113]

Not favoring dinosaurs by any means, Knight commented in 1935 on the contemporary state of popular paleontology. Knight conspicuously acknowledged paleoimagery's power, particularly in the form of "excellent articles and illustrations in books and magazines" for generating the sustained high level of interest then.

> So recently as thirty or forty years ago, in these United States, the average young man obliged to take a course in either geology or paleontology was looked upon as a species of martyr by his fellow students., and usually regarded himself as a most unfortunate human being. The very idea of entering the local museum with its cases of badly stuffed animals and birds, its dusty fossils and dreary-looking rows of sea shells, was enough to send a nervous chill up the spine of even the hardiest youth.
>
> Today, however, and indeed for some years past this balky mental attitude has completely changed; and now we see hundreds of our red-blooded young men (and women too for that matter) crowding zoology and geology classes, thronging to the museum exhibits and lectures, and beseeching the college and museum authorities to let them join various scientific expeditions. This reversal of feeling on the part of the student body is remarkable and complete; but it may be accounted for, partly at least, by the equally changed attitude of the professors and teachers themselves toward the whole subject. These gentlemen have at last come to realize that what they had to offer in the way of information and interest must not be confined to conversations among themselves or hidden away in bulky volumes. On the contrary, they now see that it should be brought out, dusted off, as it were, and spread before the fascinated gaze of a hitherto totally uninformed public.
>
> Many and various agencies have been at work, of course, to bring about this truly enviable state of affairs—newspaper publicity, for one thing, which has placed the latest ideas on the subject before the eyes of millions of readers; excellent articles and illustrations in books and magazines; lectures with colored slides; and the close cooperation of great museums in placing on exhibition splendidly mounted specimens of modern animals and elaborate series of fossil types, set up with great care and expense in halls devoted to the purpose. All these forces, and many more, have combined to make the world in general, and the United States in particular, supremely conscious of the vast sources of knowledge and interest by which we are surrounded, but of whose latent possibilities for entertainment and culture we have hitherto been so supremely unaware.
>
> Having thus indicated in a very general way the place held by paleontology in the world today, it may be of interest to go back a long, long time in world history.[114]

Near the end of Knight's triumphant prehistoric time tour, readers' eyes feasted on his 1923 restoration of "Early Man" confronting a Woolly Mammoth mired in a bog, perhaps the most symbolic of all in *Before the Dawn*.

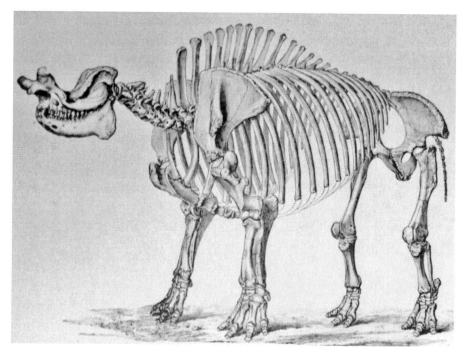

Othniel Marsh's skeletal restoration of a Brontothere (or "Titanothere"), as reprinted in Lankester's *Extinct Animals* (1905 ed.; collection of Allen G. Debus, photographed by the author).

Wild Neandertal men savagely hurl spears, rocks and lever a massive boulder upon the girth of the unfortunate behemoth. This is the spiritual battle modern savants and later scientists had struggled with on so many levels for so many decades. Man versus prehistory; man defeating the primeval monster! It was clear by then that prehistoric men had coexisted with great monsters of the past. Seeing through Knight's eyes was surely believing. Viewers were coaxed into accepting not only our murky heritage from savage ancestors, but our perceived killing instincts needed for survival, then as in modernity. Such a restoration could have never been conceived by Jefferson, the illustrious Cuvier, or the haughty Richard Owen. For, as Knight visually reinforced, a monstrous version of Man had indeed waged war upon Prehistoric Beasts, just like in the early Hollywood films and published tales of fantastic fiction. In the paleontological arena, fiction was becoming melded, if not conflated, with irrefutable fact.[115]

    While W. Maxwell Reed's and Jannette M. Lucas' *Animals on the March* (1937) was dedicated to fossil mammalogy,[116] by then dinosaurs had supplanted fossil mammals in the 20th century's pop-cultural battle for survival. There are *several* reasons and factors for this result. These *were* books dedicated to dinosaurs, such as W.E. Swinton's well illustrated *The Dinosaurs: A Short History of a Great Group of Reptiles* (1934) (London: Thomas Mufby & Co.). However, two primary factors will be elucidated here. First, dinosaurs had bested the mammals at the 1933/34 Chicago World's Fair (A Century of Progress). Secondly, by the early 1930s, dinosaurs captivated the media largely due to *filmic* representations of dinosaurs and other sauria recreated chiefly by Willis O'Brien and Marcel Delgado for two movies—*The Lost World* (1925) and *King Kong* (1933).

Dinosaurs on Public View

Charles Knight's artistry soon wended its way into the movie industry, as Willis O'Brien and Marcel Delgado relied on Knight's visions of dinosaurs for their own special effects, stop-motion puppets, animated for *The Lost World* (1925). Most might not recall that Conan Doyle's novel cast a rich assortment of prehistoria, including a giant flightless "Terror Bird," *Phororhacos*, which chases Professor Challenger up a rocky slope, and the huge rodent *Toxodon*. However, aside from a pair of apes, for the film version O'Brien and Delgado populated their lost plateau solely with dinosaurs and pterodactyls— no fossil mammals appeared. Rather than Conan Doyle's (leaping) megalosaur and *Iguanodon*, instead they animated *Triceratops, Agathaumas, Allosaurus, Tyrannosaurus*, "Brontosaurus," and a duckbill genus allied to *Hadrosaurus*. *Stegosaurus* was common to both film and novel versions. Some scenes portrayed carnivorous theropods battling horned dinosaurs, a Knight-inspired theme. Another scene showing the *Allosaurus* attacking the duckbill genus (i.e., "Trachodon"—related to *Hadrosaurus*) is thematically derived from 1860s imagery. This state-of-the-art movie, of which more will be said in chapter nine, proved a sellout sensation; although a silent picture it remains a dinophile favorite. By the early 1930s, the impact of Charles Knight's 28 mural paintings depicting geological history, completed for Chicago's Field Museum of Natural History during the late 1920s, soon supplanted that of his famous *Century Magazine* portrayals, or even his American Museum paintings. Knight's Field Museum paintings were reproduced seemingly everywhere — books, postcards, articles. Rival artists found Knight's artistry so irresistible that they couldn't refrain from copying his designs and prehistoric themes for their own use, emulating Knight. Consequently, Chicago certainly seemed to be *the* place for experiencing dinosaurs during the early 1930s. Beginning in May 1933, prehistoric animals invaded the Chicago World's Fair grounds, within easy walking distance of the museum where Knight's murals resided. Only five of Knight's 28 murals featured dinosaurs; while twelve murals showcased fossil mammals. At the Fair, however, three distinct exhibits presented sculptural, three-dimensional paleoimagery forms:

Erwin S. Christman's life-sized sculptures of (left) *Brontotherium*, and (right) *Eotitanops*, mammals of the early Tertiary Period. Christman's sculpture was intended to exemplify the majesty of evolutionary processes, capable of (orthogenetically) transforming a smaller ancestral variety *Eotitanops* along a "driven" course of evolution into the larger, more impressive (i.e., to human paleontologists' eyes) great-horned *Brontotherium*. In advising Christman, herein Osborn implied that a "wired" ladder of evolution directed transformations in this lineage, as opposed to a more "Darwinian," bushy, branching phylogenetic "tree of life" (from H. F. Osborn's *The Origin and Evolution of Life on the Theory of Action, Reaction and Interaction*. New York: Scribner's, 1918 ed.).

George H. Messmore's and Joseph Damon's "The World a Million Years Ago," the Sinclair Dinosaur Exhibit, and the Century Dioramas Studios. And so it was in Chicago, where the dinosaurs finally emerged, in a sense, "victorious" over rival fossil mammals. Knight's Field Museum murals and "The World a Million Years Ago" exhibit would be their last hurrah.

In a sense, "The World a Million Years Ago" culminated much of what paleontologists and paleoartists had been striving for over a century: popularizing prehistoric life. Messmore's and Damon's recreations were life-sized, like Hawkins' Crystal Palace sculptures. And they were temporally ordered, informatively, according to geological succession, as the elaborate visual time tour wound through a hemispherical, dome-shaped metal building representing planet Earth, with its principal land masses and oceans painted on the building's shiny exterior. Outside, representations of a *Stegosaurus*, sauropod and huge gorilla adorned a shelf overlying the entrance, luring audience lines. Although the building looked "futuristic" from the outside, inside everything was very ancient indeed. Visitors were conveyed along a mobile beltway from one display to the next. Most of the thirty-seven recreations were mechanized with tape-recorded growls and screeches added for sound effects. To a 1930s crowd, the monsters must have seemed verily alive! It was as if patrons had actually gone back into prehistory to witness the prehistoric monsters, one after another from the Paleozoic Era through the recent Ice Age. For its time, "The World a Million Years Ago" was the most realistic prehistoric experience imaginable. During its 1934 Chicago run, the name of the display was changed to "Down the Lost River — Thru the World a Million Years Ago."

While there are scant records left today indicating how the display was organized at the Fair and what it contained exactly, several records indicate that fin-backed *Dimetrodon* and fossil mammals played key roles in the prehistoric experience. Besides dinosaurs, winged pterodactyls and marine reptiles, evidently also on display were a Woolly Rhinoceros, an armored Glyptodont, Woolly Mammoths, shovel-tusked mastodonts *Platybelodon* and *Trilophodon*, a saber-tooth lion situated in a tar pit scene representing the Rancho La Brea locality, a Cave bear, a giant ground sloth, an Irish Elk, prehistoric horses, Titanotheres, and a ten-foot tall "Kong-like" prehistoric gorilla. Messmore and Damon capped off the ambience of their "million-year"-old recreations with a number of mechanized prehistoric humans, including "Java Man," "Piltdown Man," "Neanderthal Man," and "Cro-Magnon Man." Through exhibits and other similarly themed promotions, on a popular level, fossil mammals, dinosaurs and cavemen could still collectively represent the planet's geological history, just as Osborn decreed.

American Museum paleontologists Roy Chapman Andrews (1884–1960) and Walter Granger (1872–1941) lauded the display, stating on December 14, 1932, with the models nearing completion,

> Both of us were delighted with the ingenious reconstructions of prehistoric animals.... We were glad to see that they were doing it in a really serious way and making great efforts to have the models as accurate as possible. Dr. Granger and I feel that this has great possibilities educationally. It almost makes fossil animals live and I look forward to the time when our scientific museums will have exhibits of this sort. I have been amazed to find the great interest which children, as well as adults, take in the past life of the earth. I can well believe that the proposed exhibit at the World's Fair will be one of the most effective installations from the standpoint of education and popular interest.[117]

While "The World a Million Years Ago" was the most ambitious prehistoric display at the Fair, and despite the fact that it was also very well publicized and promoted, in retrospect, it was overshadowed by Sinclair's outdoor display of mechanized Mesozoic recreations. Due to their public accessibility Sinclair's dinosaurs may have instilled more lasting impressions. Both exhibits were situated adjacent to one another across a traffic midway, then at a location approximately where, today, the I-55 ramp stems from Chicago's Lake Shore Drive, near the current Merchandise Mart. Whereas Messmore and Damon's most impressive recreations were tucked away within the hemi-spherical building, Sinclair Refining Company's dinosaurs could be observed for free, as they were placed outside in prominent view. Although, arguably, Messmore and Damon's geological time exhibit represented a more "global" and inspired undertaking, Sinclair's startling Jurassic and Cretaceous menagerie, borrowing plainly from Charles R. Knight's artistry—or seem-ingly from *King Kong*, became the more memorable attraction.

The American Museum also lent its expertise to this competing display. This time it was Barnum Brown supervising the slightly larger-than-life dinosaur restorations constructed by contractors guided by P. G. Alen. Full-scale sculpting commenced following Brown's approval of Alen's miniature dinosaur designs, made in clay. There is a fair amount of photography and footage available today indicating how these animals were displayed. Alen's breathtaking models of *Tyrannosaurus*, "Brontosaurus," *Stegosaurus*, *Proto-ceratops* (one of the 1920s most heralded dinosaurs), a swimming "Tracho-don" (actually a crested *Corythosaurus*), and *Triceratops* looked sensational. The horned dinosaur and toothy "tyrant king" were posed as colossal adver-saries advancing toward one another as in Knight's by then already famous Field Museum mural. To match "The World a Million Years Ago's" engineer-ing sophistication, Alen incorporated motors into huge steel reinforced bod-ies (covered with rubber hide for skin) that swiveled heads of tyrannosaur and three-horn, and turned the sauropod's twenty-foot-long neck while curl-ing its thirty-foot-long tail. According to souvenir "Picture News" published by Sinclair for Fair distribution, the models also "growled." Foliage situated about the display area was designed to recreate that existing during the dinosaur age.[118]

As Don Glut opined in 1980, "The Sinclair Dinosaur Exhibit was cer-tainly one of the best-remembered highlights of the Century of Progress exhi-bition."[119] While numerous fossil mammalian forms were nestled within Messmore and Damon's "World a Million Years Ago" tour, Sinclair chose not to display any prehistoric mammals, or those marine reptiles commanding center stage in Verne's then 70-year-old *Journey* novel. The aforementioned, collectible "Picture News" publications reveal how delightful and popular were Sinclair's dinosaurs. Pretty bathing-suited gals posed next to the tyran-nosaur, or in other photos pretended to give monstrous "Rex" a manicure. A caption to a photo printed in one of these papers boasts, "Most pho-tographed exhibit—Attendance at the Sinclair Dinosaur Exhibit is running over 1,000,000 people monthly. In a single day, 79,000 persons have 'met' these fantastic creatures who occupy a World's Fair spot that is a 'throwback' in time to the remote ages of the world when life prevailed on a gigantic scale. The dinosaurs' lair not only recreates the strangest animal life of all time, but also reproduces a landscape typical of that time. Thousands of photographs are being taken weekly of these weird dinosaurs, the Sinclair Exhibit providing a favorite background for picture-taking by amateur photographers who want striking 'shots.'" While dinosaurs then could still be regarded as "weird," they were also rapidly becoming prehistory's most familiar fauna.

Although "Brontosaurus" was the largest animal in full public view, the influential *Tyrannosaurus* (lauded as "killer of all monsters") versus *Triceratops* (billed as "Nature's bad dream") pairing commanded a greater sense of awe, mystery, and perhaps horror. Surely, Fair-goers would have agreed this was the ultimate, "Battle of the Centuries! Each day at the Sinclair World's Fair Exhibit a battle scene is re-enacted between *Tyrannosaurus rex*, the most ferocious animal ever known, and his ancient foe *Triceratops*. These animals actually lunge at each other while visitors gaze in awe from the 'ringside.'"

While, today, several "World a Million Years Ago" monsters are owned by a private collector, the whereabouts of thirty or more are unknown.[120] Sinclair's dinosaurs decayed following a curtain call at the 1936 Texas Centennial exhibition. But through its promotion of Knight-inspired dinosaurs, Sinclair arguably prevailed over its rival "World a Million Years Ago" display (with its many mammalian faunal recreations). Knight's Field Museum depictions offered the most celebrated geological time tour of all. However, in staging its titanic dinosaur battle, Sinclair plucked out of his grand display one scene in particular for the ages. Sinclair's visions, derived from Knight's, would prove everlasting, resulting in the most influential "island" oasis of life-sized prehistoria constructed since Benjamin Waterhouse Hawkins' heyday. Thanks to Barnum Brown, P.G. Alen and Sinclair (in a manner left unexploited by "The World a Million Years Ago's" creators—and therefore lost to posterity), dinosaurs finally triumphed over rival prehistoric mammals, especially when they could be imaginatively joined in combat with a huge hairy Ape who joined the fray! Yes, RKO's *King Kong*, or that "mastodonic miracle of the movies,"[121] with its startling cinematic effects, had opened in March 1933; the tragically symbolic brute, a fictional evolutionary cousin to humanity, battled several gigantic saurians "live" before the camera, including that Late Cretaceous "killer of all monsters."

Between Kong's unforgettable, dramatic exploits and Sinclair's handiwork, from this time forward there was little room left in everyday popular imagination for seemingly lesser animals such as Cope's obscure Loxolophodon, Buckland's century-old cave hyenas, or Osborn's idiosyncratic and doctrinaire evolutionary theories founded upon the fossil records of prehistoric horses and titanotheres. By 1928, Jefferson's once foreboding Great-Claw had degenerated into H. G. Wells' "dreary megatheria."[122] On the strength of their spectacular paleoimagery placed on show, dinosaurs, particularly the biggest and most combative kinds, had captivated public imaginations beyond that of the less consuming mammalian paleo-monsters![123]

During harsh historical times, such as in times of oppressive war or during the years of America's Great Depression, people turned increasingly to

Heralded dinosaur collector and writer Donald F. Glut stands adjacent to several pieces of his "World a Million Years Ago" memorabilia, with the mechanical Ptero-dactyl to his right. Amazingly, the "robot" still squawks when plugged into an electrical socket (for more on Don's outstanding collection, see his website posted at: www. donaldfglut.com).

"escapist" fantasy — movies, novels, comic books. Indeed, the 1930s was a time when Hollywood's movie monsters were introduced; when Franken-stein, Dracula, the Invisible Man and many others ruled the silver screen, commanding box offices. It would appear too that dinosaurs were a viable alternative form of escapism for the public too, as the 1930s fostered an unprecedented interest in dinosaurs.

During early decades of the 20th century, paleontologists would increas-ingly discover, reconstruct and restore new genera and species of dinosaurs allied to the 19th century archetypical forms. Thus plated varieties such as *Kentrosaurus*, a South African analogue to North American *Stegosaurus* and British "*Omosaurus*," joined the quadrupedal ranks. North American *Styra-cosaurus* and *Centrosaurus* were clearly variations on the horned dinosaur theme, already so cemented in public consciousness by *Triceratops*' bizarre form and near equally by Knight's famous painting of Cope's poorly known *Agathaumas*. Armored North American *Ankylosaurus* shed light on its British evolutionary cousin, Gideon Mantell's *Hylaeosaurus*. Astoundingly, *Brachio-*

saurus exceeded weight limitations already set by immense sauropoda such as "Brontosaurus."

And a certain North American, Late Cretaceous carnivorous genus, that "killer of all monsters," would surpass the 19th century's considerably frightening theropod trio—"Laelaps," *Megalosaurus* and *Ceratosaurus,* thus becoming *the* most iconic real dinosaur of all time ... which will be the subject of chapter eight.

SEVEN

Cretaceous Sanctuary

In order to make lucid interpretations from eroded bedrock formations *and* to comprehend organisms and past environments which no longer exist, paleontologists see through the fourth dimension — *time*. With acuity, paleontologists are trained to restore and reconstruct the murky past from weathered strata, geochemical markers and bits of fossilized bone or shell. In order to embellish an imaginative living link to the past, paleontologists often resort to verbal or written explanations, much of which, however, is sheer speculation based on a modicum of fact. While in an earlier day of the science, paleontologists were cautioned to avoid introducing such speculation, during the late 19th and early 20th centuries the lure of speculating proved irresistible.

Scholars rarely consider the theme of paleontology reflected in literature.[1] Few paleontologists craft original science fiction stories based on their subject matter expertise. Charles H. Sternberg's (1850–1943) imaginative science fictional writings, in particular, are usually overlooked. Paleontologist Sternberg penned the earliest imaginary descriptive tyrannosaurid-horned dinosaur (i.e., *Gorgosaurus* versus *Styracosaurus*) battle and became vertebrate paleontology's most illustrative pioneer of paleontologically-themed science fiction. Sternberg wrote literary descriptions of marine sauria recreated in his mind's eye, and recorded personal observations of a science fictional encounter between a large flesh-eating tyrannosaurid and a herbivorous "duckbill" dinosaur. Such vignettes, later filmed in stop-motion animation for dramatic scenes in *The Lost World* (1925), were fancied by American Museum scientists and display artisans during the early 20th century.

Sternberg's creative writing avocation evidently spanned a mid life crisis between 1898 to 1917. He began writing fantasy vignettes and poetry — speculative paleontological "pen-pictures" — which ultimately evolved into an overlooked autobiographical, paleo-science fiction *story* framed within his book, *Hunting Dinosaurs in the Bad Lands of the Red Deer River, Alberta, Canada: A Sequel to the Life of a Fossil Hunter* (1917). His *A Story of the Past, or the Romance of Science* (1911) is transitional from the earlier pen-picture style.[2] Although he wouldn't have recognized his unusual contribution, Sternberg may be viewed as a significant pioneer in the writing of paleo-themed fantastic fiction.

What is the significance of his fictional contributions? What could have been the motive? And was the objective (simply) to entertain readers with gripping tales of battling prehistoric monsters, analogously as would today's Hollywood producers through the visual medium?

Sternberg's science fictional style evolved from a 19th century tradition of "pen-picture"[3] popular science writing. By 1911, however the paces he put his fictionalized beasts through had become rather incidental to an ulterior, personalized purpose for writing. While, as acknowledged by historians of science today, during the late 19th early 20th centuries, Sternberg contributed most significantly to rising popularity of dinosaurs and other sauria through discovery and distribution of many remarkable fossils and dinosaur skeletons to natural history museum institutions across the United States and abroad, his vivid pen-picture and science fiction literary skills have gone relatively unnoticed. Therefore, Sternberg, who experimented with literary techniques (e.g., verses and speculative verbal recreations) honed by 19th century paleontologist forbears, before publishing a genuine science fiction story involving time travel and prehistoria, may also be regarded as a "paleoartist" of his time. With respect to dinosaurology, his writings both contributed to as well as reflected contemporary popular culture.

In popular culture, paleontological pen-picture re-creations may be regarded as scientific bench marks, points of departure for crafting more elaborate fictional accounts— genuine *stories*— involving interesting characters staged in prehistory. Sternberg's imaginative writings and science fiction evolved as an extension of his natural talent for crafting dinosaurian pen-pictures—"living" prehistoric animals recreated with *words*, rather than relying on *visual* illustration.

Sternberg's sf and fantasy writing fulfilled important needs.[4] Beyond construction of simple pen-pictures, in the science fiction tale component of his 1917, *Hunting Dinosaurs*, Sternberg established a nostalgic, artistic means to grieve and reconnect with his recently deceased, beloved daughter Maude. In a personalized fantasy cyberspace, Sternberg reassured the importance of his life's work for posterity, identifying with fossilized creatures and organisms he'd dug from outcrops for forty years. Readers were thus advised of the validity of his scientific contributions and accomplishments.

Lacking formal paleontological training, Sternberg more than made up for this deficiency by resurrecting fossil bones from "God's great cemeteries." Not content to be labeled as a consummate collector or discoverer, Sternberg contemplated the nature of the strange creatures he spent a lifetime disinterring from outcrops. Through sheer diligence he became a revered paleontologist.

No single professional scientist impressed him more than the brilliant, indefatigable Cope who first hired him to collect fossils in 1876. Plagued by an all too vivid imagination, dreaming nightmarishly of dinosaurs trampling him underfoot, during daylight hours Cope delighted in eloquently relating his opinions about the strange animals they dug from the rocks, how they appeared and lived, and of their importance to science.[5] Sternberg listened intently.

Because Cope harbored erroneous Darwinian ideals, Sternberg's understanding of evolutionary principles must have been likewise distorted. Cope, and presumably Sternberg, espoused orthogenetic evolutionary principles. Cope's opinions on wildlife conservation evidently made a stronger impression on Sternberg. From his family base, re-established in Kansas, Sternberg gained expertise in the old Cretaceous sea and its former inhabitants, preserved in Kansas chalk beds. Sternberg's first professional publication — on "Pliocene Man" (1878) — was ghost-written, erroneously attributed to Cope.[6]

Perhaps discouraged by Cuvier's shade,[7] throughout the 19th century men with paleontological training avoided introducing speculation into geology, and such a ban would have extended to the writing of prehistoric fiction. So early paleontologists' literary efforts were, at best, crude examples of simplified fiction — verbal restorations inspired by fossils, founded on science. Then during a relatively short interval, circa 1909 to 1926, a handful of scientifically minded men, several of whom we may properly regard as paleontologists, fictionalized creatures out of prehistory.[8] Their stories (defined rather broadly here as, early exercises in fantastic writing) ranged from simple, paleontological pen-pictures to a full length novel, offering means of time travel, in the sense that modern human beings bridged time to stand alongside living mammoths, dinosaurs and prehistoric birds.

Sternberg began composing a number of paleontological pen-pictures as early as 1898.[9] And he wrote several for an autobiographical account, *The Life of a Fossil Hunter* (1909). His verbal flights of fancy restored prehistoric environments and fauna, extrapolated from science. These are perhaps not simple stories, per se, but, rather, vignettes or scenes composed from knowledge of fossils and deep time. These passages unveil a motif that would be considered unusual in dinosaur books of today — Sternberg's efforts to rationalize fossils with his providential sense of divinity infused with ecological, conservatory flavor.

Picture Sternberg peeping, through a key-hole into time's recesses, glimpsing a living prehistory, then telling us what he saw —*his* interpretation, based on contemporary science or perhaps what he has learned on the wagon trail astride the like-minded, pious visionary, Cope. In an introductory, imaginary encounter with the past, as printed in Sternberg's *Life of a Fossil Hunter,* we're whisked back to the Cretaceous— a period unsullied by man, but where his mind so often dwelled.

How often in imagination I have rolled back the years and picture central Kansas, now raised two thousand feet above sea level; as a group of islands scattered about in a semi-tropical sea! There are no frosts and few insect pests to mar the foliage of the great forests that grow along its shores, and the ripe leaves fall gently into the sand, to be covered up by the incoming tide and to form impressions and counterparts of themselves as perfect as if a Divine hand had stamped them in yielding wax.

Go back with me, dear reader, and see the treeless plains of today covered with forests. Here rises the stately column of a redwood; there a magnolia opens it's [sic] fragrant blossoms; and yonder stands a fig tree. There is no human hand to gather its luscious fruit, but we can imagine that the Creator walked among the trees in the cool of the evening, inhaling the incense wafted to Him as a thank-offering for their being. All His works magnify Him.... Many other beautiful plant forms grace the landscape, but the glorious picture is only for him who gathers the remains of these forests, and by the power of his imagination puts life into them; for it is some five million years ... since the trees of this Kansas forest lifted their mighty trunks to the sun.[10]

Not content to quote dry vital statistics in encyclopedic fashion for prehistoric animals he dug from outcrops (such as their skeletal measurements, deducing from their preserved teeth whether they ate plants or flesh, whether they were quadrupedal, bipedal or amphibious, and how long ago they lived), Sternberg *showed* readers this information. He offers elaborate descriptions, making it seem as if we're standing there with him (and Him), watching reanimated creatures slide through foamy waves, or slink along misty jungle trails.

In another passage, Sternberg invites us on another brief foray into prehistory, "will my readers go with me on another expedition to these Kansas chalk beds? 'How fleet is a glance of the mind!'"[11] In other words, just *think* about it — if you'll let him guide — and you're suddenly there in a living prehistory. Then, only pages after describing how his team had excavated a chalky 30-foot-long fossil marine tylosaur, he introduces a living specimen which quickly engages another marine saurian in battle. "The great creature strikes its opponent with the impact of a racing yacht and piercing heart and lungs with its powerful ram, leaves a bleeding wreck upon the water. Then raising its head and fore paddles into the air, it bids defiance to the whole brute creation, of which it is monarch."[12]

Later, readers are treated to Sternberg's visions of mosasaurs, as restored in his mind's eye from "those old cemeteries of creation." Ultimately, however, Sternberg apologizes for his literary limitations. "That past life, at least a very small fraction of it, I have sought to bring before my readers with pen pictures.... The restorations of Mr. (Charles R.) Knight's restorations of many of the extinct animals brighten my pages ... if I have failed in my pen pictures to take my readers into the misty past, these brilliant restorations will certainly have the desired effect."[13]

So, if he's failed in his endeavor to write convincing pen-pictures, there are still Charles R. Knight's excellent restorations to look at. In a sense, Sternberg's pen-pictures served as elaborate captions to Knight's inspirational paintings. Here, the ultimate literary challenge must be to describe in words, while dispensing with visual aids. Indeed, by the 1900s, visual forms of paleoart were gaining in status as a means of portraying prehistoric worlds and their inhabitants.

True, Sternberg's 1909 glimpses of living prehistory don't quite make it as science fiction stories, per Robert Silverberg's helpful definition,[14] but in meeting his own objectives he took matters as far as he then felt comfortable. Instead, Sternberg's dazzling fantasy passages seem mere, bare bones special effects, or stepping stones that could enhance a science fiction tale. Sternberg had yet to introduce a protagonist who fully interacts with the Cretaceous environment, or a character who surmounts hardship or resolves an insurmountable problem. His vignettes had yet to achieve critical mass of crucial story land ingredients in a remote, forbidding prehistory.

Sternberg's subsequent book, *A Story of the Past, or the Romance of Science* isn't technical or scientific. Rather, it's a book of poetry and verse, not unlike other volumes published then. There are 14 chapter headings; ten are theological, three others are paleontological, and another is simply titled "Maude." These last four take up nearly 70 percent of the book's 87 printed pages. *Story* is an incongruous merging of topical ideas: a hodgepodge.

And yet we see connections bridging 1909's *Life of a Fossil Hunter* and subsequent *Hunting Dinosaurs*. Sternberg's first chapter, "A Story of the Past," is autobiographical; he leads to a familiar beacon — the old Cretaceous seaway.

> But they are not dry bones alone;
> I see them as they were...
> So will you go with me, dear friend
> To the Cretaceous shore,
> That sea that seems to have no end,
> Its mysteries explore?[15]

And so we wake up 5 million years before, magically, daydreaming. "Presto change! We're back once more." Thereafter, in a sequence of recreated scenes we spy a fishing fleet of long-necked plesiosaurs, and a mosasaur; these two engage in bloody battle, "And so we watch with bated breath." Then we're treated to pen-picture imagery of winged pterodactyls soaring from above into the waves, fishing.

Passages reminiscent of Jules Verne's (1828–1905) *Journey to the Center of the Earth* (1864) are also encountered as we read how the past and its troubled waters are supernaturally interwoven. As waters rush incessantly toward the Earth's center, a river called Styx,

Charles H. Sternberg drew inspiration from Charles Knight's restorations for "pen-picture" writing, as in the case of this *Nectoportheus pririger* (a mosasaur), which grew, as William H. Ballou believed, up to 50 feet long (from *Century Magazine*, vol. 55, 1897).

> bears me on through gates of death
> Where I shall all forgotten be,
> As voyager on life's troubled sea
> But on the great eternal shore
> I wake to LIFE forevermore.

In "The Permian Beds of Texas," Sternberg claims that "God gave us minds to see the past." Much of this poem relies on river of time metaphor, as Sternberg bids his readers to sail with him on the Tide of Time, "My oar-beats keeping gentle rhyme." As his speed rushes him along faster into the past where "man's destructive power's not known," more or less as did Verne in his *Journey* half a century earlier through Axel's waking dream, Sternberg takes us from the day of living *Mastodons*, past the Miocene and Eocene (epochs) with their characteristic vertebrata, then on into the age of three-horned *Triceratops* (the Cretaceous), swiftly past the "Jura" age and eventually into the Permian Period, where he collected many specimens for Cope.

But rather than fossils, here we spy living specimens such as the amphibian *Eryops* and sail-backed cotylosaur *Dimetrodon*.

Finally, "In the Laramie" conveys in verse the story of the famous "Trachodon" mummy's discovery. A brief fantasy scene wherein we watch the aquatic herbivore "stem the rising tide" and eat, ends after Sternberg directs that he'll "dream no more." There are also references to another beast known from these deposits, Trachodon's tyrannosaurid devourer. But here are tears for things as well.

In his first chapter, Sternberg bares his soul referencing discovery of the Maastricht mosasaur specimen examined by Cuvier; Sternberg laments that its bones would be "Admired by the Savant wise," but not such as he, "The unlettered," who must be content to be merely "lost in surprise."

And his sorrowful poem concerning his daughter "Maude," references Lazarus's tomb; indeed she later is resurrected in Sternberg's later *Hunting Dinosaurs*, where Maude is first seen exiting a cave along the Cretaceous shore. His love for her, "The joy of my life, my comfort and pride," is heartfelt.[16]

Adding his *Story of the Past* to his portfolio of imaginative writings, we see how since the 1890s Sternberg cleverly experimented with fiction, using different stylistic approaches. Sternberg was obsessed with the Cretaceous sea and its glorious inhabitants, the aquatic Trachodon and its menacing tyrannosaurid hunter, Maude's resurrection, the Permian beds, and providential, prehistoric settings. Why this unusual, if not profound, aggregation?

Sternberg's descriptive powers had grown by the time of his next full length publication, *Hunting Dinosaurs: In the Bad Lands of the Red Deer River, Alberta Canada* (March 1917), the first edition of which was self-published, with an embossed "Trachodon" on the green cover. Sternberg lures us into prehistory once more for his unveiling of "Trachodon," a herbivorous "duck-billed" dinosaur he was highly familiar with from numerous excavations and American Museum of Natural History exhibits. Painting a rich picture of how the Red Deer Valley of Alberta in Cretaceous times resembled the Florida Everglades, Sternberg beckons,

> Often in my day dreams [*sic*] some stately dinosaur has passed before my mental vision! The forests, the rivers, the lakes and the oceans of those ancient days have appeared in imagination as though they actually existed. So I ask the reader to put on my glasses.... Perhaps our imagination has carried us back to a bayou of the Edmonton Cretaceous. Yes![17]

Encouraged to don special time portal glasses, we soon spy a living Trachodon, which in a vignette is quickly dispatched by a three-toed carnivore. Rather than true story, this fictional passage is not much more than an embellished scientific description of a typical, representative day in the life of a typical, generic Trachodon. Other than Sternberg's simple narration, as before, the tale lacks an interacting human protagonist. The scene is extinguished by a display of primeval savagery.

Twice, we are urged backward in time, first to the Belly River stage. First, readers "witness" the meanderings of an armored dinosaur (possibly *Euoplocephalus*), and an aquatic plesiosaur. Next, we "find ourselves sitting in the shade of a giant redwood." We are shown a living horned dinosaur, *Styracosaurus* quickly coaxed into battle with the Tyrant of the Everglades— carnivorous *Gorgosaurus*, of the tyrannosaur guild; perhaps the first such encounter staged in literature. Now in this passage readers do seem more interactively involved than in previous passages, as if Sternberg is building sufficient confidence with his craft of writing scenes from deep time. We aren't simply peeping through a keyhole or peering through special glasses into the past. No, this time—although perhaps in a ghostly sense—we're actually there, walking through a jungle and fully interacting with this carefully recreated cyberspace; Sternberg's vision of the Cretaceous environment based on fossils. And once again, we're treated to a Knightian battle-royal between the two giant animals. As Sternberg instructs, "We watch the combat with bated breath."

Artistic restorations featuring tyrannosaur-ceratopsid combatants (Charles R. Knight–inspired) arrest our attention; their titanic struggles are cliché in *literary* fictional reconstructions stemming from science. Yet Sternberg still suppresses an urge to let his engaging story completely "out of the vault." Referring to Silverberg's story definition (see note 14), this is still pen-picture territory, not yet a full-fledged story.

Finally, in a chapter titled "There Were Giants in Those Days" Sternberg takes us back to the Cretaceous Period again, a journey we're accustomed to. However, this time there's no portal to peer through. Relying on a waking dream, more reminiscent of "Rip van Winkle" than Verne's majestic use of the device in *Journey*, Sternberg simply falls asleep in a mine in the Red Deer River valley, waking up three million years ago in the Cretaceous. No longer practice or a simple vignette, this is genuine story-time, per Silverberg's science fictional definition.[18]

> The day had been hot and sultry ... I came upon a coal miner's tunnel ... I found relief by going in some distance. The floor was deeply covered with fine dust, making a restful place; and it is little wonder that I fell asleep. I never knew how long I slept, but when I awoke, I was overcome with surprise, I could not tell whether I had awakened in eternity, or Time had turned back his dial, and carried me back to the old Cretaceous Ocean.

Thus, Sternberg's genuine sf story begins, sans technology and without fanfare. However, story developments were foreshadowed several times already in *Hunting Dinosaurs* through prior peeking into the Cretaceous keyhole; the element of surprise has been compromised. Next, Sternberg describes the terrestrial, coastal setting, the salty air, reiterating (while justifying) that we shouldn't be overly surprised at his ability to recreate the past, because after all, "How fleet is a glance of the mind, compared with the speed of its flight, the lightning itself lags behind, and the swift arrows of light." Further foreshadowing suggests another "Tiger of the Everglades" will reappear to upset this tranquil setting, but only after we spy another meticulously described living Trachodon. It's as if Sternberg's prior pen-pictures were warm-up trials for this more elaborate, confidently written verbal recreation.

Sternberg's movements in the recreated environment, his building of a raft to observe the swimming dinosaur up closely and his sense of perspective is often amateurishly interrupted by references to his former fossil digs of the past 40 years, the numerous skeletons he'd shipped to various museums, strange poetry, and frequent allusions to a deity. Rather like Jules Verne's *Journey*, Sternberg's tale is laced with providential overtones. Sternberg's allusions to real science aren't harmoniously spliced into the story's flow. Sternberg strove to build readers' (as well as perhaps, initially, his own) confidence in his ability to craft realism in the effects of a story by relying on scientific facts, which (as opposed to Verne's artistry in *Journey*) has a tendency to jolt readers out of any welcome sense of reverie.

The fierce tyrannosaur is introduced through a page of rather archaic rhyming verse. Before our eyes, the tyrannosaur "conqueror" rushes in to devour poor Trachodon, while an accepting Sternberg gazes at the awful terror. Then our intrepid protagonist procures dinosaur skin from the cadaver, tanning it to make sails, ropes and tarpaulins, as if he intends to stay awhile in the Cretaceous. Surprisingly, Sternberg is never distressed about his woeful situation, and lacks inclination to return to the present-day.

We note that Sternberg's tale is a backward life-through-time tale (opposite in temporal direction from Verne's *Journey* which moves upwards in time from Primeval toward the Present), because in the next segment Sternberg awakens in a slightly older Cretaceous setting corresponding to the Niobraran stage, principally known from western Kansas. Next, Sternberg composed dialogue between himself and his deceased daughter Maude, who miraculously appears alive in a cave. Together they eke out an existence along the shore of the old Cretaceous seaway, named "Mosasaurian Bay," gathering seafood and supplies, preparing makeshift tools, and building shelters and a raft with which to explore the marine life.

Yet Maude and Sternberg merely talk shop, as, to Maude's delight, he describes the various fish and saurians swimming by. But he also mentions the institutions he'd shipped such specimens to—millions of years later. Such dialogue is far inferior to, for example, the enlightening and amusing exchanges between Axel and Lidenbrock peppered throughout Verne's *Journey*. Sternberg's devout sense is apparent as he repeatedly praises the Deity for His glorious organic works, observed alive in the recreation—many of which (as we're reminded) Sternberg will eventually dig from solid rock millions of years hence.

Sternberg and Maude have been miraculously projected into the past, while knowing in future ages—although in his personal past—he will discover the plant and animal life witnessed alive, although fossilized. Yet Sternberg doesn't wrestle with temporal difficulties of his transit. Disappointingly and unwittingly, Sternberg never considers how any of his deliberate actions might affect his ordinary present; nor do the two time travelers ever ponder whether their presence in the past might cause a butterfly effect rippling down through time's corridors, altering the course of future events. No, they're ostensibly on a grand sight-seeing trip. Sternberg exultingly describes many great organisms he has dug from outcrops. Despite evident hardship they never suffer but, incongruously, *rejoice*. The problem Sternberg-as-protagonist resolves is still not readily apparent. So after falling asleep (again), Sternberg reawakens, although this time re-materializing in the geologically older Texas Permian Period swampland, "twelve million years" ago.[19]

In this final segment of Sternberg's time travel trilogy, a family reunion takes place in the Permian. Over the next handful of pages, the family, joyously reunited in a nostalgic prehistory, discover means of survival, cooking, foraging, camping while Charles H. delightedly observes the curious batrachian animals preserved in the beautiful primeval forest. Was it all just a dream? It's more than simply that.

Sternberg's detailed pen-pictures and fantasy writing represented, for posterity, a means of personally identifying with — putting his stamp on — prehistoria he could not describe with sufficient technical formality for scientific journals. Sternberg's "fear of oblivion," as noted by Spalding, is analogous to George Gaylord Simpson's (1902–1984) growing awareness "of his own mortality," as noted by Laporte. Simpson had become "anxious about posterity's judgment of his scientific contributions."[20]

Sternberg craved recognition for his accomplishments gained through countless hours of lonely, physically demanding and often death-defying fieldwork, perhaps inwardly regretting that fruit of his labors and perseverance was often handed over to other men who would scientifically describe "his" fossils.

While the art of writing paleontological pen-picture restorations was introduced during the 19th century, they were intended as popular — not necessarily scientific — interpretations, and were written usually for public edification, certainly not for scholarly peer review purposes. Sternberg's several first-person "voyages of the mind" expressed in his *Hunting Dinosaurs* has the ambience of a personalized, autobiographically founded journey. He has paid obeisance to the almighty Creator of all things primeval in "God's Cemetery," he has through the Lord's grace been permitted to see living creatures he'd otherwise only been able to handle as dusty fossils in modernity.

Marcel Delgado's and Willis O'Brien's *Lost World* battling dinosaurs — the Jurassic *Allosaurus* (left) and Cretaceous "Trachodon" (right). While this film was released eight years following private printing of Sternberg's *Hunting Dinosaurs*, this filmic scene matches a similar scene described in a chapter titled "There Were Giants in Those Days." This vignette also captures the spirit of ideas surrounding *Tyrannosaurus* and their unfortunate duckbill dinosaur victims, as entertained by American Museum paleontologists during the 1900s and 1910s.

Caught up in nostalgia, he is joined by his dear family who are all living in the recesses of a glorious Permian. For only in the past can he, along with deceased Maude and other cherished family members commingle with God's sublime creatures of another day. It is as if Sternberg yearned to sort out the problem of his own mortality and the meaning of his life and career by invoking the past, merging happy bygone days with more illustrious, ancient and providential prehistory — the essence of his inner joy sharing an unspoiled past.

Evidently this was Sternberg's ultimate aim in crafting imaginative segments of his *Hunting Dinosaurs*, appealing for its lost world-ish/travel-through-time nature. At this latter stage of Sternberg's life, prehistoric ages, where his mind so often dwelled while pondering fossils, acquired nostalgic qualities. Here, Sternberg effectively coupled both the business of describing his living scientific specimens with the pleasure of greeting a loving family in personalized, prehistoric "cyber-space." Sternberg achieved prayerful salvation in a prehistoric sanctuary, and thus the purpose of his peculiar autobiographical story was fulfilled.

Sternberg never made an impression on admired science fiction writers such as Ray Bradbury, who wrote the 20th century's most enduring dinosaur time travel tale, "A Sound of Thunder" (1952). Although he probably didn't realize the impact of his work at the time, vertebrate paleontologist Sternberg may be viewed as a significant pioneer in the writing of paleo-themed fantastic fiction.

His pen-pictures and stories of the past are unique for their reverential undertones.[21] How quaint yet utterly characteristic of Sternberg to employ famous dino-*monsters* in this vein to flesh out a utopian setting of the past; primeval places not of horror but of healing and solace.

By the end of the 1910s, dino-monsters of every persuasion had been introduced: through museum skeletal displays; painted and sculptural visual restorations; popular science literature; fantasy writing; and early filmic efforts. But the greatest, most influential prehistoria of all would materialize flesh-on-the-bone beginning (especially) during the 1920s, leaving their marks indelibly in the hearts and souls of men.

EIGHT

Rex Battles

Increasingly since 1915, as acknowledged by Donald F. Glut in 2000, *Tyrannosaurus rex*'s "gigantic head, very long teeth, and diminutive, two-fingered forelimbs are familiar images to scientist and layman alike, with the animal itself having taken on mythic status in our collective consciousness."[1] Paleontologist Christopher A. Brochu, who studied "Sue's" skeleton at the Field Museum, stated, "Nothing evokes prehistory more than *Tyrannosaurus rex*. Nearly any five-year old in the industrialized world knows what it is.... One could ... argue that tyrannosaurids fulfill a cultural need in the sciences; tyrannosaurids are no more relevant to phylogenetics or comparative biology than any other group of organisms, but they are extremely popular. When we do science with *Tyrannosaurus*, we do it with a broader audience than if it were done with almost any other animal."[2]

So what is a *T. rex*, what has it come to symbolize over the past century, how has it been portrayed both literally and metaphorically, and why is *this* of all the real prehistoric creatures known to science commonly regarded as the most monstrous of all?

There are a handful of additional significant discoveries as well contributing to our knowledge of America's grandest prehistoric symbol — perhaps the most significant *real* prehistoric monster, spawning many fictional, pop-cultural morpho-types since (including the dramatic larger-than-life Kong movie version).

One of the first individual fossil specimens later assigned to the *Tyrannosaurus* genus name was found by paleontologist Edward D. Cope (1840–1897), in 1892. Cope had assigned the genus name, *Manospondylus*, to a single dorsal vertebrae, (specimen number AMNH-3982). But decades later came murmurings that AMNH-3982 belonged to another dinosaur instead, one which hadn't been described on the basis of more complete remains until 1905.[3] Circumstances surrounding the intriguing possibility of Cope as *T. rex*'s (first, unheralded, i.e., during his lifetime) discoverer were publically consumed in 1999 when a Black Hills Institute dinosaur bone collector named

Bucky Derflinger happened across another South Dakota *Tyrannosaurus* specimen, then the 30th such specimen known to science. The 1999 remains were already *stacked* in a pile, as if "buried, placed carefully by someone with opposable thumbs," in the vicinity where Cope was known to have collected.[4]

The evidently late 19th century, archaeological dig site suggested to Peter Larson of the Black Hills Institute that this must have been the very locale where Cope had found his *Manospondylus gigas*. Are *Manospondylus* and the 1999 Rex one and the same dinosaur specimen? Larson exclaimed, "To find out if there is any possibility that this dinosaur might actually be 'the one,' I'll need to compare trace elements from the site with Cope's original vertebra.... If they match, we can be pretty sure they're the same. In that event, the rules of nomenclature may put the name *Tyrannosaurus rex* in danger. As unimaginative as *Manospondylus gigas* is, it was there first — and would take precedence."[5] Appropriately, this new specimen — thought to be approximately 10 percent complete — was informally named E. D. Cope. However, as Adrienne Mayor suggests, "Cope was not the only pioneer paleontologist to explore that area." Other paleontologists explored the terrain as early as 1874, during George Armstrong Custer's cavalry expeditions to the Black Hills region.[6] One scientifically trained officer, George Bird Grinnell (b. 1849), reported the discovery of fossil bones in the area, and so it is possible that he, rather than Cope, stacked the *Tyrannosaurus* bones, two decades before Cope's arrival!

Then between 1900 and 1916, a flurry of events transpired, publicizing a trio of partially complete skeletons found by Barnum Brown (1873–1963), later named *Tyrannosaurus* by Henry Fairfield Osborn (1857–1935). *T. rex* and *Manospondylus* weren't the only scientific names this dinosaur of all dinosaurs went by during its first half century of scientific intrigue! Alerted to the possibility of unexplored fossiliferous deposits in the Hell Creek formation through a paperweight (a section of *Triceratops* horn) presented by William T. Hornaday, director of the New York Zoological Society,[7] in 1900, Brown found a partial (about 13 percent complete) dinosaur specimen in Wyoming. Initially described as *Dynamosaurus imperiosus* by Osborn, this was later shipped to London's British Museum (specimen BM-R7995) in 1960. Next, in 1902 he uncovered another specimen (10 percent complete, no. CM-9380) in Montana, which was sold much later in 1941 to Pittsburgh's Carnegie Museum partly for precautionary reasons (in case the American Museum was bombed during World War II).[8] Now it was left to Osborn, who was then racing against the Carnegie Museum's Olof Peterson, to write the definitive

Opposite: Two views of the American Museum's famous *Tyrannosaurus* skeleton — America's reigning paleontological icon, AMNH-5027, as fully mounted by 1910 (from H. F. Osborn's *The Origin and Evolution of Life on the Theory of Action, Reaction and Interaction.* New York: Scribner's, 1918 ed.).

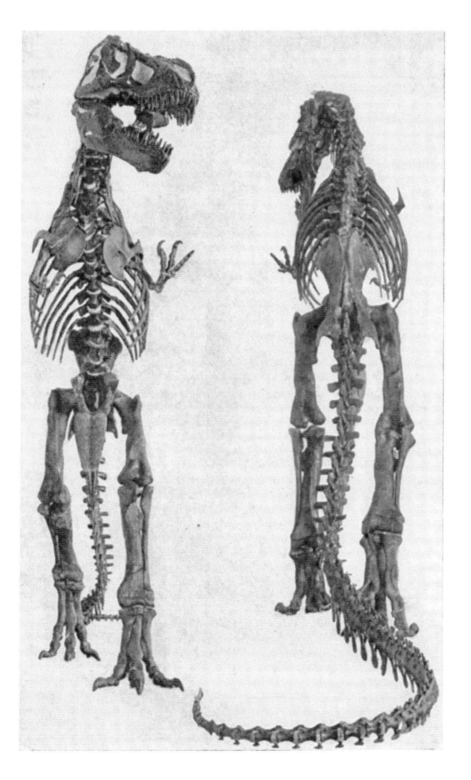

paper on tyrannosaurs. But thanks to Brown's field efforts, it would be Osborn and the American Museum, not Peterson or the Carnegie, who would define what sort of monster the assemblage of bones represented.

Osborn initially believed significant anatomical differences existed between specimens, hence his concomitant erection of two genera — *Dynamosaurus* and *Tyrannosaurus* — in 1905. A year later, however, Osborn recanted this conclusion, claiming rather that "The separation of these forms now proves to have been an error. Continued excavations in the summer of 1905 brought to light many additional parts of the type of *Tyrannosaurus*; and during the past winter the remains of the type of *Dynamosaurus* have been carefully worked up. The two animals are found to be generically if not specifically identical." It seems as if remains attributed to *Dynamosaurus* had included some bony armor fragments (osteoderms) which were ankylosaurian.[9] To memorialize the glorious monster king, Osborn defined a new family — Tyrannosauridae, replacing Cope's former Deinodontidae founded in 1886 "upon type species *Deinodon horridus* (*nomen dubium*)."[10]

Brown's grandest discovery, offering America its finest view of the Monster came from Montana's Hell Creek deposits in 1908. This is *the* American Museum specimen (no. 5027) which has captivated the souls of several generations. Or as Brown claimed, the "most formidable fighting machine ever devised by Nature."[11] Brown also collected bones of three other large specimens which were far less preserved.[12] With Brown's inspiration and a supporting team of skilled museum artists, Osborn also began visualizing dramatic museum display possibilities for the American Museum's new wealth of dinosaur specimens. In the case of *Tyrannosaurus rex*, three themes swiftly presented themselves — *T. rex* battling another of its own kind, *T. rex* stalking (or scavenging) duckbilled dinosaurs, and — from posterity's perspective — most influentially, *T. rex* menacing *Triceratops*. Or Rex versus Tops, as I informally refer to it — the most widely known and frequently played dinosaur theme of all!

Erwin S. Christman's early fascinating artwork showing *T. rex*'s reconstructed skull and paleontologist William Diller Matthew's (1871–1930) and artist L. M. Stirling's more dramatic reconstructions (respectively, 1905 and 1906) showing entire skeletons composited from the several specimens, in each towering above a diminutive looking human skeleton, accompanied Osborn's American Museum's *Tyrannosaurus* descriptions published in 1905 and 1906.[13] But few museum patrons would be expected to purchase let alone *digest* dry scientific publications.

More dramatic portrayals of dinosaurs such as the *Allosaurus* and *Tyrannosaurus* were needed in order to, as W. D. Matthew stated in 1908, provide an edifying, "vivid picture of a characteristic scene of that bygone age, millions of years ago, when reptiles were the lords of creation, when 'Nature, red in tooth and claw' had lost none of her primitive savagery, and the era of brute force and ferocity showed little sign of the gradual amelioration, which was to come to pass in future ages through the predominance of superior intelligence."[14] (One may wonder whether mankind has returned to a vaunted age of brute force and ferocity.)

In an early (circa 1909) vision, two tyrannosaurs glower over the "mummified carcass of a Trachodon, a unique specimen which was purchased last year from Mr. Charles H. Sternberg.... This group will make a very effective and striking centerpiece for the Hall of Cretaceous Dinosaurs."[15] More or less concurrently, by 1908, Barnum Brown was envisioning a pair of "Trachodon" skeletons mounted as if feeding in the marsh, "when one is startled by the approach of a carnivorous dinosaur, *Tyrannosaurus*, their enemy, and rises on tiptoe to look over the surrounding plants and determine the direction from which it is coming. The other Trachodon, unaware of danger, continues peacefully to crop the foliage." Speculating further, Brown suggested, "Perhaps the erect member of the group had already had unpleasant experiences with hostile beasts, for a bone of its left hind foot bears three sharp gashes which were made by the teeth of some carnivorous dinosaur."[16] By 1915, meanwhile, Osborn had noted *Triceratops* frill fragments associated with the "Dynamosaurus" specimen deposit, and Matthew was commenting how the "great horns and bony neck-frill of *Triceratops* ... were developed no doubt to resist the attacks of the huge Tyrannosaur."[17] *Triceratops'* three-horn fate evidently rested in *T. rex's* cruel jaws. Such visualizations led to Christman's striking *Tyrannosaurus* miniature scale test models, which few patrons ever saw, and eventually a magnificent painting, since haunting generations of paleo-enthusiasts!

In 1913, Osborn reported an early concept for the *Tyrannosaurus* skeletal display, a configuration suggested by Raymond L. Ditmars, Curator for Reptiles at the Bronx Zoo, which admittedly presented "mechanical problems of very great difficulty."[18] To gauge the feasibility of the design, museum staff recreated every bone of two tyrannosaurs in miniature, at one sixth scale. Even so, the erect model stood 3 feet tall. Christman actually tried four approaches, three of which weren't satisfactory. In the remarkable pose Osborn described, one *T. rex* is posed crouching over bones of a prey animal ("represented by a portion of a skeleton"), and the second is shown closing in to contest ownership of the corpse. Osborn believed that museum patrons would then be able to see the mighty pelvis and "perfectly preserved skull" of the crouching individual close to eye level, while observers could also admire the height of its towering attacker. Osborn stated that this pose represented the animals "just prior to the convulsive single spring and tooth grip which distinguishes the combat of reptile from that of all mammals."[19] Furthermore it was hinted that a fifth study will "embody some further changes," due to avoid precarious balancing of the more erect figure. In one of the possibly earlier diorama studies, the two *T. Rexes* are shown grappling with one another, their skulls close, the foot claws of one forcing the lower jaws of the other open.[20]

Through 1915, only individual, separated bones and the reconstructed skull, "beyond question the most impressive dinosaur skull ever found,"[21] of *T. rex* were on public display. That is until the world's first *Tyrannosaurus* skeleton was conservatively mounted, rather anthropomorphically, in an upright walking pose, with a lighter weight cast of its skull — replete with massive toothy jaws — suspended 18 feet above the museum floor. (The actual fossilized skull and one of the small ⅙ scale models were displayed in a separate cases nearby.) Adjacent to *Tyrannosaurus* were skeletons of the Trachodon group and its chief perceived adversary — *Triceratops* — collectively suggesting "the awesome, terrifying power of nature."[22]

Barnum Brown described the planned exhibit most avidly, as well as the 47-foot-long feature creature itself, in a popular *Scientific American* article (October 9, 1915), titled "Tyrannosaurus, a Cretaceous Carnivorous Dinosaur: The Largest Flesh-Eater That Ever Lived." Emblazoned on the magazine cover was Vincent Lynch's striking painting offering a life restoration glimpse of two recreated tyrannosaurs fighting over the corpse of a duckbill dinosaur — as it would have appeared to observers three million years ago. Noting there was presently insufficient museum floor space for Osborn's fully planned skeletal diorama, Brown stated, "The completed group, 54 feet long and 12 feet wide, will comprise three skeletons posed similar to the models to represent a scene of daily occurrence in the dim distant past."[23] Brown even offered a short pen-picture describing inspiration behind the scene.

> It is early morning along the shore of a Cretaceous lake three millions of years ago. A herbivorous dinosaur 'Trachodon' venturing from the water for a breakfast of succulent vegetation has been caught and partly devoured by a giant flesh-eating *Tyrannosaurus*. As this monster crouches over the carcass, busily dismembering it, another *Tyrannosaurus* is attracted to the scene. Approaching, it rises nearly to its full height to grapple the more fortunate hunter and dispute the prey. The crouching figure reluctantly stops eating and accepts the challenge partly rising to spring on its adversary.

While according to Ronald Rainger this display never came to fruition, its spirit lives on, perhaps captured in Czech artist Zdenek Burian's (1905–1981) 1938 painting of *Tyrannosaurus* approaching a cowering pair of Trachodon.[24]

Brown concluded his 1915 article with a rather precocious, influential description of the tyrant king lizard.

> *Tyrannosaurus* was a powerful creature, active and swift of movement when occasion arose. Its anatomical characters show distant relationship with lizards, crocodiles and birds. Like those of birds, the bones are hollow and the hind limbs in contour and construction closely resemble those of birds. Long, powerful hind legs carried the body upright, balanced by a long tail, and the front limbs, no longer a means of locomotion, had become rudimentary and

restricted for use only in grasping and holding. The massive head was armed with 13 dagger-like saw-bladed teeth in each jaw, the largest 5 inches long.... *Tyrannosaurus* was capable of destroying any of the contemporary creatures and was easily king of the period and monarch of its race.[25]

Not only was the artistic idea of a pairing of *T. rex* combatants anticipated and preceded by Cope's vision of his leaping, carnivorous dinosaur — Laelaps— as sculpted and painted by Charles R. Knight in 1897, but afterward, such a dramatic theme would also not be long neglected by America's most brilliant illustrator of prehistoric life — Knight again!

In 1866, paleontologist Cope had described a carnivorous dinosaur from Cretaceous deposits in New Jersey named Laelaps (now referred to genus *Dryptosaurus*). Cope had already sketched this animal, perceived as an adversary to his herbivorous *Hadrosaurus foulkii*, both viewed as leaping dinosaurs due to their long, muscular hind limbs. But now he sought a professional artist to bring his monster to life. Osborn offered Knight's considerable talents; Knight succeeded beyond Cope's prayers with a painting which has fascinated many dinosaur lovers through the decades. Knight's colorful study shows one scaly Laelaps individual springing onto a second laying on its back, arms and foot talons outstretched defensively. A sense of accuracy instilled in the restored fiercely fighting dinosaurs is falsely suggested by the painting's vitality, charm and detail, because so little fossil material actually exists with which to describe *Dryptosaurus* on firm scientific basis. However fanciful or imaginary the restoration, Knight restored Cope's Laelaps vividly from remote prehistory, as he would employ his extraordinary talents in the case of *Tyrannosaurus* within the next decade.

In 1906, Knight painted an Upper Cretaceous scene which has inspired so many others since. This time *T. rex* was paired against a family of *Triceratops*. As the scaly, upright-walking tyrannosaur advances confidently with tail elevated slightly above ground level, adult *Triceratopses* angle their sharp horns menacingly. Knight only suggested the titanic struggle about to commence, sparing viewers the inevitable, imagined bloodshed. Thanks to Knight, from this point forward, in our mind's eye *Tyrannosaurus* and *Triceratops* would share a special bond, forever united in Mesozoic Valhalla.

In another publication,[26] I suggested that *T. rex* has transformed through three paleoimagery Eras, with Knight's artistry dominating the first, known as the Era of "Savage Rex" (circa 1902–1942). *Tyrannosaurus* was one of Knight's favorite and frequent subjects, as he sculpted and painted *T. rex* in various memorable poses. In two cases, analogous to his "Laelaps," Knight painted a pairing of tussling tyrannosaurs. But his most popular and perhaps influential painting was completed as part of a magnificent series of murals showing Life's history through geological ages for Chicago's Field Museum of Natural History in the late 1920s. Here, out of misty, metaphoric "Nature red in tooth and claw" primeval time, *Tyrannosaurus rex* once more engages

Charles R. Knight's 1897 "Laelaps" restoration was one of the most influential dinosaur restorations of the 20th century (from *Century Magazine*, vol. 55, 1897). Note smoldering volcano at upper left.

Triceratops in battle waged on a forgotten primeval North American plain — the most terrible and titanic confrontation imaginable between two land vertebrates!

Through the gloom, an elephant-sized three-horned lizard grudgingly squares off against the most fearsome predator to ever walk the Earth, *Tyrannosaurus rex*. Only one of these mighty beasts may have survived the struggle, and yet the fate of both victim and victor was inevitably sealed through the last rites of extinction. Huge muscles strain. Thick dinosaur hide bleeds. Deafening cries bellow from parched throats. A battle to the death! The viewer sees the beasts ambling toward one another, nearly hearing the din: toothgnashing, turf-raising sounds. The death throes of an imagined victim (which one — a gored carnivore, or a slashed bull?) fade into fantasy. While so many of Knight's paintings entertain the savage struggle for survival in prehistory, a thematic war among vertebrates waged for supremacy through a succession of geological ages leading to our own ascendancy, arguably no single painting of prehistory has symbolized Darwinian survival of the fittest more fittingly than the Field Museum's Rex versus Tops mural.

Of this painting, and its featured monsters, Knight asserted,

Charles R. Knight's famous *Rex vs. Tops* painting of 1906 (from H. F. Osborn's *The Origin and Evolution of Life on the Theory of Action, Reaction and Interaction.* New York: Scribner's, 1918 ed.).

Tyrannosaurus (king, tyrant-lizard) with its massive head and long jaws armed with rows of serrated teeth, its gigantic hind limbs and tiny forelegs, must have presented a most awesome appearance as it strode stiffly about in search of prey. A flesh-eater, it could have swallowed a man at one gulp, had there been any on earth at the time. Fortunately, for our early ancestors, they did not appear until much later in world history. On the other hand, *Triceratops* (three-horned face), in spite of its formidable mien, was perfectly harmless, a vegetable feeder exclusively, and not aggressive in character for all its great horns and armored bony frill. The spiny projections on the head were merely for protection in case of attack from the ferocious carnivores of the period, and no doubt saved the peaceful, clumsy brute from early extinction. The sword of fate, however, hung over both these vast creatures, for the climate of their world was slowly changing.... Nature itself seemed weary of the great variety of reptiles which had dominated the planet for so many long ages, and which had now seemingly paved the way for the advent of a new type of being — small in size but possessing a hairy instead of a scaly body, and warm, fast-flowing blood in place of the sluggish reptilian stream.[27]

Through the next two decades, *Tyrannosaurus'* image became well publicized (usually posed in conflict with *Triceratops*) through a number of artistic renditions, chiefly owing to Knight's striking visions of the Late Cretaceous. We see Knight's remarkable impact on the dinosaur "industry" of the 1920s and 1930s (and later!) enlivened through a host of early restorations and classic dinosaur films. Unlike the enormous grip on popular consciousness held by mighty Mastodon and *Megatherium* over a century before, now aspiring dinosaur enthusiasts witnessed the most impressive restorations and reconstructions imaginable of two even more impressive beasts. The names *Tyrannosaurus* and *Triceratops* also swiftly entered the popular lexicon during this period through a number of published, memorable scientifiction magazine stories and dime store novels. By 1933, thanks especially to the Chicago World's Fair, *Tyrannosaurus* and *Triceratops*— two genuine

Charles R. Knight's mid–1920s small test painting for his later, grander Chicago Natural History Museum Rex vs. Tops mural. This painting was displayed near the entrance to Princeton University's Guyot Hall in 1993, when it was photographed by the author.

Mesozoic monsters—had vaulted to prominence in paleontology's subculture!

Not only was Knight's Rex versus Tops theme being reproduced in a number of popular magazine articles as well as both popular and geology text books, but other popular artists were moved to create their own arguably derivative works often published in booklet form. For instance at the 1933/34 World's Fair patrons could purchase copies of Sinclair's dinosaur booklet souvenir featuring James E. Allen's artwork, with its eye-opening centerpiece Rex versus Tops scene splashed across pages 6 and 7. (On its front cover readers were drawn to a Jurassic scene equivalent, *Allosaurus* mortally wounding an ornately plated *Stegosaurus*.) Allen also contributed a Knight-like Rex for the first, eventually, of three, Sinclair Dinosaur Stamp Albums. Meanwhile, another Fair exhibit called The World a Million Years Ago offered its own booklet souvenir written by Leon Morgan and illustrated by H. G. Arbo. A recreation mainly inspired by Knight's 1906 Rex versus Tops painting graced page 5 of this publication, Morgan's caption stated, "Though not as large as some of the great herbivorous creatures the Tyrannosaurus was carnivorous, feeding off his less agile neighbors. Standing twenty feet high and armed with huge teeth and sharp claws he was indeed a terrifying object, but of his great bulk only one pound was allotted to brain. With his powerful hind legs he could run or jump but his forelegs were of little use save for holding his prey." (Based on more recent skull endocast studies. Rex's brain certainly weighed more than one pound.) Notably, while Arbo's Rex's hands came each equipped with three fingers—like its Jurassic cousin *Allosaurus*—thanks to Barnum Brown's consultation Allen's Rexes were correctly restored with two-fingered hands.

By then *T. rex* had already been witnessed in a number of silent film productions. Yet the most dramatic display ever mounted outside museum grounds was hosted at the Chicago World's Fair, namely the Sinclair Dinosaur Exhibit. Here, a three-fingered *Tyrannosaurus* versus *Triceratops* display dominated a number of five other dinosaur statues each designed by P. G. Alen with Brown's help, situated in awesomely landscaped primeval settings. Not only due to their impressive size, but because their motorized heads rotated slightly forward as if in deadly combat, this display was the most thrilling popular dinosaur display until that time. According to Donald F. Glut, these dinosaurs reappeared at the 1936 Texas Centennial exhibition.[28] And an eighteen-foot-long *Triceratops* interacted with its deadly Rex foe, both constructed by the Messmore & Damon Company for the Fair's *second* major paleo-spectacle —"The World a Million Years Ago." This extravaganza was situated inside a hemispherical rotunda.[29] To the delight of many young paleo-philes of the time, lifelike, hollow lead-cast dinosaur toys offered in conjunction with this "World a Million Years Ago" included Rex and Tops replicas. These were developed chiefly with Roy Chapman Andrews' (1884–1960) advice. Messmore and Damon's life-sized dinosaurs proved more durable than Sinclair's, being showcased in a number of other settings and venues throughout the next four decades. Meanwhile, another Rex versus Tops study, based on Knight's Field Museum mural—this time a miniature sculptural display—was on view in the Century Diorama Studies for the Century of Progress exhibit.

By 1936, Works Progress Administration Project no. 960 commenced on a Rapid City, South Dakota, hilltop with the building of several life-size concrete dinosaur statues designed by sculptor Emmett A. Sullivan (1895–1936). Included among the Mesozoic menagerie were a Knight-inspired sculpted pairing— Tyrannosaurus vs. Triceratops.[30] Also during the 1930s, a team of sculptors led by John Kanerva (b. 1883) was busily constructing a grouping of prehistoric animals for the Calgary Zoo, many of which were native to Mesozoic Canada. However, in this setting Calgary's long demolished Rex statue was dwarfed by the truly immense "Dinny" the brontosaur (which is still standing).[31]

By the early 1940s, popularity of the Rex versus Tops theme still hadn't played itself out. Booklets such as Bertha Morris Parker's *Animals of Yesterday* (1941) incorporated Frederick E. Seyfarth's dramatic such pairing into a geological time chart; the theme was evidently prospering through a number portrayals published in foreign publications too.[32] By the 1940s, perhaps because the Rex versus Tops concept was becoming cliché, artists were increasingly turning to other genera (e.g., imagined battles between *Ceratosaurus* and *Stegosaurus*, a pairing of which was first sculpted for the Hagenbeck Zoological Gardens during the 1910s) with which to artistically convey Nature's struggle for existence, then popularly perceived as unmercifully savage during primeval times.

Dinosaurs-in-movies has evolved into a highly favored subculture of its own merit. While *Tyrannosaurus* is often and routinely cast among the saurian stars in such films from the beginning through (and surely beyond) *Jurassic Park III*, we find there were relatively few films prior to 1940 incorporating *T. rex* or themes inspired by Knight's tyrannosaur restorations. Perhaps the earliest such categorical film was a 15-minute-silent, *The Ghost of Slumber Mountain* (1919), featuring the stop-motion talents of Willis O'Brien. Here, following a deadly sparring match between two *Triceratops*, an *Allosaurus* (not a Rex) strides defiantly onto the scene, killing the surviving three-horn. When, threatened by the meat-eater, protagonist Jack begins shooting bullets at the allosaur; at the last moment Jack is awakened from his nightmare. Predating Knight's Field Museum murals, the climactic dinosaur battle must have been partly inspired by Charles R. Knight's 1906 American Museum Rex versus Tops restoration. *Tyrannosaurus* first appeared in film in 1923 —*twice*. First, as Mark Berry relates, in *Monsters of the Past*, Virginia May is shown sculpting a clay tyrannosaur, "by filming her tearing a finished sculpture apart and running the footage backwards! The animated scenes feature a Brontosaurus and a rather fanciful T. rex–Triceratops fight."[33] But Rex also appeared, although erroneously identified, in Max Fleischer's *Evolution* (1923). In this film a life-size *Iguanodon* statue then on display in Germany's Hagenbeck Zoo is misidentified as the "tyrant giant lizard (Tyrannosaurus)." And so in a later sequence where dinosaur battle scenes from *Ghost of Slumber Mountain* are spliced in, it was certainly implied to the audience that *Triceratops'* allosaur adversary was the *Tyrannosaurus*.

Perhaps contrary to popular perception, Arthur Conan Doyle (1859–1930) included no *T. rexes* in his classic novel, *The Lost World* (1912). However, *Tyrannosaurus* did make quite a splash in another silent, *The Lost World* (1925), being featured in several battle scenes animated by Marcel Delgado (1901–1976) and Willis O'Brien (1886–1962). Rex battles and then devours the horned *Agathaumas*—the latter one of Cope's dinosaur genera erected from fossil scrap, although (like Laelaps) made famous through another of Knight's superb 1897 paintings. *Lost World* even borrows a favored scene with a *Triceratops* family, replete with youngster, attacked by a fitting Rex substitute—the Jurassic carnivore *Allosaurus*—dramatic footage obviously inspired by Knight's 1906 painting.

While carnivorous dinos witnessed in *Lost World*'s stop-motion sequences might appear scrawny based on today's standards, that would not be the case with O'Brien's and Delgado's next beefy Rex, also certainly inspired by Knight's artistry. Their new Rex was pitted against pugnacious ape Kong in RKO's *King Kong* (1933). While based on today's science the appearance of Kong's Rex must be judged inaccurate, nonetheless this was the "no quarter" movie battle-to-the-death waged between prehistoria setting the stage and standard for all others to come! Several years later, in 1938–1939, such film fare inspired a young sculptor and animator Ray Harryhausen to create his own test footage, produced for an uncompleted film, *Evolution*. His objective was to relate "The history of life on Earth, told entirely in animation." Harryhausen's *T. rex* versus *Triceratops* fight sequence is one of the finest ever done, although as he later commented resignedly following his viewing of Disney's *Fantasia* (1940), "Oh, they're covering the same ground, and they did it so beautifully, I might as well abandon it."[34] A savagely compelling yet scientifically inaccurate animated sequence in *Fantasia* staged Jurassic herbivore *Stegosaurus* in a losing fight against a three-fingered *Tyrannosaurus rex*; the two genera were separated geologically in time by over 90 million years.

The year 1940 also brought the first of several suitmation tyrannosaurs to the silver screen in Hal Roach's *One Million B.C.* As Berry noted whimsically, "This quickly fabricated beast was, after all, only man-sized, since no optical enlargement effects were employed — nor, unfortunately, did stuntman Paul Stader happen to be 40 feet long and 18 feet high — and the scene's direction and editing wisely make every effort to ensure that the viewer never gets a clear look."[35] However, despite this inauspicious beginning as we shall see, over a decade later, the popularity of *T. rex* (or Rex-derived) costumes of filmdom had attained cult status idolatry. Clearly by 1940, in the arena of *T. rex* animation for cinema, we find that all the tricks had seemingly been tried. Was there nothing new under the sun?

Well, by the late 1940s, mankind *had* learned how to generate the power of artificial suns, brilliantly gleaming over southwestern sands, triggered by military testing.

Apart from movies but perhaps to lesser degree, mankind feasted on fantasy literature presenting tyrannosaur tales as well. Nevertheless, there were relatively few examples of *T. rex* in fantasy literature during the pre-atomic period. Usually, whenever Rex appeared *Triceratops* did too: the usual suspects. Thus, between 1917 and 1941, *T. rex* (or tyrannosaurids either conflated with or readily equated to Rex) roared and bellowed their way defiantly through a host of short stories and novels written beginning in 1917 with a novella segment — "There Were Giants in Those Days"— incorporated into Charles H. Sternberg's (1850–1943) *Hunting Dinosaurs in the Bad Lands of the Red Deer River, Alberta Canada* (no doubt inspired by Osborn's and Brown's researches as well as their American Museum Rex displays), Edgar Rice Burroughs (*Out of Time's Abyss*, 1918), Harley S. Aldinger ("The Way of a Dinosaur," 1928), Lester Dent (1904–1959, aka Kenneth Robeson); *The Land of Terror — Doc Savage*, 1933), Delos W. Lovelace (1894–1967) (*King Kong*, 1933), Alexander M. Phillips (1907–1991) ("The Death of the Moon," 1929), John York Cabot's ("Blitzkrieg in the Past," 1942), Robert Moore Williams (1907–1977) ("The Lost Warship,"1943), and Oskar Lebeck and Gaylord DuBois (b. 1899) (*The Hurricane Kids on the Lost Islands*, 1941).[36]

Here, three examples shall suffice to illustrate the image popular writers typically projected of their newfound real monster toy — which had supplanted the mastodon (and sauropods such as Brontosaurus and *Diplodocus* besides) as a revered national symbol of antiquity, ferocity and prehistoric identity.

In Edgar Rice Burroughs' (1875–1950) *Out of Time's Abyss* (1918) protagonist Bradley's encounter with a tyrannosaur — then, a species thought to be six million years old — is the novel's only graphic scene involving a saurian which storms onto the scene, rearing "up on its enormous hind legs until its head towered a full twenty-five feet above the ground. From the cavernous jaws issued a hissing sound of a volume equal to the escaping steam from the safety — valves of half a dozen locomotives." The *T. rex* devours character John Tippett. Its "sharp, three toed talons of the forelimbs seized poor Tippett, and Bradley saw the unfortunate fellow lifted high above the ground as the creature again reared up on its hind legs, immediately transferring Tippett's body to its gaping jaws, which closed with a sickening, crunching sound as Tippett's bones crackled beneath the great teeth."[37] The remaining members of Bradley's party kill the Titan with bullets. "It was an arduous and gruesome job extricating Tippett's mangled remains from the powerful jaws, the men working for the most part silently." In Robeson's *The Land of Terror*, rather incongruously, *Tyrannosaurus* hops menacingly toward the novel's heroes who collectively regard the toothy monster as a "cross between a crocodile, the Empire State Building and a kangaroo."[38] One of the Doc Savage characters named Renny even bears the misfortune of parachuting down onto the lost island — horrors — right smack into the middle of a nasty scrap between — you guessed it! — *T. rex* and *Triceratops*.

And in Lovelace's *King Kong* novelization, we read, "Out from the bushes ... below Ann's perch came a grotesque *hopping* creature [my italics] of very little less bulk than Kong's.... Its long slender neck scouted hungrily in every direction as it progressed upon powerful hind legs. Of forelegs it had almost none, only frail, clawlike members good for nothing save to lift food to its mouth. It was the mouth that caused Ann to scream again. Nothing she had seen was quite so horrible as that red aperture filled with pointed carnivorous teeth."[39] Evidently, the popular 19th century dinosaur-kangaroo paradigm (culminating in Knight's 1897 Laelaps and *Hadrosaurus* paintings for the *Century* magazine) of how monstrous bipedal dinosaurs moved certainly persisted into the filmic stop-motion animation period. And in 1931 (or 1932) O'Brien and Byron Crabbe completed a detailed pre-production drawing of a three-fingered *Tyrannosaurus* bounding like a kangaroo toward a belligerent, fist-waving Kong.[40] Fortunately for their viewing audience, in *King Kong*, sculptor Maracel Delgado and Willis O'Brien chose to have their *T. rex* puppet triumphantly *stride* rather than hop on screen into its mythic battle with Kong. True, if you watch closely enough at their battle sequence, Rex does take one short jump toward Kong, but after 1935, perhaps partially due to

This illustration by *Prehistoric Times* editor Michael Fredericks captures the spirit of Willis O'Brien's and Byron Crabbe's impressions of how the bloody, titanic battle between Kong and Rex would be orchestrated on film. Here, Fredericks has documented the original concept of a leaping Rex that was considered by Kong's creators and script writers. *King Kong* (1933) represents a significant transition in *T. rex* locomotion. Prior to release of this film, factoring in Lovelace's 1932 novelization of the script, and pre-production art staging this scene, Rexes, like all other theropods of fantastic fiction and film, leaped or hopped — perhaps a consequence of Charles R. Knight's striking 1897 "Laelaps" painting for *Century Magazine*. But afterward, despite Rex's short, subdued hop toward Kong visible in the movie, a vestige manifestation of former ideology, beginning with this breathtaking scene and inconsequence to Knight's 1935 "stiffly striding" interpretation for the American Museum and Chicago Natural History Museum, the idea of hopping tyrannosaurs fell by the wayside (courtesy Michael Fredericks).

Knight's inference that Rexes "strode stiffly," we read (and see) less of those incongruously hopping Rexes.

In my *Paleoimagery* book I referred to *Tyrannosaurus'* rule during the period, circa 1947 to 1975, as the "Lordly Rex" era. By this time, Rex's savage mastery over prehistoric nature was more prominently conveyed in restorations than ever before, such as through the exceptional artistry of Burian, Neave Parker (1910–1961), Louis Paul Jonas (1894–1971) and, especially, Rudolph Zallinger (1919–1995), whom we'll come to shortly.[41]

But before we consider *T. rex* in all its latter 20th century phases of glory and recognition, its main generic adversary—*Triceratops*—should now be properly introduced. By 1918, Osborn believed that *Tyrannosaurus rex* and *Triceratops* co-represented a pinnacle of dinosaurian evolution.

> This evolution took place stage by stage with the evolution of the predatory mechanism of the carnivorous dinosaurs, so that the climax of ceratopsian defense [*Triceratops*] was reached simultaneously with the climax of *Tyrannosaurus* offense. This is an example of the counteracting evolution of offensive and defensive adaptations, analogous to that which we observe today in the evolution of the lions, tigers, and leopards.... It is a case where the struggle for existence is very severe at every stage of development and where advantageous or disadvantageous chromatin predispositions in evolution come constantly under the operation of the law of selection.[42]

In other words, the evolutionary forces resulting in the brute *T. rex* essentially became, intricately and rather mystically according to Osborn, manifested in *Triceratops* as well. The two animals were not just associated in geological time and space, they were *genetically* linked. As evident through their common geological association in Hell Creek deposits, they are inseparable; one would not have evolved without the other's concomitant evolution. (However, Osborn's orthogenetic views were *not* entirely consistent with Charles Darwin's evolutionary premise.)

Considering its historical association with *Tyrannosaurus*, Tops may be the second ranking iconic dinosaur of all time. But, from an American pop-cultural perspective, whereas good-mother lizard *Maiasaura* climbed to the number one position on the charts for several years during the decade of the dinosaur (i.e., the 1980s), *Triceratops* has never been *numero uno*. In the annals of the history of paleontology, discoveries of the ever popular plated dinosaur *Stegosaurus*, and Great Britain's *Iguanodon* predate *Triceratops*' discovery and description. However, in our pantheon of prehistoric vertebrates, neither of these has risen to as prominent a place occupied by the dynamic duo Rex and Tops. *Triceratops* as *the* real runner-up dino-icon of all time? Yes and here's why. More than any other art form or written scientific description, it is the imagery of prehistoric life, especially life restorations, which kindles the fires of imagination. Such images fueled the public's acceptance of Tops, a genus inherently representing both prehistoric savagery and deep time, as a leading pop-cultural dinosaur decades ago. Mainly, of course it's all those restorations where Tops is posed fending off Rex's attacks, beginning with Charles Knight's influential 1906 American Museum painting. Tops became well known to science before its association with T. rex became traditionalized.

Triceratops was described in 1889, thirteen years prior to T. rex's discovery, although later, because of their geological association, scientists imagined the titans often engaging in deadly struggle. Especially thanks to the efforts of field collector John Bell Hatcher, *Triceratops* was known from fairly complete, composited remains, leading to Othniel C. Marsh's 1891 skeletal reconstruction of the three-horned beast.[43]

In 1892, wildlife artist Joseph Smit (1836–1929) painted the first life restoration, published as the frontispiece to the Rev. H. N. Hutchinson's (1856–1927) classic popular text, *Extinct Monsters: A Popular Account of Some of the Larger Forms of Ancient Animal Life* (Chapman and Hall).[44] Nevertheless, *Triceratops'* real bones, and eventually casts of those bones, wouldn't be assembled three dimensionally for several years. Additionally, drawings of *Triceratops'* skull and a mounted reconstructed skeleton appeared on pages 116 to 117 in Hutchinson's book. Also, besides Knight's later restorations, another early striking restoration was completed by Alice B. Woodward (1862–1951) for Henry R. Knipe's (1855–1918) *Evolution in the Past.*[45]

One of the earliest exhibitions of an assembled *Triceratops* skeleton, if not the earliest, was at Paris' Universal Exposition of 1898, inaugurated on July 21. During the first two days, 21,000 patrons ascended a staircase to the fossil animal gallery where they spied a mounted *Triceratops* standing amidst a variety of prehistoric mammals. Then, a year before *T. rex*'s discovery, artisans had produced a plaster cast of *Triceratops* for the Smithsonian's exhibit at the 1901 Pan-American Exposition, held in Buffalo, New York. Three years later, a cast (perhaps the same exhibit), appeared at the 1904 World's Fair, or the Louisiana Purchase Exposition, held in St. Louis, Missouri. Stereographic slides of the mounted skeletons displayed at both expositions, most likely produced as souvenirs, no doubt cemented *Triceratops'* formidable if not bizarre skeletal form in the minds of Fair patrons. During the 1910s, German sculptor Josef Pallenberg (1882–1946) designed striking life-sized *Triceratops* statues for Hamburg's Hagenbeck Zoological Gardens. Thereafter, by 1923, a life-sized *Triceratops* head restoration was sculpted by Lang and Otto Falkenbach for the American Museum of Natural History, a photograph of which was recently reproduced on the book jacket cover to Steven M. Stanley's *Extinction* (1987).[46] By this time, however, *Triceratops* was already on its way to becoming a sci-fi movie monster star.

Triceratops' chief *ceratopsian*, pop-cultural competitors through the period of its cinematic debut, however, were (1) the fanciful *Agathaumas*, first sketched as a living animal in the late 19th century by paleontologist Edward Cope and later more fully restored through artist Charles R. Knight's 1897 sculpture and painting, and (2) following its discovery at Mongolia's Flaming Cliffs—the unhorned, horned dinosaur *Protoceratops.*[47] *Protoceratops'* popularity stemmed from its association with fossil egg nests, at the time believed to be the first unequivocal proof that dinosaur young were born as hatchlings. But *Protoceratops* wasn't as large as its apparent evolutionary descendants, nor was it as powerful as either of the two larger genera, *Agathaumas* and *Triceratops*. And so, *Triceratops'* glory would surpass that of its evolutionary cousins, *Protoceratops* and *Agauthaumas*.

Charles R. Knight's most impressive early restoration of a horned dinosaur was not of the *Triceratops*, but in fact its screen rival — *Agathaumas*. Both horned dinosaurs made significant appearances in *The Lost World* (1925). This restoration is perhaps one of his least accurate, anatomically speaking, given the paucity of fossil remains attributed to this genus. Yet it remains one of the most beautiful artistic renderings made to date of any of the horned dinosaur species (from *Century Magazine*, vol. 55, 1897).

While *Agathaumas* was described from scanty remains, entire *Triceratops* skeletons had been assembled as museum displays, which no doubt added to the growing popularity of the horned genus. The World's Fair plaster casts, for instance, had been based on Charles Gilmore's assembly in the U.S. National Museum (Smithsonian, which he described in a 1905 paper). Tops' skeletons were also placed on display both at the British Museum of Natural History and the American Museum in New York. Furthermore, museum patrons purchased *Triceratops* memorabilia available in gift shops of the time, fostering impressions that *Triceratops* was the Mesozoic Era's most outstanding horned reptile.

Certainly since the late 1960s, marking the beginning of the Dinosaur Renaissance, *Triceratops* and other horned dinosaurs have been well represented in public view. However, even back during the 1920s, while most dinosaur genera known to contemporary science remained obscure to the public, Tops was a familiar figure, conventionalized through appearances in a number of early sci-fi and horror movies. For example, its co-starring role with *Agathaumas* in *The Lost World* (1925) was rather symbolic. A publicity photo framed using the *Agathaumas* and *Triceratops* puppets in a *Lost World* set (not a filmed sequence), shows the two horned creatures squaring off versus each other.[48] While *Triceratops* remained in the public eye, *Agathaumas*' on-screen performance was to be its last hurrah! Despite Knight's sensational 1897 restorations of this genus, cursed by a lack of fossil material or promotional backing, *Agathaumas* faded from the mainstream.

Triceratops became a usual suspect in early dinosaur filmography. For instance, besides *The Lost World* (1925), *Triceratops* imagery could be seen in *The Ghost of Slumber Mountain* (1919), *Adam's Rib* (1923), and *Monsters of the Past* (1923). In his *Dinosaur Scrapbook* (1980), Don Glut mentions a Laurel and Hardy comedy short, *Flying Elephants* (1927, produced by Hal Roach) which "included a briefly viewed *Triceratops* that was, like the numerous horses and donkeys of so many vaudeville sketches, two men in a costume who had perhaps earlier argued about which one would portray the back end."[49]

Delos W. Lovelace described an action packed scene involving King Kong's sparring match with *Triceratops* in his 1932 novelization of the movie script which unfortunately never made it into the jungle chase scene in RKO's *King Kong* (1933). Lovelace's 1932 dramatic descriptions of the never filmed Tops scenes went like this:

These obviously were still of the more gigantic creatures which had survived on Skull Mountain Island from a forgotten age. Huge, four-legged things they were, with thick short necks and short heavy heads ending in horns. There were three horns on each head, short pointed weapons which shook implacably after Kong.... "What are those brutes?" Driscoll asked.... Denham hesitated... "...I have it. Triceratopses... Just another of Nature's mistakes." ... Kong ... stood now on a dry mound in the center of the morass. He had put Ann down on the far side of the mound, away from the Triceratopses.... Already Kong was carrying on a long distance fire. Great slabs of the hardened asphalt swung up over his snarling face and went hurtling down upon the triceratopses' horny heads. "No!" Denham said ... "I won't believe it. There was never a beast as strong as that." What amazed him ... was the power with which Kong cast his huge projectiles. One, striking fairly, broke off a horn. The triceratops staggered, obviously hurt, and Kong redoubled his attack. The second of the two beasts swung grudgingly off to the flank and retreated slowly toward the watching group on the crest. The first also tried to retreat but another missile hit it again on the head and it fell. Kong roared in triumph and beat his chest. "We'll have to get out of this," Driscoll said. Off to the right, through a fringe of trees, could be seen the rocky edge of a narrow, stark ravine; at one point what looked like a fallen log led to the seeming safety of the far side. Driscoll pointed, and they all began sliding away. The ravine invited for a second reason. Kong, still roaring his triumph, had picked Ann up and was moving off. His course bent at an angle which, it seemed to Driscoll, would carry the beast-god around to the far side of the ravine. Only by crossing on the log bridge could they keep in touch."[50]

Later, when Driscoll's men crossing along the log bridge in pursuit of Kong and Ann, a *Triceratops* lumbers out of the trees:

Kong came lumbering up a slope, and at sight of the men on the log roared out and beat his chest.... Still angry from his earlier fight with the tricer-atopses, he was doubly enraged now by the men. And at the further sight of the three-horned beast charging toward the ravine his rage broke all bounds.... The men on the log could do nothing. To advance against Kong was impossi-ble. To retreat was no less so, for the triceratops, sighting his old foe, rushed up to the end of the log, and bellowed a challenge. Denham and Driscoll, from their caves, watched the tragedy helplessly.[51]

Footage from an unfinished film, the Kong precursor *Creation* (1930/31), shows a dramatic sequence featuring a charging mother *Triceratops*. Tops' worst cameo, however, was in Hal Roach's *One Million B.C.*, where the famed three-horn suit was incongruously worn by a live pig![52] From the sublime to the nadir, such films and lost scenes do exemplify *Triceratops*' contemporary, pop-cultural if not immortal appeal.

Triceratops was also memorialized by author Edgar Rice Burroughs. With over a dozen novels delving into fictionalized prehistoric settings and themes to his credit, Edgar Rice Burroughs may be the most prolific of all dinosaur writers. Word of Burroughs's fictional African realm, Pal-ul-don, first reached a delighted readership in 1921, through a novel *Tarzan the Terrible*, a sequel to his well-received *Tarzan the Untamed* (1919, 1920).[53] In the latter, ape-man Tarzan finds himself embroiled in World War I's African campaign, venge-fully waging a single-handed war on the Germans, who have destroyed his home and evidently murdered his wife, Jane Greystoke. At its conclusion Tarzan learns the awful truth, however, that a German officer has taken Jane hostage into wilder, more primitive and uncharted territory. And so in *Tarzan the Terrible*, Tarzan sets upon his undying, Homeric quest to rescue Jane. His path takes him to a prehistoric setting replete with sabertooth cats and lost cities populated by factions of a hostile civilization — descendants of *Pithecan-thropus* (which, unlike their real fossil ancestor, have prehensile tails). Mon-sters lurk about in the jungle too— such as an aquatic carnivorous reptile, a "frightful survivor of some extinct progenitor," as well as immense evolu-tionary descendants of the North American genus *Triceratops*, known to the natives as *gryfs*.

From his visits to London's Natural History museum, Tarzan instantly recognizes the *gryf*'s triceratopsian features (even though artist J. Allen St. John's (1872–1957) paintings of this conjectural dinosaur for the 1921 novel edition rather differ from Burroughs's description). As explained in *Tarzan the Terrible*'s Glossary, "The *gryf* of Palu-ul-don is similar [i.e., to *Triceratops*] except that it is omnivorous, has strong, powerfully armed jaws and talons instead of hoofs. Coloration: face yellow with blue bands encircling the eyes ... bony protuberances yellow except along the spine — these are red." The *Gryf*'s three red-colored parallel rows of osteoderms cover their spines, and ivory horn-like projections adorn their heads as in fossilized three-horns. These vicious animals are 20 feet tall at the shoulder and grow to about 70 feet long. Encased in their armored hides, *Gryfs* are invulnerable to spears and clubs. They're fairly uncontrollable too, but they can be (somewhat) domesticated, to the extent they may be ridden into battle — as Tarzan discovers. While J. Allen St. John's impressions of a (bipedal, theropodous-looking) *gryf* barely resemble the *Triceratops* known to science, other, more recent artists have elected to restore *gryfs*, as a more of a stegosaur/ceratops cross, highlighting the ceratopsian features.

A decade later, in Burroughs's 1930 novel, *Tarzan at the Earth's Core*, Tarzan spied a herd of *gyor*, a variety of horned dinosaur, much closer morphologically to the extinct *Triceratops*. This time, although in Pellucidar, Tarzan is "impressed by its remarkable likeness to the Gryfs of Palu-ul-don." Except *gyor* are herbivorous and lack osteoderms, and in fact, Tarzan was "convinced ... that he was looking upon an almost unaltered type of the gigantic triceratop that had, with its fellow dinosaurs, ruled the ancient Jurassic world." Yes, he (or rather Burroughs) probably meant to say Cretaceous instead.[54]

Tops' chief competition for the runner-up iconographic slot (after Rex) would perhaps be of the sauropod persuasion. While *Triceratops* was charging in the early movie footage and battling with carnosaurs a decade before release of RKO's King Kong, sauropods Brontosaurus and *Diplodocus* were also creating quite a stir. While industrialist Andrew Carnegie (1835–1919) had sent casts of his magnificent *Diplodocus* to museums around the world, underscoring America's industrial might and mastery over the natural world, the Sinclair Refining Company had adopted a Brontosaurus for their corporate logo. Not only that, but scary and destructive Brontosauruses starred in scenes of *The Lost World* and *King Kong*. And few realize that a Bronto-monster made it into Frank Saville's engaging 1901 novel *Beyond the Great South Wall: The Secret of the Antarctic*.[55] So, how does one reconcile Tops versus Bronto?

Throughout the 20th century, both long necks and Tops made repeated appearances in movies as well as in favored museum and World's Fair displays and dinosaur theme parks. While Bronto was cut from the *Jurassic Park* lineup, Tops was featured in *Jurassic Park* (1993 movie and 1990 novel). Tops has also appeared on over two dozen postage stamps, ranking only behind *T. rex* and *Stegosaurus*. For many of you, it may well be a tossup between Tops and Bronto for the coveted number two spot, but, to me, the more exotic-looking *Triceratops* outscores "Brontosaurus" on inherent "coolness" factors, due to its menacing demeanor. And I'd suspect that my psychological reception to Tops is rather like that of any dinosaur lover of the past century.

During the 1950s and 1960s, from the world's perspective, *Tyrannosaurus,* perhaps not unlike the spirit of the contemporary United States' geopolitical domination, was popularly perceived as a most commanding figure! Adopting the American Museum's specimen no. 5027 skeletal stance, tyrannosaurid reconstructions and restorations were usually displayed standing in upright, anthropomorphized poses. In this erect stance, they were lords of all they surveyed, capable of readily defeating nearly every living thing they encountered, including *Triceratops*! For a time, Rex was personified as the ultimate alpha male dinosaur. Yet while capable of defeating all other prehistoria, Rex was also cast as perpetual loser in its battles with mankind, or Nature itself. On the heels of Ernest Untermann's (1864–1957) 1930s and mid–1940s Rex versus Tops theme paintings for the Utah field House of Natural History, *Tyrannosaurus-Triceratops* imagery still prospered during the Lordly Rex era, chief examples being children's books, museum murals and another World's Fair display. By then youngsters were acquainted with Darlene Geis' *Dinosaurs and Other Prehistoric Animals* (1959) with R. F. Peterson's illustration of a Knight-inspired battle, as well as Jane Werner Watson's *The Giant Golden Book of Dinosaurs and Other Prehistoric Reptiles* (1965), featuring Rudolph F. Zallinger's (1919–1945) Rex-Tops battle scene.[56]

Zallinger's magnificent, simply jaw-dropping 110-foot-long, 16-foot-tall "The Age of Reptiles" mural, completed on June 6, 1947, for Yale University's Dinosaur Hall in the Peabody Museum, earned a Pulitzer Prize in 1949, was championed by *Life* magazine in 1955 and underscored nationalistic pride when featured as a block of four set of U.S. postage stamps issued in 1970. At the Cretaceous end of this life-through-geological-time portrayal, one spies a pot-bellied *T. rex* with its tail end turned defiantly toward a pair of *Triceratopses* (one seen in the distance). Confident king-pin Rex knows that not even three-horns would dare challenge his supremacy. But if Rex's reign may be dubbed lordly then, in retrospect, it was also certainly the most ponderous of its conceptual periods. Dating back to 1925, especially due to Richard Swann Lull's (1867–1957) assessment, *Tyrannosaurus* was considered dull-witted and slow of movement, a wretched tail-dragger. This perception pervaded for decades, through the late 1960s when Rex's clumsy and awkward ways, as for example espoused by William Swinton (1900–1994), predominated dinosaurian popular culture.[57]

Two other tyrannosaurid recreations dating from the Lordly Rex period represented the glorified Late Cretaceous view, when Rexes reigned over prehistoric nature. First was the centerpiece fossil in the Chicago Natural History

For many years, this is how "Gorgosaurus" (with lambeosaurine "prey") appeared in the Field Museum's Stanley Field Hall. Maide Wiebi's small, no longer displayed, sculptural restoration (right) of the scene accentuated the skeletal reconstructions. The author and his younger brother, Richard — later a professional geologist — stand at left in front of "Gorgy & Prey" (these photographs were taken by Allen G. Debus in 1965).

Museum's (now the Field Museum of Natural History) skeletal *Gorgosaurus libratus* mount, titled "Dinosaurs, Predator and Prey," posed standing triumphantly over its Lambeosaurine prey. Technically, our 30-foot-long *Gorgosaurus* wasn't a *T. rex*, although it highly resembled one. It stood fully erect as the museum's mascot, the Stanley Field Hall's main attraction; one had to venture upstairs to see the disembodied *Triceratops* skull, as well as Knight's famous Rex versus Tops mural. Later, *Gorgosaurus*— described in 1914 — became synonymized or technically conflated with another generic name, *Albertosaurus*, defined by Osborn in the same 1905 paper where he described *T. rex*.

Then, as of 2006, the Field Museum's marvelous "Gorgosaurus" specimen was referred to tyrannosaurid genus, *Daspletosaurus*. Although the paleontologist splitters and lumpers have flip-flopped on these names over the past three decades, for our purposes here, we'll refer to the Field Museum's mounted specimen as Gorgosaurus (even though as just stated you'll read the name *Daspletosaurus* instead on museum signage). For more on these dinosaurs, see Donald F. Glut's succession of entries for these genera in his

Dinosaurs: The Encyclopedia volumes.[58] Likewise, Russian paleontologists had discovered several well preserved specimens of another tyrannosaurid in Mongolia, to which they assigned the name *Tarbosaurus bataar* in 1955. Through the years since, paleontologists have noted skeletal similarities between North American *Tyrannosaurus* and geologically older, Asian *Tarbosaurus*. In fact at times it has been suggested that the latter genus should be instead referred to as a second *Tyrannosaurus* species—*Tyrannosaurus bataar*. (Herein, however, we'll refer to it as Tarbosaurus.)

If not for Gorgosaurus, fairly complete specimens of which had been found by Barnum Brown, Charles H. Sternberg, and Lawrence Lambe, by the 1930s paleontologists wouldn't have been able to make educated guesses as to how the forearms of *T. rex* appeared—before specimens were found, or how many fingers Rex had on each hand.[59] While *T. rex* was often conventionally figured in a powerful, relentlessly stalking pose, because Gorgosaurus was—skeletally—a more gracile form, paleontologists mounted its bones at the American Museum in a running pose, inspiring Christman's 1921 life restoration of the carnivore dashing after a pair of *Saurolophuses*.

The Field Museum's beautiful specimen was collected by Barnum Brown[60] in 1914 and was mounted by Field Museum staff by 1956. As Rainer Zangerl related in 1956, the composition, perhaps then the world's *second* most impressive tyrannosaurid skeletal mount, intended to show Gorgosaurus "as it has just come upon the carcass of another animal [*Lambeosaurus*] that had recently died. While the predator was looking it over, something disturbed him: he has reared up, startled."[61] (Shades of Barnum Brown's and Osborn's original American Museum *T. rex* display design!) In any case, "Gorgy's" magnificent pose resembled Zallinger's masterful *T. rex* rendition, as well as one sculpted several years later by fellow paleoartist, Louis Paul Jonas.

Sinclair featured dinosaurs for its display at the 1964/65 World's Fair held in New York City. And both Barnum Brown as well as Yale's John Ostrom (1928–2005) were called upon to advise Jonas as to the dinosaurs' proper appearance. Their life-sized *T. rex* was regarded as the most scientifically accurate restoration ever completed, a fitting climax to the Lordly Rex era! Sinclair's Rex symbolically battled Tops once more. Rex's lower jaw moved up and down, while Tops' neck swiveled to and fro. Not to be outdone by Sinclair, the Walt Disney Studios also projected their own dinosaur visions for the Fair's Ford Motor Rotunda Magic Skyway life-through-geological-time primeval exhibit. This time though, while in one diorama *Triceratops* guarded her hatchlings (as in Knight's 1906 American Museum painting), Disney's 22-foot-high, growling *T. rex* model had found another worthy adversary, borrowing from its famous, bloody vignette in *Fantasia*—a spiky-tail-thrashing *Stegosaurus*. This incredible savagely depicted segment of the display was enlivened through Disney's audio-animatronics artist team. Disney's dinosaurs were moved to Disneyland in 1966 where this author had the pleasure of seeing them a decade later.

Disney's scientifically *verboten* battle between a giant Rex-like carnivore and a geologically 85-million-year older *Stegosaurus* was more correctly envisioned in Karel Zeman's (1910–1989) 1955 Czech film, *Journey to the Beginning of Time*, where a hungry non-tyrannosaurid, *Ceratosaurus*, dramatically dispatches its North American contemporary, *Stegosaurus*, through the use of stop-motion puppetry. Zeman's film was released in the United States in 1966. However, such artistry was anticipated by (both) Josef Pallenberg's life-sized *Ceratosaurus* attacking *Stegosaurus* (circa 1910) for Germany's Hagenbeck Zoological Park, and an atmospheric 1942 painting by American Museum artist George Geselschap.[62]

Speculations as to the natural history of the peculiar morphological form represented by tyrannosaurids (gigantic bipeds with shortened arms and equipped with massive toothy heads and hind limbs), introduced by Brown and Osborn, were extended by Lawrence Lambe (1863–1919) who in 1917 pondered simple questions— in what stance would tyrannosaurs most comfortably rest and were they scavengers? Such contemplations led to further related matters such as their posture, maximum running speed and their probable physiology (were tyrannosaurs warm- or cold-blooded animals?)— questions which in one form or another have been debated ever since, *especially* during the current phase of tyrannosaurid hype and love which I have dubbed the era of "Renaissance Rex."[63]

Lambe, convinced that animals such as 30-foot-long Gorgosaurus were cold-blooded reptilians, championed the idea that such meat-eaters scavenged food rather than actively preying on huge contemporary herbivores. To him, Gorgosaur teeth seemed relatively free of wear, a result which could only signal its presumed scavenging lifestyle.[64] In other words, Lambe would not have accepted the authenticity of a Knightian *Tyrannosaurus-Triceratops* battle. "It is believed, therefore, that Gorgosaurus confined itself to feeding upon carcases of animals that had not been freshly killed, that it was not as an intrepid hunter but as a scavenger that it played its useful part in nature, and no doubt its services were fully required when we consider the immense numbers of trachodonts, ceratopsians, stegosaurs, and other dinosaurs and reptiles that lived and died in this particular time of the Cretaceous period."[65]

Lambe attributed the gastralia (belly) ribs and broadened pubic bones found in Gorgosaurus as special adaptations allowing gorgosaurs (and presumably all tyrannosaurs) to snooze on their bloated bellies through most of the day. Definitely not an inspiring picture for those of us who prefer imagining roaring Rexes charging across the prehistoric plains in search of danger, mayhem and Nature red in tooth and claw bloody combat with others of its kind over the spoils of war, along the way battling packs of sickle-clawed raptors, three-horns and club-wielding ankylosaurs! However, Lambe's theory wasn't so popular even among his peers, as Matthew oversaw the mounting of the American Museum's gorgosaur in an active running pose, while Christman illustrated the terrible lizard charging (without a credit card) after a dashing duckbill dinner. By 1933, with *King Kong*'s active fighting Rex in public consciousness, Lambe's boring, slovenly, scavenging (and definitely uncool) tyrannosaurids were left snoozing by the wayside, that is until 1970 when such questions resurfaced with a vengeance. Then, contemplation of Gorgosaurus led to reconsideration of *Tyrannosaurus rex*'s natural history.

In spite of how the Field Museum's Gorgosaurus was much later mounted — emulating the American Museum's skeletal mount — Lambe envisioned a different posture for the former. In 1917, "Lambe depicted this beast walking with a definite stoop rather than adopting the imposing upright stance always accorded to *Tyrannosaurus*. He also found in the leg joints signs that it had been bowlegged in life.... *Gorgosaurus*' stooping walk with tail raised of the ground as a counterbalance must have been exhausting."[66] In 1970, British Museum paleontologist Barney Newman reinterpreted the contemporary view of *T. rex*, although rather aligning his thoughts with Lambe's views on Gorgosaurus' posture. Newman thought *T. rex* walked with a stoop, like Gorgosaurus, and that it used its relatively tiny forearms to push itself from the ground in raising from a reclined resting position. Newman also quite correctly chopped twelve feet of tail from *T. rex*, most of which had been added conjecturally by Osborn over half a century before, thereby providing proper balance to the now more horizontally aligned animal. According to Newman, Rex no longer dragged, but waved its abbreviated tail triumphantly behind, while in front the head was also held horizontally aloft capping a flexible swan neck. In fact, the British Museum's *T. rex* 1960s skeletal mount reflected Newman's then radical views.

But now, given Newman's musculo-skeletal realignment of the hind limbs, Rex appeared pigeon-toed, perhaps a metaphorical sign of the times that were then dawning, as ever since the early 1970s birds were increasingly viewed as dinosaur descendants.[67] As Glut outlined, "With an *Albertosaurus* for comparison ... Newman ... stated that the hind legs could not have borne the stresses in walking if the body were oriented so that the tail dragged, and speculated that the weight of the walking animal must have been carried by one leg at a time, the gait most likely sinuous, producing a pigeon-toed waddling as in birds."[68] No longer a Gorgo-like tail-dragger, or a Godzilla-esque ground stomping creature, *T. rex*'s science was catching up to Gorgosaurus, taking on novel proportions that were unfamiliar to most!

Meanwhile, ever since those earlier days of Lovelace's Kong and Robeson's *Land of Terror* — with their kangaroo-like, hopping Rexes — *T. rex* had evolved considerably in pop-cultural representations as well.[69] For example, a favored 1970s Frank Frazetta painting reproduced as a wall poster featured a Knightian battle between a pair of musclebound dinosaurs—(three-fingered) Rex and Tops; in the foreground we spy a spear-wielding Tarzanian savage. Among other examples, and around the same period, Elbert H. Porter's (b. 1917) magnificent dinosaur statues were placed on public display in Orderville, Utah, one such denizen of his acclaimed Dinosaur Gardens being a life-sized posed fiberglass Rex, for its time every bit as accurate as Sinclair's 1964 World's Fair statue.

Because principal examples of science fiction, horror and other fantastic literature involving *Tyrannosaurus* dating from this later, modern period (circa 1950 to 2005) are well examined in my *Dinosaurs in Fantastic Fiction*,[70] it will suffice here to note that Rex versus Tops imagery has persisted into the Renaissance Rex age. Avid readers may encounter one such climactic battle in an Edgar Rice Burroughs' spinoff—John Eric Homes' 1976 novel *Mahars of Pellucidar*. Here, on the edge of our seats and with our heroes trapped in treetops above, we vicariously view the roaring contestants from above. Holmes thrills us with this description. "I saw the triceratops lower its head, paw the ground like a bull and then rush the tyrannosaurus. The big man-eater was too quick to be caught off guard, however, and he dodged out of the way, snarling and clashing his jaws."[71] Let the battle begin!

Undeniably though, *T. rex*'s primeval exploits were more prominently (yet sometimes ridiculously) conveyed to the public through science fiction and horror *movies* of the 1960s, '70s and '80s. Here we find that the dramatic presentation of special effect Rexes of filmdom often lingered behind the science. This was especially true in instances of suitmation saurs masquerading as *T. rexes*, and other creatures of the animal kingdom selected by producers to be adorned with accouterments received by the viewing audience as sufficiently Rexian. (This latter category will be covered more thoroughly in a later section of this book.) Besides *One Million B.C.*'s suitmation *T. rex* cameo, there were at least two others that readers may recall.

In Universal-International's *The Land Unknown* (1957), scientists intently investigating Antarctica's foreboding warm-water oasis mysteriously find themselves in a tropical, primeval setting populated by one-each-of-a-kind Mesozoic creatures. In one harrowing scene, a gigantic *Tyrannosaurus*, described by Mark Berry as a "zoologist's nightmare," menaces the helicopter crew. The upright, tail-dragging, twelve-foot tall suit equipped with hydraulic face controls cost a king's ransom to build. Interestingly, the retired head later became the pet dragon prop known as "Spot" on *The Munsters* show.[72] A second Rex entry appeared in Tsuburaya Productions' *The Last Dinosaur* (1977). While *The Land Unknown*'s Rex costume was created during the Lordly Rex era, *Last Dinosaur*'s Rex suitmation was filmed at a time when paleontologists were reassessing Rex's posture. So while the former special effects artist team can be excused for merely reflecting contemporary popularized scientific opinions, *Last Dinosaur*'s suitmation artists really should have known better. In *The Last Dinosaur,* the dinosaurs are found at the opposite end of the Earth in an Arctic oasis, accessed from the Arctic Ocean. Encountering a host of prehistoria, heroic character Masten Thrust, played by Richard Boone, covets a Rex trophy. Adding to his difficulty, the chunky, clunky suit-a-saur passing itself as Rex is practically Godzilla-sized! Furthermore, if this strange Rex can defeat its gigantic and feisty adversary, a suit-mation *Triceratops* resembling one of the Aurora Models' 1970s toys, then how can a mere mortal such as the obsessed Thrust ever defeat it? While studying a handful of western-made theropod dino-monster suits of the 1940s may have facilitated Japanese special effects artists in designing the various *daikaiju* monsters of later Japanese cinema, certainly *The Last Dinosaur*'s Rex and Tops together resulted in derivative fashion and reciprocally from creatures such as Toho's Gigantis and horned Anguirus.[73]

However, film critics collectively have viewed stop-motion animation sequences, so painstakingly achieved, such as witnessed in *The Beast of Hollow Mountain* (1956), *Dinosaurus!* (1960), *The Animal World* (1956), *One Million Years B.C.* (1966), *The Valley of* Gwangi (1969), *Dinosaurs ... Terrible Lizards* (1971), *Planet of Dinosaurs* (1978), *Caveman* (1981), *Prehistoric Beast* (1985) and a 1970s televised series—*Land of the Lost,* which featured *Tyrannosaurus* scenes (or in certain cases bloody battles between vicious theropods and horned dinosaurs, inspired by Charles Knight) more favorably than those relying on suitmation gimmicks.[74] Here, thanks to stop-motion wizards Wah Chang, Ray Harryhausen, Marcel Delgado, Jim Danforth, Phil Tippett, Stephen Czerkas and Jim Aupperle, the accuracy of the dinosaurs is remarkably enhanced; their animated Rexes seemingly instilled with improved accuracy compared to those resulting from oft impugned suitmation techniques. But most of this sense of technical accuracy so appreciated by the viewing audience owes simply to the fact that sculptors designing small scale Rex movie model armatures are not limited to the parameters and confines of human anatomy (in contrast to suitmation's inherent limitations).

The 1970s ushered in significant changes for the paleontological sciences. With respect to dinosaurs, much of the impetus behind a then mushrooming ideological revolution has been attributed to Robert T. Bakker's warmblooded dinosaur theory.[75] However, in the early years of the period referred to as the Dinosaur Renaissance, *Tyrannosaurus* re-invention didn't figure so prominently in furthering Bakker's revolutionary cause, as scientists initially focused on genera other than *T. rex.* At first there were few new Rex depictions. It was as if nobody wanted to challenge Knight's masterful yet dated visions, while at the same time artists perhaps felt uncertain about how to portray the newly embroidered king. In a behind-the-scenes sort of way, spirited by Barney Newman's inferences, despite what the movie industry was promoting in films such as *The Last Dinosaur* (1977) with its upright walking, tail-dragging Rex versus Tops suitmation battle, undeniably, *Tyrannosaurus* (like Gorgosaurus) was evolving.

Accordingly, Chicagoans can be proud of a local company, the Richard Rush Studios, where the first life-sized sculptural rendition of a *Tyrannosaurus* was made that departed scientifically with conventional views, now considered less accurate, based on paleontologist Donald Baird's consultation. In 1983, Rush's Rex was placed in a life-like, atmospheric diorama inside Milwaukee's Public Museum, poised over a massive *Triceratops* carcass. Replicas of these dinosaurs have been sent to other institutions as well, such as Indianapolis' Children's Museum. Rush's Rex was more horizontally posed than Elbert H. Porter's (which retained possibly for stability reasons a tail-dragging character and generally upright posture).[76]

But most paleo-philes had their first glimpses of Rex's new, more accurate looking stance and proportions not from Midwest museum restorations, but instead through the illustrations and paintings by artists Jessica Gwynne, whose pair of tyrannosaurs startled viewers in Desmond's *The Hot-Blooded Dinosaurs* (1975),[77] Ely Kish, who restored horizontally postured tyrannosaurs under advisement of paleontologist Dale S. Russell, and — perhaps especially — Gregory S. Paul, who championed Bakker's fresh warm-blooded ideals through lively portrayals of sprinting ceratopsians and prancing tyrannosaurids. Kish's and Paul's delightful restorations began appearing in a number of popular paleontological books and magazine articles, enhancing Renaissance Rex's appeal. As Bakker might have said at the time, Out with the (C-)old and in with the New, that is, Renaissance view, for as heralded in his 1975 *Scientific American* article we had entered a Dinosaur Renaissance.

The new, ubiquitous Rex was also a more engaging fighter, but then so was the new, refined and (according to Bakker) warm-blooded, nimble-limbed *Triceratops*. So naturally their imagined battles were fiercer, bloodier and more concerted than envisioned before. Thus, two influential 1980s books — Bakker's *The Dinosaur Heresies* (1986), followed by Paul's *Predatory Dinosaurs of the World: A Complete Illustrated Guide* (1988) — more so than any other publications of the period instituted and conventionalized views of warm-blooded, energetic tyrannosaurs, illustrating their mythic battles with spry, horned dinosaurs in modern light. Both were very artistic men who captured their heretical views brilliantly through life restorations. With such imagery on the loose it was becoming increasingly difficult to think of tyrannosaurs as being anything other than quick and light on their feet and capable of efficiently dispatching any form of fleeing or opposing prey.

Bakker and Paul led the movement toward convincingly captivating readers both in words and through their original drawings how nimble Rexes and other Late Cretaceous tyrannosaurs surveyed and effectively ruled their world as super-predators. Bakker's views of *T. rex* became celebrated through his 1986 volume, *The Dinosaur Heresies*. Emblazoned on its cover was John

Gurche's 1986 painting showing a battle between tyrannosaur *Daspletosaurus* versus its perceived nemesis, multi-horned *Styracosaurus*. Intriguingly, *Daspletosaurus'* left taloned foot can be seen upraised, fending off a swipe from *Styracosaurus'* sharpened shield and nasal horns. Were tyrannosaurs so limber and could they engage in battle so energetically? Yes, claimed Bakker, who also mounted a cast of AMNH-5027 in the Denver Museum of Natural History's main hall in this pose, below Gurche's dramatic mural. Accordingly, as if about to be crushed, now delighted museum patrons may pose for camera shots standing below Rex's threatening paw!

Casting aside conventionalized mid–20th century views of dinosaurs and, rather, reverting to Osborn's earlier visions of Rexes and Topses as more lively animals (as well as mystical, genetic links between them), in a chapter titled "Mesozoic Arms Race," Bakker lectured on nature's "co-evolutionary link" between dinosaurs who relied on armor and other defensive body gear for survival, versus their gigantic carnivorous, predatory foes. Bakker's illustrations of "big-plate stegosaur 'Diracodon'" battling *Ceratosaurus* and a confrontation between Rex and a pair of Tops (as well as a host of other restorations showing active, predatory theropods sprinkled throughout his book) underscored Bakker's convictions that gigantic tyrannosaurs were adapted to bring down even the most ornately defended prey animals of the time.[78] Yet no matter how loudly their stenchy, empty bellies may have rumbled, dispatching horned dinosaurs would have been no easy task for tyrannosaurids!

Since 1971 Bakker had envisioned horned dinosaurs as being fast-paced, able to gallop and therefore supremely dangerous for attacking predators. Bakker's skeletal reconstructions and life restorations of horned dinosaurs in *Dinosaur Heresies* as well as in another 1986 publication revealingly titled "The Return of the Dancing Dinosaurs," illustriously proved just how savage must have been their frequent Late Cretaceous struggles.[79] So, logically, this meant that Rexes, obviously adapted for a predatory lifestyle, had been perfected as Nature's ultimate killer monster machines! Fortifying this view, in *Dinosaur Heresies* Bakker pictorially restored Rexes ripping through a *Triceratops* corpse. And in homage to Cope, Bakker's book was also graced by a restored pairing of feathered *Deinonychus* leaping in combat as in Knight's latter's 1897 painting. Clearly, to Bakker the prehistoric world could be characterized as through its self-evident titanic ecological battles.[80]

Gregory S. Paul simply wrested the ball from Bakker's grasp and ran with it, scoring numerous points for active Rexes. In fact most modern paleoartists would proclaim they've been more inspired by Paul's visions of tyrannosaurids and other predatory dinosaurs than any others.' Throughout the latter 1970s and into the mid 1980s, Paul had produced a number of extraordinary dinosaur restorations—both paintings and drawings. Paul preached that tyrannosaurids were active, warm-blooded animals capable of charging toward or after one another, and that the racehorse-swiftest theropods were predators, not scavengers. Their cruel confrontations with sprinting horned dinosaurs were savage and far more brutal than Charles R. Knight would have dared to dream or capture on canvas. Paul also affirmed his artistic visions of the major dinosaurian groups in a 1987 publication, "The Science and Art of Restoring the Life Appearance of Dinosaurs and their Relatives: A Rigorous How-to Guide."[81] Paul's dramatic visions of how tyrannosaurids and horned dinosaurs interacted on the fertile plains of yesteryear are worthy of contemplation.

> Perhaps *Tyrannosaurus* used the power of its stouter limbs to equal the running performance of *Albertosaurus*. Or perhaps the former were better sprinters and the latter better long distance runners. The very size of *T. rex* may have made it the fastest tyrannosaur; there is no way to be certain. Stout *Tyrannosaurus* was well built for ceratopsian killing. To safely and successfully hunt ceratopsians, tyrannosaurs probably had to surprise them, or panic them into a run in which they could be approached from the rear. Otherwise the powerful horned dinosaurs may have reared like enraged bears to try and intimidate the tyrannosaurs. If that failed, a running charge was the horned dinosaurs' answer, and then the tyrannosaur often did the fleeing![82]

Paul's vivid portrayal of a Rex slicing off a hunk of flesh from the rump of a fleeing Tops was printed on page 28 of his 1988 book.

The idea of sprinting, warm-blooded Rexes stoked the fertile imaginations of science fiction and horror writers. Although there were no cliché Rex versus Tops battles described in Harry Adam Knight's (i.e., John Brosnan's) 1984 chiller novel *Carnosaur*, tyrannosaurs steal the show. Not only does a terrifying *Tarbosaurus* chase after a fleeing vehicle, racing at speeds sustained up to 35 miles per hour, but a mad scientist named Penward, who evidently has far too much time and money on his hands, sadistically murders his wife, whom he's chained securely in the presence of two hungry *T. rex* hatchlings. "Something pitter-patted over the newspaper towards her. She looked round and saw what it was. As she feared, it wasn't a rat — it was a baby *Tyrannosaurus rex* ... it only weighed seven pounds — but already its jaws were powerful and its teeth very sharp. Nature had designed the baby *Tyrannosaurus* to enter the world as a fully operational killing and eating machine. It bit Lady Jane in the calf then ran off, taking a small chunk of her leg with it. The reptile's brother, embroidered by his sister's success, followed suit.... Lady Jane's screams carried for a long way across the untended farmland, and continued for almost 12 hours, but no one heard them."[83] The dinosaurs, including a vicious *Deinonychus* which Penward gathers on his British estate, have been ingeniously restored from dino–DNA. As absorbing as Brosnan's book was, however, a trilogy of movies inspired from the novel featuring *Tyrannosaurus* and raptor dinosaur props are rather forgettable.[84]

While relatively few were aware of novel or movie versions inspired by

At left, the Burpee Museum's juvenile tyrannosaur, "Jane." At right, the actual skull of the Field
Museum's tyrannosaur "Sue" (photographs by
the author).

the novel *Carnosaur*, Rex's popularity as a cinematic monster was reestablished following publication of Michael Crichton's phenomenal 1990 novel,
Jurassic Park. Once again extracting fossil DNA, although this time from bugs
sufficiently preserved in amber, a team of mad scientists isolated on an island
(instead of a secured British estate as in Brosnan's *Carnosaur*) discover the
means of growing entire packs of dinosaurs—including a *pair* of tyrannosaurids. While the adult Rex obsessively chases Dr. Grant and John Hammond's grandchildren throughout the doomed Isla Nublar park setting, the
deceptive 8-foot-tall juvenile scores prey as well, after toying with a human
victim horrifically in cat-like fashion.[85] Rexes also monstrously devoured
expedition members on Isla Sorna in Crichton's 1995 sequel *The Lost World*.
Here, Brosnan's scene where a live prey victim is fed to Rex nestlings is fright-
eningly replayed. While both novels were best-sellers, Rex's real impact was
most keenly felt through the Jurassic Park film trilogy—*Jurassic Park* (1993),
The Lost World: Jurassic Park (1997), and *Jurassic Park III* (2001), inspired by
Crichton's novels.[86]

Creative use of computer graphics imaging, not words alone, would
complete Rex's regal transformation. For many viewers, *Tyrannosaurus'* image
seemed entirely transformed on film — an absolute paradigm shift — as it
strode magnificently into its electrifying *Jurassic Park* debut. Bellowing at the
puny, superior humans shivering in their stalled rail ride cars (and at the
trembling movie audience presumably as well), Rex triumphantly announced
its new age on film, catalyzed through computerized special effects (so analogous to the significance of computers used to revive dinosaurs in the *Jurassic Park* novel). Unlike its tail-dragging filmic ancestors, *Tyrannosaurus* moves

with gravity and grace. This is the perfected Bakkerian Renaissance Rex. It is the CGI-embodiment of Paul's restorations caught alive — an ideologue conceived collectively from the insights of Matthew, Newman, Donald Baird, Bakker, and John Horner.

In *The Lost World: Jurassic Park*, a Rex searching frantically and vengefully for its confiscated baby plunders is way through downtown San Diego. According to Berry, "The San Diego sequence pays affectionate tribute to decades of rampaging movie-saurs from the 1925 *Lost World* and the radioactive beasts and behemoths of the fifties, to *Gorgo* (also a parent searching for its captive infant) in the sixties." In another vignette, one Japanese businessman yells comically to his fleeing friend, "I left Tokyo to get away from all this!"[87] Move over Godzilla, the new improved Rex has arrived! But while mad Rex menaces humans time and again throughout Spielberg's trilogy of terror, an Isla Sorna specimen finally succumbs in battle to an even more powerful rival — a fin-backed *Spinosaurus*. Then in 2005, Rex-like theropod monsters reprised a legendary battle with Kong in Universal's dramatically staged CGI sequence.

Lately, based on enhanced, modern understanding of Rex, fortified on the strength of over 40 partially complete specimens (as of 2004), 80 percent of which have been collected within the past 30 years, *Tyrannosaurus*' role in prehistoric natural history is more accurately known than in Osborn's and Brown's day. Despite this, debates over whether Rex lived its life primarily as a scavenger or a hunter, how fast it could run, how quickly it grew into adulthood, whether it lived as an endothermic (warm-blooded) animal throughout its lifetime, how strong its forelimbs were, whether it sported feathers during any growth stage, the nature of its familial and social sense of kinship, and the extent to which it was related in an evolutionary sense to other tyrannosaur types such as *Tarbosaurus* and other proposed genera, have redoubled in the past decade and a half. While such scientific matters are intriguingly interrelated they might merely seem like hair-splitting (as opposed to hair-raising) details to monster film fans and science fiction readers. Indeed, perhaps especially due to its dramatic portrayals in the *Jurassic Park* films and *King Kong* (2005), a popularized consensus of Rex as a bellowing super-predatory beast has emerged. Today, even paleontologists find themselves likening battle-scarred Rex to pop-cultural superstars such as Michael Jordan, James Dean and Arnold Schwarzenegger.[88]

Increasingly since the early 1980s, Rex and Tops were symbolically cast as virtual witnesses to the K-T asteroid or comet which according to popularized theoretical scenarios wiped out the dinosaur clan (minus the avian lineage) 65 million years ago. Two of the many instances merit acknowledgment here. For the cover of its May 6, 1985, issue, featuring an exposé posing the question, "Did Comets Kill the Dinosaurs?" *Time* magazine's editors selected Braldt Brald's painting of a burly looking Rex baring its teeth as comets streak across the night sky — a mushroom cloud rises menacingly over the mountains beyond. Especially thanks to *Time*, from this time forward it wasn't difficult to associate Rex's image with its *ultimate* destroyer — not man, but Nature's fiery sword.[89]

For example, in Will Hubbell's science fiction novel, *Cretaceous Sea* (2002), a human female named Con, who has time-traveled backward to witness the extinction event, finds herself clutching at the bloated corpse of a *Tyrannosaurus* for survival in the Hades-like impact event's aftermath. Thus, in an ironic and ghastly sense, we understand that monstrous Rex is necessary for survival. "She ... discovered that there was a cavity between the corpse and the ground. The Tyrannosaurus had collapsed on its side with its back turned uphill. Its lower abdomen, stiffened by projecting bones that formed the front of its pelvis, made a shelter from the rain. Con dug in the rain-softened earth ... to enlarge the cavity. Eventually, she excavated a space in which to wedge her body. She crawled out of the rain, curled up, and fell into the merciful arms of sleep."[90] Poor starving Con even forces herself to *eat* the *T. rex*'s rotting flesh — surviving without grace just as our furry mammalian ancestors managed. Even the mighty shall fall, as the puny and pathetic shall reign! Another marvelous visual example of the impacting K-T boundary bolide's association to Rex is at the IMAX film's climax, *T-Rex: Back to the Cretaceous* (1998). Here, just as a young girl (who in a psychological sense has traveled into the Late Cretaceous Period) is about to be devoured by a Rex, the comet collides with the Earth, causing pressure blast waves to jettison her — presumably — through time back into her present.

The timeless theme of Rex versus Tops (or versus Rex's other worthy adversaries), originated through the imaginative interplay of Charles R. Knight and Osborn a century ago lives on resplendently through a host of life restorations (sculptures, paintings and drawings). Paleophiles revel in the color covers and pages of *Prehistoric Times* magazine, as well as a host of other visual displays paying homage to, as well as celebrating Charles R. Knight's prototypical visions.[91]

This chapter was first conceived as a historical, pop-cultural piece summarizing the "Rex battles" of paleoart, literature and film. That is, until I realized there was considerably more to the story. According to Paul Semonin, by the early 20th century, "the American monster [i.e., the American mastodon] was no longer the ruling species in prehistoric nature.... Osborn put the skeleton of *T. rex* on display ... inaugurating the reign of a new race of monstrous carnivores whose massive jaws were destined to symbolize the savagery of prehistoric nature."[92] While Osborn's preparators and technicians mounted skeletons of Rex and Tops, Charles Knight captivated our souls and psyches with his awesome *T. rex* versus *Triceratops* portrayals.

In a host of prehistoric versus imagery of the 20th century, depictions melding "nature red in tooth and claw" with the violence-in-prehistory theme, Knight's enduring, symbolic *T. rex* and *Triceratops* figures still live in spirit, although often transformed into other prehistoria. Cinematic distortions (e.g., footage of battling behemoths, Gigantis and horned Anguirus as in Toho's *Gigantis the Fire Monster*, 1959) merely borrow the idea's essence. In fact, one may classify the diversity of "versus" depictions into a number of categories—such as by geological period, historical episode, media, contestant genera (if Rex isn't a participant), or by Rex adversary (when *Triceratops* is absent), but for simplification let's survey a few examples.

In just how many ways can Rex and Tops be redistilled, preserving any modicum of originality? Well, many are the ways, even with subject matter (i.e., as *T. rex* versus *Triceratops*) which, by now should rightfully seem prosaic. Reviewing a century's worth of these rexy restorations can make one Tops-y-turvy, reeling with excitement! Besides Knight, notable paleoartists— James E. Allen, Mark Hallett, Gregory Paul, Robert Bakker, Michael Skrepnick, Paul Domke, Rudolph Zallinger, Wayne Barlowe, Serge and Paul Dupuis, Douglas Van Howd and James Gurney have all produced distinctive variations on the time-honored *T. rex* versus *Triceratops* theme. Besides the Milwaukee Public Museum which has a thrilling life-sized diorama showing *T. rex* preparing to feed from a Tops cadaver, within the LA County Museum visitors may see a dramatically reconstructed Rex versus Tops skeletal display (while the aforementioned Van Howd's life-sized sculptural restoration lures patrons inside from the museum grounds). Similarly, across the Atlantic at Germany's Munchehagen Dinosaurierpark, a *T. rex* versus Tops sculptural display lures tourists past the turnstiles into the main exhibition.[93]

Within this grouping of striking renditions, several sub-themes are evident. In the earliest examples, artists strove to emulate Knight's brilliant face-off scene, with Rex and Tops striding toward one another, yet witnessed prior to the inevitable impact and bloodshed. In Zallinger's mid 20th century "Age of Reptiles" mural, *T. rex*'s backside is turned toward the three-horned face.

By the 1980s, renaissance dinosaur paleontologists viewed *T. rex* as a high metabolic, lively creature. As if to prohibit pending nuclear inferno, in Cold War fashion paleoartist Gregory Paul doubted Rex would ever challenge Tops head-on in a face-off duel to the death, as Knight had envisioned over sixty years earlier. Therefore, as in Paul's 1988 restorations published in his *Predatory Dinosaurs of the World*, Rex's attacks are instead characterized as skirmish chases. For if Rex bites through hide of a fleeing Tops, then three-horn will eventually die from blood loss and disease spread by microbes festering in Rex's unflossed, Colgate-unclean six inch sabers. Only then can Rex scavenge. In Michael Skrepnick's 1998 stylistic restoration seen on the jacket cover of the Lanzendorf paleoart collection volume (see note 91), Rex feasts on a dead Tops, a scenario suggesting the tyrant king's possible role as scavenger.

In a Mark Hallett 1984 painting, *Triceratops* individuals round themselves up into a protective ring (like covered wagon trains are often shown in westerns whenever the Indians attack). Hallett's *Triceratops* band is shown brandishing their horns outward against the enraged pack of encircling *T. rexes*.[94]

Many rock-bands may be able to play "Stairway to Heaven," but, let's face it, there was only one original, the classic version. Similarly, Knight's original Rex versus Tops pairings are definitive, even though others have copied his designs. New discoveries, such as the early 1970s discovery of the giant Texas pterosaur *Quetzalcoatlus*, enlivens the possibilities for savagery-in-prehistory themed restorations.[95]

So, paleoartists have portrayed savagery-in-prehistory by introducing an alternate cast of dinosaurian contestants, an array which may be broken down into a phylogeny, a conceptual "family tree" of derivation. For example, if you will, instead of Rex versus Tops, Rex is often paired with another contemporary horned dinosaur such as a styracosaur. Then, in the next stem, Rex is often depicted combating (or having defeated) a non-ceratopsid animal such as *Quetzalcoatlus*, an armored dinosaur such as *Euoplocephalus*, or, anachronistically, versus *Stegosaurus*, or even the fictitious prehistoric ape — King Kong. Next, Rex imposters, theropods most related to *T. rex* such as *Daspletosaurus*, are portrayed in mortal combat with horned dinosaurs (*Styracosaurus* as in John Gurche's 1986 painting, or *Albertosaurus* vs. *Pentaceratops* as in David A. Thomas's life-sized bronze sculptural displays outside the New Mexico Museum of Natural History).

Rex has encountered other theropods as well. Aforementioned *Jurassic Park* imagery — Rex versus Raptors, and Rex versus *Spinosaurus* battles, for instance — have chilled us to the marrow! Or, finally and before leaving the Cretaceous period, how about Rex's worst nightmare — Rex versus itself — another Rex, or a *Giganotosaurus*, or *Daspletosaurus* versus *Daspletosaurus* which have been done by Knight, James Gurney, and Gregory S. Paul (or, for good measure, reconsider even Knight's famous Laelaps versus Laelaps 1897 pairing)?

Paleoartists and image-makers have frequently turned to the Jurassic period for inspiration too. The previously mentioned Jurassic classic example is *Ceratosaurus* versus *Stegosaurus*, or less commonly *Allosaurus* battling *Stegosaurus*. We've also seen Ceratosaurus battle *Triceratops* in *One Million Years B.C.* (1966), and the allosaurid Gwangi fight *Styracosaurus* in *The Valley of Gwangi* (1969). However, we may be straying a little off the main topic here as these portrayals cannot rival the Cretaceous savagery of Rex and Tops. And in fantastic literature and film, Savage, Lordly and Renaissances Rexes have of course battled *mankind*, at first for supremacy but increasingly for mere survival![96]

There are far too many examples, especially stemming from dinosaur filmography, to cite here. So then, arguably, Rex's most *symbolic* battle occurred in a relatively obscure 1992 film, *Doctor Mordrid,* where in one fascinating stop-motion animated sequence a tyrannosaur skeleton fights a Mastodon skeleton for supremacy.[97] And there's little question as to whether Rex actually did battle with Tops, for recent paleontological evidence does so indicate that the two confronted one another on the Mesozoic plains. "Ambush hunter" Rex wasn't simply a scrounging scavenger, and Tops did indeed need those horns to defend itself! Rex possessed the means to knock over a Tops, "before it could get away; then, as the prey animal struggled on the ground, the tyrannosaur could easily have killed its victim and eaten it."[98]

So, whenever we see *T. rex* in conflict, at the movies, in a painted or sculptural restoration, or even transformed into comical aberration —*know the beast!* Recall its derivation from the once-feared American mastodon, culminating in its 20th century reincarnation. Shape-shifting Rex, or should I say the mighty mastodon, knows many disguises.

Nine

Modern Prometheus

Just as *Tyrannosaurus* aspired to reigning king of the dinosaurs, surely Godzilla was crowned King of the Monsters. Godzilla was an unprecedented prehistoric monster of our own design, a dinosaurian harbinger of doom suited to our psyche. But Godzilla decidedly isn't a genuine species. To what extent can Godzilla be considered an iconic *dinosaur*? Rather ironically, Godzilla may now be the most iconic prehistoric animal of all. Pervasive Godzilla visuals and associated imagetext form a basis for the prehistoric monster's iconic, neo-mythological status.

Despite Godzilla's popular treatments as a movie dinosaur, Godzilla's relevancy to real dinosaurs and the science of paleontology is rarely regarded. That's because to experience Godzilla, we turn on our televisions, not visit natural history museums. To most paleontologists of the 1960s, Godzilla was something of an embarrassment; terrestrial vertebrates can't grow *that* huge, lift train cars in their mouths or hibernate for millions of years. Such a creature isn't known in the fossil record and its proportions defy scientific scaling laws. No alleged dinosaur, let alone any animal living for very long, could exhale radioactive particles. A single blast of cannon fire would mortally wound any real dinosaur. But not Godzilla!

Therefore, it follows that if Godzilla is a *dinosaurian* creature, then it should have prehistoric origins. Its stock should be rooted in Mesozoic prehistory. Accordingly, efforts are often made to reconcile Godzilla's dinosaurian origins and evolutionary ties to both real and questionable dinosaurians. In 1998, dinosaur paleontologist Dr. Kenneth Carpenter, who in 1997 dignified Godzilla by naming a real 225-million-year-old dinosaur specimen in its honor—*Gojirasaurus quayi*—published a serious-sounding, tongue in cheek article about Godzilla from a paleontologist's view.[1] So, if real paleontologists are giving the Big Guy the nod, then, especially given its pop-cultural status, how can Godzilla *not* be *the* iconic dinosaur? Is it time for real dinosaur icons, like that time-honored triumvirate of *Tyrannosaurus rex*, *Triceratops* and Brontosaurus especially, to step aside? Yes—and here's why.

According to monster lore, Godzilla was the result of Tomoyuki Tanaka's (1910–1997) "stroke of genius." While aboard a flight in the spring of 1954 he conceived a film about a dinosaur sleeping in the Southern Hemisphere, which had been "awakened and transformed into a giant by the Bomb."[2] Although, today, Godzilla scholars question the authenticity of Tanaka's lightning stroke Eureka moment, key facets of the monster's origination are captured in his recollection.

Here we have yet another, not entirely original case of an amphibious sleeping dinosaur. This sleeping condition implies that a dinosaur of some unfathomable species has been hibernating for considerable ages, or whose species has mysteriously survived in a hidden part of the globe. This fantastic scenario recalls Ray Bradbury's classic 1951 short story, "The Foghorn," where a dinosaurian survivor, haunting the oceanic deeps, is lured to a mariner's fog horn sounding atop a doomed lighthouse. It also summons a memorable 1942 *Superman* cartoon episode, *The Arctic Giant*, where a Rex-like dinosaur is accidentally thawed from an icy imprisonment. It is uncertain whether Tanaka was familiar with either of these productions.[3] But, for *Gojira*'s sleeping *dinosaurian* aspect, Tanaka was admittedly influenced by Eugene Lourie's *The Beast from 20,000 Fathoms* (1953), which drew inspiration from Bradbury's tale.[4]

Next, the body of the sleeping giant is resurrected. Unlike in Bradbury's tale, Tanaka viewed the transformed (i.e., mutated) monster *Gojira* as "the son of the atomic bomb. He is a nightmare created out of the darkness of the human soul. He is the sacred beast of the apocalypse." But Godzilla was more than a bomb metaphor. Ishiro Honda viewed the monster as a "physical manifestation of it."[5] According to Steve Ryfle in his 2006 article, "Godzilla's Footprint," "Godzilla's impossible size lends the film added thematic weight: the creature is as massive and unstoppable as the technology that created it, as if nature itself were retaliating against man for his foolish tinkering with the laws of physics."[6]

So, essentially, we have the case of, perhaps, a more ordinary sort of relic dinosaurian that—until the fateful event of its deadly exposure to atomic radiation—was unobtrusively slumbering or otherwise surviving in an out of the way environ. Then, intense rays from the nuclear blast must have killed or destroyed this original, progenitor creature before its awful transformation and resurrection into a corpse-like radioactive, mutated form. (Thus, as shall be further explained later, the overall effect is analogous to the Frankenstein monster's reanimation.)

The altered Godzilla, salvaged from extinction, haunts mankind relentlessly. The coming of Gojira, seemingly more dead than alive, is a ghost-like manifestation. Its vengeful Tokyo stomp becomes the original "night of the living dead."

The aforementioned Dr. Carpenter's welcome contributions came long after those times when young baby boomers avidly watched the original Godzilla series as the films came out, relentlessly, one after another. Back then, it was relatively easy for impressionable young dinosaur enthusiasts to conflate fantasy movie-land dinosaurs of the 1950s and 1960s as being somehow more serious than fanciful. Through the 1960s, during a time of societal angst and world chaos, despite the attraction to Godzilla among a younger viewing audience, Godzilla was ridiculed, not revered, by American establishment. During the late 1960s, however, things were rapidly changing on many fronts. By then, paleontologists were reconsidering *status quo* opinions as to how real dinosaurs lived and appeared in life.

The term *Dinosaur Renaissance* wasn't printed until April 1975, when *Scientific American* coined that phrase to describe the new lively views of (warm-blooded) dinosaurs perceived by paleontologists such as Yale University's John Ostrom and, especially, the "young turk" himself—Robert Bakker.[7] But few realize that the Dinosaur Renaissance had its cultural beginnings a decade earlier, in 1965, when paleontologist Loris Russell published speculations as to whether dinosaurs' warm-blooded condition may have led to their extinction. Also in 1965, a new raptor dinosaur, later described as *Deinonychus* in Ostrom's 1969 monograph, was discovered in Montana.[8] By then, five Godzilla movies had been produced by Toho Productions, four of which had reached U.S. cinemas. By the end of 1965, from a contemporary intelligencia perspective, there seemed to be no relevancy between Godzilla's lowly B-class status and discovery of a then obscure middle Cretaceous reptile bearing sickle-shaped claws on its big toes.

Bakker, Ostrom and other converts published radical ideas about dinosaurs throughout the early 1970s. In retrospect, their heresies salvaged a dead science — paleontology—from the scrap heap. The artistic Bakker portrayed dinosaurs in poses which seemed politically incorrect or downright wrong to nearly everyone, even though these illustrations are now revered as revolutionary. Almost overnight, springing on the heels of the new intelligent, lively and possibly warm-blooded dinosaur's (i.e., *Deinonychus'*) discovery, paleontology had become fun, lively and strangely relevant! How curious then that Godzilla's popularity rose coincidentally during a time when paleontologists re-engineered dinosaurs!

While the most popular monsters of previous decades had been creatures like Frankenstein's Monster, Dracula, werewolves and King Kong, in 1973 Godzilla won a poll conducted by *The Monster Times*. By then, eleven Godzilla films had been released in the U.S., including his recent winning, and then socially relevant, bout over Hedorah the Smog Monster. As Donald Glut declared in 1978,

> To the editors' astonishment, the winner of the poll in the category "your favorite monsters of all time" was not the immortal Kong. Runnersup in the poll were, in descending order; the ever popular Dracula, Frankenstein

Monster, King Kong, and Wolf Man, with the Mummy and Black Lagoon Creature making impressive showings in the horror race. But the creature who took in the majority of votes was Japan's awesome reptilian superstar Godzilla, the so-called (and appropriately so in the seventies) King of the Monsters.[9]

That would also make Godzilla the most popular *prehistoric* movie monster ever.

Godzilla's magazine cover debut coincided with a rapidly emerging rebirth in dinosaur science. Godzilla's first official magazine front cover appearance was in *Monsters of the Movies* no. 5; this February 1975 issue came out just weeks before *Scientific American*'s landmark Dinosaur Renaissance issue. In March 1975, *Famous Monsters of Filmland* followed suit with their first Godzilla cover.[10]

Since the early 1990s, as paleoartists (i.e., artists who create imagery of prehistoric life) came forth in droves, Godzilla and his Japanese monster nemeses have formed an exclusive subset of paleoimagery, an offshoot of dinosaur imagery, or *Godz-imagery*. Godzilla's imagetext (written or unwritten but sometimes vaguely understood meaning reflected by visuals), may be interpreted in relation to other dinosaur imagery. Godzilla denotes prehistoric savagery, although projected into modernity. Even though Godzilla isn't really prehistoric because he lives in our present, people still think he (or she or it) is. (According to pseudo-science spun by actor-paleontologists in the older film series, Godzilla would then qualify as a sort of living fossil, like the *Coelacanth* fish.) Furthermore, Godzilla's enormity, strength and invulnerability symbolizes the grandeur of prehistory compared to our puny, enervated era on earth. By the late 1960s Godzilla, a powerful godlike, drago-dinosauroid, had become a savior, protecting us from powerful foes, calamitous circumstances and aliens.[11]

In his article, "Wardrobe! The Many Suits of Godzilla," John D. Lees describes the evolution of Godzilla's appearance from the beginning in 1954 through his 22nd film, "*Godzilla vs. Destroyah*" (1995).[12] Herein, the observant Lees noted progressive trends in "suitmation" design. However, these changes in Godzilla's appearance would be considered relatively subtle considering how different restorations of real iconic dinosaurs known to contemporary science such as *T. rex* appear today, that is, when compared to how paleoartists formerly restored them during the year of *Gojira*'s release—1954. To a certain degree, even real dinosaurs are fictional, like Godzilla, because of the evolving way in which human intellect has recreated and interpreted them.

After a few false starts for Toho's 1954 masterpiece, *Gojira*, technicians and artists arrived at a suit design that was sufficiently dinosaur-like for their prehistoric monster, borrowing heavily from

two upright-walking reptiles, the *Tyrannosaurus rex* and the *Iguanodon*. They also borrowed a distinctive feature of the quadrupedal *Stegosaurus*: three erect rows of dorsal fins lining the spine, from neck to tail, with no function other than to give the creature a unique look (it was decided, after the fact, that the fins would glow when Godzilla emits his radioactive breath). In taking artistic license with evolutionary history, the designers were clearly more interested in creating something fantastic rather than realistic or logical, a spirit that marked the beginning of *kaiju* history.[13]

Additionally, Ryfle noted Godzilla's "rough-hewn pleats" of its alligator-like hide representing radiation scarring, not noted in real dinosaur mummy fossils, where skin impressions are sometimes remarkably preserved.

While, as Lees noted, Godzilla's suitmation design evolved slightly through the 1970s (some might say decayed) over many movie performances, by the 1990s artists began visually re-conveying Godzilla's origins as a genuine prehistorian. In fortifying the monster's prehistorical associations, their designs, based on modern Dinosaur Renaissance views of how dinosaurs appeared, departed significantly from Toho's original concept. In 1998, for instance, using up-to-date science, vertebrate paleontologist Dr. Carpenter published "A Dinosaur Paleontologist's View of Godzilla," in which he reconstructed Godzilla's skeleton, in the process inferring many of its supposed life habits and phylogeny.[14] No doubt chuckling to himself, Carpenter — who in 1997 had named a genuine Triassic fossil carnivore *Gojirasaurus quayi* — claimed that movieland Godzilla was related to a carnivorous, horned theropod variety of dinosaurs known as ceratosaurs.[15]

Throughout the early 20th century, real dinosaurs such as *Tyrannosaurus*, *Triceratops* and Brontosaurus increasingly became pop-cultural icons. These dinosaur genera, especially, symbolized deep geological time, extinction and the savagery of prehistoric times. Following the onset of the Dinosaur Renaissance in America, however, by the 1990s a host of new real dinosaurs gained iconic status, particularly *Deinonychus*, *Maiasaura*, and *Jurassic Park*'s vicious raptors—*Velociraptor*.[16] A flock of feathered dinosaur species captured late 1990s media attention. Synchronously, Godzilla came out of semi-retirement, smashing his way through new, more highly urbanized movie sets. Until the dawn of the Dinosaur Renaissance, Godzilla rarely received much good press and review of Godzilla movies were often far from rave. However, once the Dinosaur Renaissance was in full swing, Godzilla, like the new, pop-cultural challengers—*Deinonychus* and *Jurassic Park*'s raptors—increasingly became an American pop-cultural icon. (Comical references to Godzilla on *Saturday Night Live* and a cartoon short *Bambi Meets Godzilla*, 1969, fortified the monster's rising public appeal.)

Let's place things in cinematic context, focusing on three pseudo-dinosaur icons from the movie industry. At a young, impressionable age, 45 years ago, this writer experienced three fantasy dinosaurs, all of which at one time or another were revered by monster movie fans. Four decades later, only one of these has attained iconic status in America. The movies are *King Kong*

(1933); *Godzilla, King of the Monsters* (1956), the American version of *Gojira*; and *Gorgo* (1961). Of course, while Kong remains a bastion of the movie industry, he wasn't a dinosaur. Yet for many years the individual *T. rex* which Kong battled and killed on Skull Island *was* rather considered a cinematic dinosaur icon. This dino-monster, based on paleoartist Charles R. Knight's artistry, was perhaps two to three times larger than any real *T. rex*, and was sensationally portrayed by Willis O'Brien using stop-motion animation. Whereas before, Kong's *T. rex* foe formerly epitomized what Rex is all about, today, Kong's Rex has receded in stature. Through the past century, scientific opinions of how *T. rex* appeared in life and lived have changed remarkably. Conversely, today, many new painted, sculptural, or animated *Tyrannosaurus* restorations rival or exceed O'Brien's stop-motion artistry. So, Kong's cold-blooded Rex is relegated to just another example of how *T. rex* formerly mirrored scientific ideas about this real dinosaur. Today, Chicago's Field Museum of Natural History's "Sue" specimen and her well documented imagetext is regarded as a greater iconic Rex than is Kong's Rex.

Godzilla's greatest potential dino-monster rival, more so even than fellow Toho prehistoric monsters winged Rodan, spiny Anguirus or any of the later Toho monsters, would be Gorgo. Although rarely aired today, back in 1961 dino-monster Gorgo seemed sensational. In youthful exuberance, this writer often wondered when would the movie "Godzilla versus Gorgo" finally get made? While Gorgo and its offspring mysteriously vanished beneath the waves, Godzilla kept coming back time and again, even fighting King Kong only two years after Gorgo disappeared from movieland. At the time, both Godzilla and Gorgo weren't established icons by any means. Being parental, something a general audience could sympathize with, Gorgo even held an early advantage over Godzilla. If a healthy dose of Gorgo sequels had been made spanning the time of the Dinosaur Renaissance, today Gorgo would share giant monster, pop-cultural limelight with Godzilla. After all, while over two dozen performances have shaped Godzilla's charismatic aura, Kong's Rex and Gorgo never had sufficient screen time to acquire unique, likeable personae.

Now on to the film which inspired this chapter—the Discovery Channel's *Dinosaur Planet*, aired for the first time during the week of December 14, 2003, which to me, had much to do with Godzilla as well. Viewing this amazing documentary with its realistic CGI-animated dinosaurs, made me realize that the real dinosaur icons of old are fast fading. Back in the mid–20th century, dinosaurs were typically cast in movies as solitary, statuesque, larger than life monsters on the loose. Before the Dinosaur Renaissance, dinosaurs portrayed in scientific restorations—such as those by artists Charles R. Knight, Zdenek Burian and others of their profession—were often rogue *individuals* symbolically representing the mysterious dinosaurian breed, posed in idealized paleo-landscapes. They were like the real monsters of their

Gorgo's titular monster.

respective paleo-worlds. When one heard or read the word *Tyrannosaurus* back in 1935, only a limited array of popular images sprang to mind. Back then a *T. rex* meant possibly the Kong Rex, or the American Museum's *T. rex* mounted skeleton, or either of Knight's two famous American Museum *T. rex* paintings, where the beast is shown menacing three-horn *Triceratops*. And none of these older representations look much like how *T. rex* has been restored by artists during the Dinosaur Renaissance.

However, documentaries like *Dinosaur Planet* and Tim Haines' *Walking with Dinosaurs*[17] bring us entire paleo-ecologically correct faunal assemblages, showing us, not individual dinosaurs, but entire breeding families and populations of dinosaurs interacting much like modern animals of the African veldt. These scientifically accurate portrayals de-emphasize dinosaur iconology while creating impressions that names of dinosaur genera and associated pop-cultural symbolism are of secondary importance to understanding their roles and means of survival within a contemporary food web. As scientists divulge the dizzying diversity of the fossil record, it's clear there were so many evolutionary *intermediate* dinosaur forms that it's hard to tell at a glance what genus many of these CGI-animated dinosaurs are intended to be. For instance, a herd of fabulous horned dinosaurs roving about one landscape recreated in *Dinosaur Planet* could mistakenly be considered *Styracosaurus*. Instead they turned out to be a closely related form known as einiosaurs.

The most popular *real* dinosaur icon, *T. rex*, doesn't closely resemble its former self, as it appeared decades ago. According to paleoartist cliché, isn't *T. rex* supposed to be the dinosaur which (only) attacked *Triceratops*? And didn't they always battle to the death in the shadow of smoldering, metaphorical volcanoes? Furthermore, it turns out that based on skeletal anatomy most large theropods of the Late Cretaceous period looked very much like *T. rexes*. Amidst the bewildering blur of new dinosaur species known to science, if most of us can't tell at a glance, say, whether a CGI-animated T. rex really is supposed to be a *T. rex*, versus a *Daspletosaurus*, an *Albertosaurus*, *Tarbosaurus*, *Gorgosaurus*, a Carcharodontosaur, or even a Giganotosaur, can similar-looking *T. rex* be so iconic?

Unlike circumstances 50 years ago when most dinosaur enthusiasts instantly knew from looking at iconography which dinosaur genus was which, these days one can't keep track of them all without consulting a handy dinosaur encyclopedia.[18] Today, there are so many talented paleoartists performing their mesmerizing work that it's hard to find dinosaur species that haven't been painted, sculpted and restored or artistically converted into resin model kits. Suitmation trends noted by Lees, mirroring Godzilla's evolving personality traits, are noticeable yet relatively subtle, *especially* when compared to how dramatically *T. rex*'s restored appearance has changed in half a century.

After over half a century, not only is Godzilla *basically* the same old dino-monster, but we can also always readily discern Godzilla from rival *daikaigu eiga*, or former dino-icons like Gorgo, Ray Harryhausen's Rhedosaur *Beast from 20,000 Fathoms*, and Kong's larger-than-life Rex. Whenever we see images — either filmed or derived by artists from film — of a large, bipedal dino-monster with enormous jagged plates situated along its upright spinal column, we instantly know it must be none other than Godzilla (or maybe "Gigantis")!

I'm not proposing simply that Godzilla is an iconic dinosaur. Or that *T. rex* remains our most iconic *real* dinosaur. They are! Moreover, to the intelligentsia's chagrin, Godzilla has beaten all the real dinosaurs at their own game. Recent successes in dinosaur science have arguably created an ironical icon, mightier than Rex, *Triceratops* and all the rest. Because of (1) movie sequel-itis revealing the monster's persona, spanning a significant period in the development of dinosaur science, (2) popular, endearing references to Godzilla in the media, (3) a subtly shifting, yet relatively stable basis for *Godz-imagery* and (4) a recent Dinosaur Renaissance trend toward dispensation of our older, most familiar genuine dinosaur icons, ironically, distinctive Godzilla is poised to become the most iconic dinosaur of all time.

Now let us consider what is the validity or pseudo-scientific basis of Godzilla's *prehistorical* origin, and why such a circumstance is preferred by so many fans. To answer this conundrum, we must address Godzilla pseudoscience, as well as the basis of the psychological appeal and symbolic meaning of fantastic prehistorians known as *daikaiju*.

In *Gojira* (1954), Godzilla (played by Haruo Nakajima and Katsumi Tezuka) poses the utmost dilemma for a pair of Japanese scientists, the younger Dr. Serizawa (played by Akihiko Hirata) who is mad — most honorable, yes, but quite tortured in mind — and an older distinguished professor of paleontology, Dr. Yamane (played by Takashi Shimura). Following the sinking of vessels, Japanese authorities are led to a critical panel discussion among Japan's key dignitaries and military experts where Yamane outlines his speculations about a strange creature which could be at fault. Following a hurricane-mimicking catastrophe on Odo Island, Yamane leads the expedition to find evidence for such a monster there. Evidence abounds, as Yamane notes signs of radioactivity, and a trilobite — not a fossil but a living specimen — embedded into a sizable footprint on the beach. Sand specimens obtained from Godzilla's prints on Odo Island are similar in age to Jurassic undersea deposits. Then, as if all this wasn't enough, Godzilla appears in broad daylight. Returning to Tokyo following this fruitful mission, Yamane lectures to the authorities, presenting his theory about the natural history and origin of the 50-meter tall dinosaurian relic from the Jurassic, which, as he proclaims, is a 2 million-year-old species. Yet in a pairing of scenes inserted to suspend disbelief as to his knowledge of relevant subject matter and perhaps to disguise his mistakes, Yamane is shown alongside images of genuine North American Jurassic dinosaurs, such as a small scale skeletal replica of the plated *Stegosaurus* and a restoration of Brontosaurus. (And just for the record, trilobites didn't survive into the Jurassic Period; they were extinct by about 250 million years ago!)

Why would such a living fossil as Godzilla be raising havoc in modernity? Yamane theorizes that there may be mysterious, undiscovered caverns, or pockets deep under the sea waves where such (amphibious) monsters could dwell in isolation through geological ages, until, fortuitously, disturbed by detonation of atomic warheads. Godzilla, a legendary creature known to the natives of Odo Island, occasionally made historical excursions to the island where it would accept human sacrifices. So, based on legend, Godzilla episodically leaves the abyssal pockets to explore shallower waters too. It's just that during the mid–20th century, nuclear bomb testing raised havoc with its physiology, driving the monster from its normal range of habitat.

Then, ominously, Yamane pronounces that metaphorical Godzilla isn't the last of its kind — more specimens are likely hidden away in the undersea caverns, ready to wreak havoc upon a world hellbent on nuclear totality. "I can't believe that Godzilla was the only surviving member of its species. But if we keep on conducting nuclear tests, it's possible that another Godzilla might reappear somewhere in the world again." And as Yamane speculates in antinuclear vein, Godzilla's body has uncannily absorbed massive doses of radiation. Wouldn't it be intriguing to study the monster so as to understand the effect of radiation absorption within cellular tissues? No, say his colleagues. Japan must defend itself from the powerfully destructive monster. Meanwhile, Godzilla, invulnerable to conventional weaponry and high voltage electricity, twice goes on the rampage in Tokyo, creating disasters equivalent in magnitude to nuclear destruction. Sadly, Yamane can't help his people destroy Godzilla; in this he is impotent. Can any other scientist come forward who can?

Yes, and this is where the plagued specter of Dr. Serizawa enters. Virtually isolated from the world, Serizawa has been experimenting with oxygen disintegration, resulting in a most disturbing discovery. He has developed a device no less potent than the atom bomb, a crude prototypical chemical weapon which he calls the "oxygen destroyer," that can rid the ocean of its oxygen. If this doomsday device fell into the wrong hands— well, the global, ecological consequences boggle the mind. What to do, ah yes, what *should* he do? "If the oxygen destroyer is used even once, politicians from around the world will see it. Of course, they'll want to use it as a weapon. Bombs versus bombs, missiles versus missiles, and now a new super-weapon to throw upon us all! As a scientist, as a human being, I can't allow that to happen."

But Japan needs a savior; only he can lift them at this darkest hour by defeating the most powerful prehistoric monster imaginable. Because Serizawa's knowledge cannot be usurped by other nations or madmen the price of using his weapon means self-sacrifice. Despite this, heroically, he will do what he must, and what Yamane is incapable of doing. And so (the original) Godzilla is destroyed in the depths of Tokyo Bay as tortured Serizawa bravely commits suicide alongside the dragon's remains to save the world from not one but *two* menaces, both manmade.

With Serizawa out of the picture, Yamane returned for a tussle with fire monsters in *Godzilla Raids Again* (1955). Here, the venerable paleontologist discusses the nature and origin of these monsters in this first sequel. Following observations of not one but two gigantic fighting monsters, including a new living form — an ankylosaur — zoologist Dr. Tadokoro steals the show from a repressed Yamane. As two pilots, Tsukioka and Kobayashi, shuffle through a pile of dinosaur pictures, Kobayashi identifies a picture of one of the creatures. Dr. Tadokoro claims this is our "worst fear of all," and lectures to the panel out of a dinosaur book replete with pictures. Tadokoro notes the animal Kobayashi identified is an ankylosaur genus known as the Anguirus, a 150- to 200-foot-tall carnivorous, and very aggressive horned creature having an extra pair of brains (as scientists formerly ascribed to *Stegosaurus*). However, Tadokoro never explains why an ankylosaur — an obviously terrestrial form — would be summoned from an *undersea* lair, alive and not drowned. And, here, Tadokoro corrects Yamane's former geological statement from *Gojira*, that the two dinosaurians lived at the same time not a mere 2 million, but 70 to 150 million years ago.

Yamane, who outlined a more realistic, disbelief-suspending scenario for Godzilla's mysterious geological survival in *Gojira*, can scarcely disguise his grief and anxiety over what is certain to come, given that there is no longer a weapon such as the oxygen destroyer to help. Now facing two powerful monsters, the Anguirus and a second Godzilla, Yamane claims Japan is under a greater threat than even the hydrogen bomb which awakened the sleeping monsters. Japan's paleo-scientists, impotent against the giant monster threat, can only forecast gloom.

Ill-equipped to defeat such menaces (in homage to a scene in Harryhausen's *Beast*— to be discussed in the next chapter), the second Godzilla topples a lighthouse with a twitch of its tail on its first shoreline approach before being turned away by flares— which Yamane claims will attract and anger the monsters because they'll remind the monsters of the hydrogen bomb explosion. But then, after a titanic battle to the death throughout burning Osaka, Anguirus is lying in flames along the Osaka coast, with Godzilla no. 2 (i.e., "Gigantis") roaring triumphantly. This may be viewed as flirtation with the Knightian Rex versus Tops theme, because tyrannosaurian Gigantis has dispatched the Anguirus, which has both a ceratopsian frill and horns adorning its head. Following its early successes with Godzilla, as discussed in chapter eight, Toho aimed high with its next *daikaiju eiga—Rodan the Flying Monster* (1956, 1957 — U.S. version).[19]

Beyond pseudo-science, there are also psychological reasons underlying essentiality of Godzilla's prehistoric origins. In particular, why is it so important to create (if not concoct) a *dinosaurian* origin story for Godzilla? For most of us, historical ties offer a comforting sense of continuity, lineage and meaning. And dinosaur symbolism certainly evokes a sense of the most remotely historic and mysterious heritage of all—*prehistory*. As stated in Chapter One, the story of mankind's enthrallment with prehistoric monsters began with folkloric creatures and ends in (modern) mythology.

Ultimately, the meaning and symbolism of iconic creatures—mythical monsters of the industrial age—like Godzilla and Frankenstein's Monster—extend beyond the simple pop-cultural realm of movies and books. During an age of modern technology and science, such figures have become woven into human psyche—they reflect a distinctively modern, if you will, *mythology*! And, whether heroic, dark and brooding or tragic, all mythological figures must rightfully have *origins* (that in the cases of Frankenstein's monster and Godzilla remain outside of, yet strangely explanatory of, contemporary science). Mythological figures appeal and reassure because they both reflect and assuage innate fears, and they tell us about ourselves, the way we view our world, and often in a Freudian way their exploits explain universal order. Even in highly civilized societies, the rational world may seem irrational to many who don't comprehend complexities of scientific principles, and so in the face of raw Nature the awful products of science—seen figuratively through its mythical metaphoric *monsters*—may seem irrational, yet strangely comforting, indeed.

In its 1954 success, Toho never set out to create an icon or a "god," but they were aiming for something beyond metaphor and symbolism. And because Godzilla was nightmarish, they succeeded beyond their wildest dreams. So, therefore, because Godzilla resembles a dinosaur—a group of animals known to science only since the early 19th century—then such a creature must have its unique (pseudo-rational) cosmic story that is by nature fittingly (and theoretically) pseudo-dinosaurian. Accordingly, Godzilla was vaulted to mythic status, because we—humankind—needed such a dinosaurian symbol to mask and mirror mankind's darkest period of doom, that is, in the fearful infancy of the atomic age! A monster, gaining our attention, to show us what we needed to know about ourselves—before it is too late!

At stake here, too, is desire among viewers for a sense of authenticity, even though in Godzilla's fictional case, that is, from the fossil record's perspective, there really is none. Toho's monster just looks so nifty, that don't we find ourselves wishing that such animals really had existed? But while many yearn to see Godzilla as a genuine prehistorian—Carpenter's *Gojirasaurus* aside—it is mere wishful thinking that we might someday find them in the fossil record. It has been said that much of the dinosaurs' appeal stems from the fact that they (many of them) were big, fierce and (with the exception of avians) extinct. Their fossilized bones can be safely witnessed in museum settings, or live as in televised broadcasts, or in the movies, while viewers are spared facing peril. So—naturally, if Godzilla can claim a geological history, albeit a fictitious one, then the basis of its appeal warms the heart further because it mimics that of real dinosaurs—so adored by mankind today.

Plus, in an inaccurately conveyed pop-cultural vein, dinosaurs fuel our twisted evolutionary superiority complex. After all, while they're dead, we—in our hallowed wisdom, such exalted beings—succeeded them, to conquer the planet and subjugate Nature either through ecological exploitation or by harnessing newly discovered natural forces. And then, in typical Frankensteinian fashion, we don't take full responsibility for our actions, our misdeeds—the products of such ghastly experimentation.

Godzilla is so well adored by monster movie cultists that a Japanese term, *daikaiju eiga* (daikaiju, giant monster; eiga, movie) is commonly invoked into English language to categorize fictional creatures analogous and allied to the Godzilla industry. A common misconception among fans of *Godzimagery* is that all giant reptilian creatures (*daikaiju*) are prehistorical, like Godzilla. But it isn't quite that simple when it comes to (fictitious) monsters lacking a genuine fossil record. Here we find that not all giant prehistoric monsters are *daikaiju*, while not all prehistoric *daikaiju eiga* are Japanese. Furthermore, not all prehistorical *daikaiju* are movieland monsters; there are also *daikaiju* of literature. While Godzilla and Anguirus are charter members, you might be surprised which famous giant monsters are not clear cut prehistorical *daikaiju*. We'll consider several examples to clarify this.

In 2007, John D. Lees stated in *G-Fan* that a *daikaiju* must be of sufficiently enormous size as to knock down buildings, it must present a menacing aspect, and must display some measure of purposeful intelligence beyond that of a comparable animal.[20] Surprisingly, Lees discounted King Kong from *daikaiju* ranking because Kong is insufficiently large and most noticeably fails the third criterion. However, also in *G-Fan*, author Michael Bogue[21] mounted a spirited defense for including Kong among the ranks of *daikaiju*, claiming that "Kong does seem to exhibit intelligence beyond that of an ordinary gorilla." Meanwhile, Robert Hood and Robin Pen offered their own vivid description in *Daikaiju: Giant Monster Tales* (2005):[22]

> To us, daikaiju tales require monsters of unreasonable size, impossible and outlandish dimension, relativities that border on (and sometimes cross into) the utterly absurd. Despite whatever rationalizations might be applied within the narrative, daikaiju are fantastical and provoke awe through the sheer audacity of their conception.
>
> But surely size is not all? ... A penchant for city-trashing and apocalyptic destruction. Metaphorical undercurrents. A sense that the kaiju are more than just Beasts—personality, in other words, albeit of a non-human kind. Pseudo-

scientific and metaphysical pretension. Vast scope. Incredible power. A certain cosmic inevitability. Daikaiju are not scared of Man ... classic daikaiju scorn humanity's military might. If they are often the unnatural product of human arrogance, they manage to transcend that heritage and become supra-natural. They are more like inhuman gods than unnatural beasts. They are impossible, yet they *are*.

Aside from unresolved debate over Kong's status as a *daikaiju*, larger-than-life dinosaurian fauna inhabiting Skull Mountain are decidedly not *daikaiju*. Furthermore, discordance reigns over another famous giant movie monster, Ray Harryhausen's prehistorical Rhedosaurus from *The Beast from 20,000 Fathoms*.[23] In formulating his selective definition, due to its (admittedly larger-than-life) unintelligent animalistic tendencies, J. D. Lees claimed that Rhedosaurus isn't *daikaiju*.

Conversely, Hood and Pen count Rhedosaurus among the ranks of *daikaiju* (giant monsters). In his excellent, thought-provoking article, "Man and Super-Monster: A History of Daikaiju Eiga and its Metaphorical Undercurrents,"[24] however, Hood noted that monsters such as Kong and Rhedosaurus are "relatively natural beasts, their reality deriving from a distant past or merely tweaked by a slight exaggeration in scale.... Where U.S. giant monsters tend to be direct versions of naturalistic beasts, simply grown large, the Japanese kaiju are fantastical creatures, unlikely amalgamations of divergent shapes and body parts increasingly removed from the strictures of the objective world."

But, conversely, at the outset, Godzilla's colossal proportions, near-invulnerability, self-regenerative characteristics, and so on represent "characteristics of a super-monster, something at a significant remove from the natural world."[25] So, while superficially Hood might seem at odds with Lees in the cases of Kong and Rhedosaurus, if we only reconsider Hood's and Pen's description (as previously) quoted from *Daikaiju: Giant Monster Tales*, perhaps they're not too far apart, based on definition.

Arguably, the earliest undisputed modern giant movie monster *daikaiju eiga* of any species (here melding Lees' with Hoods's and Pen's working definitions) was Godzilla in Toho's 1954 grand Tokyo performance! While the earliest 1950s giant super-monster *daikaiju* trio—Godzilla, Anguirus and Rodan[26]—were prehistorical-cryptozoological, increasingly during the later 1960s and 1970s, the purported origins of added Japanese *daikaiju* drifted away from being exclusively (or implied) prehistorical, instead shifting toward the sheer fantastical. By the mid 1960s, perhaps symbolically beginning with *King Kong vs. Godzilla,* one notes a declining emphasis on the *prehistorical* theme in Japanese *daikaiju eiga.* (However, the prehistorical theme later made a valid resurgence as a plot device in two Toho films, *Godzilla vs. King Ghidorah* and *Godzilla vs. Megaguirus*.[27])

Not all *daikaiju* have prehistoric origins (or crypto-prehistoric — because they're discovered living in the present). And, as we've seen, the prehistorical nature of such *daikaiju* is frustratingly ill-defined in film. Although no concise definition covers all cases, as a singular illustration, D. G. Valdron outlined a sensible theoretical foundation for prehistorical *daikaiju* in his fascinating short story "Fossils" (2005), published in *Daikaiju: Giant Monster Tales*. Valdron's brooding tale of a Tokyo deserted by all except for *daikaiju* "A" and a few obsessed human observers suggests dinosaurian kaiju originated in the Mesozoic. In particular, "A," described as a radioactive "god, impervious to time," sporting a glowing dorsal fin, is an allosaurian mutation — or possibly derived from plated stegosaurian, or finned spinosaurian stock. Or as Valdron professes, "The thirty-six known kaiju constitute a taxonomist nightmare. Most are clearly reptiles, relatively identifiable offshoots of archosaur lines such as dinosaurs. Their origins are placed between eighty- and two-hundred million years ago.... One is a vastly mutated coelacanth."[28]

According to Valdron, in prehistoric times, natural radiation was stronger and more prevalent than today. Depositional processes concentrated radioactive sediments millions of years ago; dead animals (dinosaurs, etc.) were buried in these toxic environments. But as radioactivity acted upon DNA in tissues, eventually these corpses transformed, arising into indestructible *daikaiju*, non-living yet "walking nuclear reactors," the titular fossils of Valdron's story which never decayed in the sterile radioactive deposits where not even decomposing bacteria could survive. Intriguingly, Valdron mentions that cyclic mass extinctions of the fossil record are caused by periodic re-awakenings of the 36 known *daikaiju*, which exterminate many other species every 17 million years. But this time, man's nuclear folly has interfered with Nature's normal course, awakening them early. Valdron's interesting examples are clearly prehistorical *daikaiju*.

For further perspective, we might consider another working definition penned by Marvin Livings, author of "Running" (published in *Daikaiju!: Giant Monster Tales*), as stated by the story's protagonist who questions intriguingly, "'Why are there no daikaiju fossils?' ... these daikaiju, leave no fossils because they're not animals, not even alive as such. They're forces of nature, like the cyclone that still roars around me, or an avalanche."[29] Uncannily, Livings' *daikaiju* bear no potential to leave traditional museum display-type fossils! To me, that clinches unnatural aspects of the *prehistorical daikaiju* formula.

Focusing on the origins and evolution of prehistorical giant mysterious beasts, let's examine how their purpose and symbolism has evolved during Godzilla's radioactive age. From a Western cultural perspective, the modern giant monster battle theme (e.g., Kong versus *Tyrannosaurus*, or Gigantis versus Anguirus) is derived from paleontological imagery.[30] And while in all the early prehistoric monster movies, paleontologists who arrive on the scene are science explainers, they're increasingly inept at offering concrete means of diverting the monster's rampage. Then, much worse, mad paleo-scientists of American fiction and film eventually evolved into ignoble Frankensteinian instigators of catastrophe, as in Michael Crichton's *Jurassic Park* (1990).[31] (However, *JP*'s cunning, crafty "raptors" are not *daikaiju* due to their mere human size.) This is a case of culturally staying the course so influentially paved by Mary Shelley nearly two centuries ago.

But while, thematically, originating in decidedly Western cultural influences (e.g., *King Kong, The Beast from 20,000 Fathoms*, the Frankenstein mythos, and a *Tyrannosaurus/Stegosaurus/Iguanodon* suitmation design), *Gojira* was most magnificently spawned in the radioactive flames of two very real, horrific atomic attacks that killed many thousands! Once sparked to life, Godzilla (another Promethean offspring and no less metaphorical than Frankenstein's Monster) and its atomic brethren proliferated, evolving into other ostentatious varieties, increasingly contorted *daikaiju*, conforming less to naturalistic design.

Beyond Frankenstein mythos, mad scientist-fueled fears of atomic holocaust are often projected in American B-movies of the 1950s. Such themes are most familiar to American audiences, but as Hood related in "Man and Super-Monster," in Japanese films, eventually,

> realism had ceased to be a priority in daikaiju eiga. The films' divergence from the pseudo-realism of U.S. giant monster films—a genre prevalent in the

Gappa was a "triphibian" *daikaiju* example.

1950s, also redolent with nuclear paranoia and apocalyptic fears—represents one of the significant differences between that subgenre and daikaiju eiga, which is much more attuned to the absurd and readily fills the scene with surreal imagery and colourfully ludicrous events.[32]

While nature-dominating Americans (who time and again refuse to learn Nature's harsh lesson—as conveyed in *Frankenstein; or, the Modern Prometheus*) become bent on destroying these monsters of their own making, in Japanese *daikaiju eiga*, the monsters are strangely embraced, accepted, and may even become ambiguous heroes, although it still proves necessary to attack them.

Discounting popular dinosaur fiction, such as Arthur Conan Doyle's *The Lost World* (1912), where its dinosaurs and other crypto-zoological prehistoria, still surviving on a lost plateau, behave more or less realistically as dinosaurian animals rightfully should, through the decades, *daikaiju* fiction continued to be published here and there, spottily, without making much of a splash. While most *daikaiju* movies are Japanese, it would seem most prehistorical *daikaiju* stories are published in English. (Here, I refer to the term *daikaiju* simply translated as giant monster, without invoking ramifications of Lees's selective definition.)

The availability of these works to the English-speaking populace meant that by the early 1990s, whether they realized it or not, fans on many levels had been treated to a variety of *daikaiju* science fictional and fantasy stories, such as Brian Aldiss' "Heresies of the Huge God" (1966), J. G. Ballard's "The Drowned Giant" (1964), Anne McCaffrey's Nebula-earning novella, "Dragonrider" (1968), Carl Dreadstone's idiosyncratic take on a dino-gigantic *Creature from the Black Lagoon* (1977), William Schoell's pair of terrifying tales *Saurian* (1988) and *The Dragon* (1989), Marc Jacobson's *Gojiro* (1991) and Joe R. Lansdale's punny-titled "Godzilla's 12-Step Program" (1994), to name a few.[33] While not all of these entries were of a prehistorical nature, one must include David Gerrold's powerfully obsessive, demonic tyrannosaur in *Deathbeast* (1978) as such.[34] Through the decades, added to the mix were novelizations of movie scripts for *daikaiju eiga* films such as *King Kong, Gorgo, Reptilicus*, and *Gargantua*.[35] Many genre stories have been printed within the pages of *G-Fan* magazine.

Schoell's *Saurian* is a splendid, illustrative example of a non–Japanese *daikaiju*, stemming from literature, not cinema. Fossils of an immense dinosaur named "Gargantosaurus" lead a manic-depressive paleontologist to a remarkable conclusion—that this crypto-genus is alive, causing recent disasters along the eastern American coastline. The animal is 250 feet long, 80 feet at the shoulder, resembling both Harryhausen's "Beast," as well as fellow sauriodinosauroid stop-motion monster, the Paleosaurus.[36] In various passages, its head shape is described as 10 percent human and 90 percent reptile, with an awful face which has a "semi-human, semi-reptilian countenance that leered and chuckled and grinned with its lipless mouth like a devil."

This surviving gargantosaur retains a strange, power, a werewolf-like, shape-shifting ability. It is remnant of a mysterious, intelligent alien race abandoned on Earth hundreds of millions of years ago. The aliens whose original bodily forms and culture have been lost to history and science, metamorphosed into dominant forms of life then existing on Earth, which during the Mesozoic Era of course meant dinosaurs. Certain aliens "refused to give up their tremendously powerful physiques; they became the dragons, sea monsters and scaly serpents of legend." One of these has evolved into a human-gargantosaur hybrid named Bronmore, an evil person who has infiltrated ranks of society. After transforming into his dino-monster Paleosaurus-like persona, Bronmore is unable to quench his horrific thirst for bloodied human flesh. Amusingly, we learn through Bronmore's racial memory flashbacks that, "If modern man only knew which of his favorite dinosaurs had actually been created by the aliens ... he would be amazed." Furthermore, real dinosaurs (both genuine forms and most individualistic alien-morphs such as gargantosaurs) were extirpated by radiation from a nearby supernova. This cosmic event also generally extinguished the aliens' shape-shifting genetic capabilities.

Saurian is a spectacularly nightmarish lead monster, which in its full reptilian form resembles 1950s famed stop-motion monsters, Rhedosaurus and Paleosaurus. However, *Saurian's* Beast is *also* instilled with *human* qualities. In fact, such a transformational merging of dinosaur with man into dino-human is a historical trend one encounters in fantastic literature, as documented in a recent book of mine.[37] *Saurian* brilliantly illustrates this theme. Schoell's "Beast," as he refers to it, is far vaster than the sum of its evil, anthroreptiloid morphology. For as the giant monster strides majestically after its victims, it also possesses crushing, godlike qualities.

In the new millennium, *daikaiju* continue to proliferate, invading from other beachheads, lately stomping through the annals of fantastic literature. Two outstanding prehistorical-type examples of *daikaiju* fiction appeared in *Daikaiju! 3: Giant Monsters vs. the World* (2007), Richard A. Becker's "The Return of Cthadron," and Robert Hood's "Flesh and Bone." In "Cthadron" the huge monster remains unchanged while the world moves on, while conversely in "Flesh and Bone" a metaphorical monster evolves with successive appearances in an otherwise unchanged world setting.[38]

Historically, culturally and socio-politically, fear is relative, as Stephen King emphasized in his acclaimed *Danse Macabre* (1981).[39] Instructively, in the case of certain *daikaiju*, what comes to be regarded as our iconic (period) monsters reflects cultural times and contemporary issues. But we can't, or won't, escape from our plight. Perhaps, ultimately, mankind's arrogant excesses and recurrent environmental mistakes are all founded on innate flaws, species imperfections and godless lapses in reasoning regardless of the historical period under consideration. Because we are who we are, deep down, genetically, we're doomed no matter what. And all things considered, for the good of the planet, wouldn't that be a good thing? We're led to such sobering affairs through Becker's "The Return of Cthadron."

Everything about Cthadron, rising for its first wave of destruction from the depths of 20,000 fathoms in 1954, reeks of death and doom. Or as Becker asserts, "Death is an impersonal force in the universe, sweeping away the good with the bad, but most of all kicking apart the embers and dust of the old in favour of the new. Death does not care what form the new may take, but clears a path for it — a herald of newness." Cthadron appears to be a 500-foot tall "heavyset Tyrannosaur, one with a bone disease that made warped spikes and ridges like vast Pacific coral reefs grow forth from his flesh." The Death Pulse — green-white nuclear fire — issues from Cthadron's maw, as "his entire face became a conduit between mankind and thermonuclear devastation." After a blissful decade-long hiatus, doomsday machine Cthadron unexpectedly returns in 1964. Why so unexpectedly? "Why doesn't a cancer victim think their tumors can ever return, after the doctors pronounce it a five-year cure?" This time with his "Ragnarok voice" (a material force whose purpose in the order of things perhaps recalls Stephen King's frightening Langoliers), invulnerable Cthadron ravages western North America. Much like undying monsters Godzilla or Frankenstein, Cthadron returns time and again. Cthadron isn't inherently evil, but, in giving us a concentrated dose of our own medicine, represents a relentless engine of entropy and a more naturalistic planetary state. One without man reigning over Nature! Change is natural and therefore good. It is time to move on.

As Hood communicated to me in Feb. 2008, "'Flesh and Bone' includes an array of monsters representing times of crisis in human history, just as Godzilla is representative of nuclear proliferation. So a Godzilla-like monster is included, but the First Monster is specifically dinosaurian."[40] Hood's gimmicks are weird physics and a Bone Computer (recalling, in a Frankensteinian-derived sense, Crichton's coupling of computers with weird or chaotic genetic forces in *Jurassic Park*). The Bone Computer allows scientists to research the succession of metaphorical "Lost Monsters" created through mankind's anxieties and foolhardiness. Each of these, including the nearly invulnerable 1950s Atomic Monster, is transitorily revivified by the Bone Computer's quantum fluctuations through a multi-layering of realities manifesting man's historical and technological shortcomings, beginning with a dinosaurian-type *daikaiju* monster existing at the dawn of man. Demented paleontologist Dr. Kumala unwisely misuses the Bone Computer to create a new *daikaiju* god that he can worship. But he's thwarted by a more responsibly-minded soul, Dalgrave, who self-sacrificially fuses with the substance of a new monster generating from the (Bone Computer–reanimated) Atomic Monster's fiery breath. Vengefully, the new winged monster destroys sinful Kumala.

By now, you might surmise how mankind's two most significant and far-reaching scientifically spawned monsters—Frankenstein's monster and Godzilla—are curiously associated. Both manifestations arrived, respectively, during pivotal historical times, illustrating how new energy forms became symbolized by the most iconic (fictional yet metaphoric) creatures of all!

As stated by science fiction master Brian W. Aldiss, the "last word" on Mary Shelley's *Frankenstein, or the Modern Prometheus* (1818 and 1831 editions) "will never be said."[41] For it is a novel far too richly woven into our culture and public consciousness to be left alone. What has not really been highlighted, though, is the extent to which Godzilla is yet another manifestation of a Frankensteinian Monster. Except that while Shelley's pivotal, allegorical tale is written from a number of perspectives—including both the monster's and its creator's—*Gojira* may be seen more as a Japanese victim impact statement, although one in which the real quantum-tinkering mad scientists are never directly accused. While in Shelley's *Frankenstein*, there are really two monsters, mad scientist Victor and his murdering abomination, in *Gojira* the tragic, radioactive resultant is prominently displayed while *its* ultimate creators remain highly suggested (i.e., via Dr. Serizawa's tortured soul) yet—before the camera—invisible.

Symbolically, when Godzilla uses fiery atomic radiation to melt and topple high tension electrical towers in a *Gojira* scene, a new era in man's wielding of science and technology is signaled. Godzilla—or the Modern Prometheus—has emerged from the radioactive ashes of irresponsible scientific experimentation (i.e., bomb testing)! Implicitly, atomic power's potential trumps the electrical, that is, in terms of its far-reaching consequences as a destructive force, and with respect to man's infernal prying into forbidden territory. At the time of *Frankenstein*'s writing, Science was moving into a distinctively modern aspect—with fundamental developments in electro-chemistry, biology and wondrous recognition of the Earth's vast prehistory at the core.

In Shelley's novel, Dr. Victor Frankenstein is fascinated by the force of lightning, when as a young man, he "beheld a stream of fire issue from an old and beautiful oak, which stood about twenty yards from our house; and so soon as the dazzling light vanished, the oak had disappeared, and nothing remained but a blasted stump.... Before this I was not unacquainted with the more obvious laws of electricity."[42] Frankenstein's monster is later witnessed on Mont Blanc amidst an incredible lightning display. Its haunting silhouette not only reinforces the Monster's laboratory origins—given that it was infused with an (electrical) "spark of being"—but also, in retrospect, symbolizes the dawning era in comprehending Earth's abyss of time. Paleontologist Martin J. S. Rudwick has expressed, for instance, that de Saussure's ascent of Mont Blanc in 1787 represents a "golden spike," or a distinctively modern aspect for the emerging earth sciences.[43]

So the monster's artificial birth in the Romantic period symbolically weds the emerging sciences of chemistry and geology. Thus, at Ingolstadt, Victor Frankenstein was urged by chemistry professor M. Waldman to rely only the latest researches of the most up to date natural philosophers. And today we know that besides assorted literary influences, Mary Shelley's imagination was certainly stoked by brilliant men of the Enlightenment such as, among others, those who experimented with electricity, Benjamin Franklin, Luigi Galvani and chemist Humphry Davy. And influential natural philosopher and physician Erasmus Darwin is cited by name in the novel's Introduction and Preface. But there were others. Perhaps, according to rumor, even an earlier physician named Konrad Dippel who, within confines of his Castle Frankenstein laboratory at Bergstrasse, practiced vivisection of animals and claimed to have discovered arcane, alchemical "secrets of life." In fact, for decades many naturalists harbored heightened curiosities concerning reanimation of the dead, using electricity, or galvanic currents.[44]

Both iconic monsters (i.e., the film *Gojira* and Shelley's novel *Frankenstein*) share many things in kind — and here I'm not merely referring to movie sequelitis persisting over the years, or the fact that there were at least two aborted attempts to bring them together in conflict on the silver screen (resulting in *King Kong vs. Godzilla, Frankenstein Conquers the World*, and a spinoff of the latter — *War of the Gargantuas*, instead).[45] True, both Godzilla and Frankenstein's Monster were assembled (literally, perhaps, stitched) from many anatomical parts (in the case of Godzilla's suit, a melding of *T. rex, Stegosaurus* and *Iguanodon* paleo-restorations), and each poses dangers to mankind with regard to potential for their awful reproduction (i.e., as in prospective progeny resulting from unholy union of Frankenstein's monster and his bride, as well as Dr. Yamane's apt suggestion at the end of *Gojira* that more Godzillas will be spawned, e.g., by the bomb).[46]

More important, while Shelley's *Frankenstein* novel carries seminal science fictional overtones, it was originally conceived as a horrific, yet visionary, ghost tale, a chiller to frighten readers. Like the later *Gojira*, Shelley's novel still succeeds magnificently on both levels. In an awful dream, Shelley foresaw that,

> supremely frightful would be the effect of any human endeavor to mock the stupendous mechanism of the Creator of the world.... He would rush away from his odious handiwork, horror-stricken. He would hope that, left to itself, the slight spark of life which he had communicated would fade, that this thing, which had received such imperfect animation, would subside into dead matter.[47]

While both monsters are powerfully destructive, vengeful manifestations, Godzilla, like Frankenstein's monster, is also a tragic character. Godzilla is a grossly mutated creature, regenerated — analogous to Frankenstein's monster — using powers that staggered the comprehensions of contemporary men. Due to their own inherent abilities to regenerate from suffered wounds, both monsters are in a sense immortal, even blasphemously godlike.[48] While the Godzilla theme is indelibly tied to dreadful misuse of atomic weaponry, Frankenstein's Monster is associated with much earlier experimentation inaugurated by scientists who pondered the lugubrious effects of electrical impulses on corpses.

Besides atomic irradiation, lightning (Zeus's thunderbolts) figures prominently throughout the Godzilla film series. Mark Justice has discussed Godzilla's religious and symbolic, reanimative "flow of life" and successive rebirths in *G-Fan*.[49] Citing the significance of volcanoes and lightning in the Godzilla theme, Justice concluded, "Godzilla's resilience, his seeming inability to truly die, his resistance to weapons of mass destruction, and his tenacity to return, all make sense when taking into consideration what Godzilla truly represents as a Shinto symbol."

Vivid lightning displays also frame *Tyrannosaurus'* dramatic appearance in Michael Crichton's 1990 novel *Jurassic Park*, symbolically linking its reanimation from fossil DNA both to the "spark of being" revivifying Frankenstein's monster, as well as Godzilla's Shinto reanimation. John Taine's 1929 novel *The Greatest Adventure* melded a dinosaurian theme with Frankenstein mythos. Both Taine and, later, Michael Crichton in *Jurassic Park* (1990) harnessed mystical *genetic* forces, which Crichton held to exceed *atomic* power's "potency." There are certainly more ways than Shelley and Tanaka dreamed to resurrect the dead ... or the extinct.[50]

So how may we benefit from knowledge of *Frankenstein* and Godzilla — the Modern Prometheus? Well, isn't it time to take responsibility for our unfolding actions, before we are vengefully rendered extinct? After all, at least since the Enlightenment, very wise people have enunciated that there's something very significant about our mythical monsters!

Depressingly, atomic fuel and the specter of nuclear weapons continue to haunt newscasts. Think of Three Mile Island, Chernobyl, North Korea and Iran. Had it not been for the memory of Hiroshima and Nagasaki, would the United States and Russia have otherwise thrown caution to the (radioactively charged) winds, exchanging volleys of nuclear missiles during that fateful period in late October 1962? In such a parallel universe, after fiery annihilation, how many would have survived the devastating nuclear winter?

Heed character Shigeru's foreboding, prophetic and metaphoric musings at the conclusion of Toho's *Rodan, the Flying Monster*:

> I realize that, by the narrowest of margins, man had proved himself the stronger. But will it always be so? May not other, and more terrible monsters, even now be stirring in the darkness? And when, at last, they spring upon us,

can we be certain we shall beat them back a second time? The answer lies in the future. Our fears, for now had gone up in flame and smoke.

From fateful dragons, through Mastodon-Behemoth, radioactive Godzilla, mightily inexorable Cthadron and now *Cloverfield*'s terrorist monster of 21st century doom — bin Laden–incarnate[51]— many of us, healthily, enjoy confronting our innermost fears in the relative safety of fiction and film. Yet with such fearsome monsters about, 'tis folly to bury our heads in the sand. Remember — prehistorical *daikaiju* mirror perceptions of *reality*.

So would we be safer eking out an existence on a deserted island, remote from civilization's trappings and sinister tendencies, symbolized by foreboding prehistorical *daikaiju*? Not if it happens to be Monster Island. Here there be dragons—"dragodinosauroid" Godzilla included![52]

TEN

"And Their Lost Worlds"

From the public's perspective, after geologists established that there were not simply one but *several* distinct periods of prehistory, each newly discovered geological period became a veritable lost world window into the past. Gentlemen of science revealed evidence for these pristine times: visions of these former worlds unfolded. Many, romantically, contemplated what such worlds were like while pondering causal astronomical or geophysical theories for why their inhabitants and planetary conditions originated, only to later vanish. It was acknowledged that those who attained proper credentials and knowledge and possessed geology's special language could be relied upon to explore, interpret and then explain prehistory's succession of worlds. By the 1850s, increasing fascination over mounting discoveries became reflected in a popular metaphoric device — the mysterious yet primitive monster island, essentially any place accessible via an imaginary, visual time tour tenanted by strange creatures out of prehistory. Rooted in geological precepts, prehistoric monster islands became a public phenomenon and a readily understood means of conveying the meaning of bewildering prehistoric recreations and scenes.

In a larger sense, the public recognized that fossils belonged to prehistory, and that prehistory was somehow a place that was rather difficult to place geographically. So for purposes of discussion and illustration, prehistoric fauna were often restored in faux naturalistic settings which, as reconstructed to augment their realism, were conveniently isolated from modernity. These anachronistic prehistoric monster settings such as Sinclair's 1933–34 Chicago World's Fair Dinosaur Exhibit, were billed as yet unexplored, or they could be re-creations of past worlds unsullied by imaginative human time travelers.

Nineteenth century paleoart comprised both literary forms and visual depictions, representing the then-unfolding panorama of geological time and its mysterious inhabitants. Lesser known today are clever theatrical recreations designed to illustrate scenes from deep time for public audiences. In a sense, these were the original paleontological stage productions, ancestral to the 20th century's dinosaur motion picture genre. For instance, O'Connor references "Geological Revolutions of the Earth" that played in 1839 at the Colosseum in Regent's Park, England. This was one of multiple sideshows included in the "Gallery of Natural Magic," also featuring electrical displays as well as microscopes and telescopes. In "Geological Revolutions," visitors were "introduced to the Caverns of the Colosseum, where the Geological Revolutions of the Earth will be exhibited in a series of scenes, viz. The Creation; Paradise; The Period of the *Iguanodon, Megalosaurus, Pterodactylus,* etc; The Deluge; and The Nile indicating the progress of civilization in the renewed world."[1] Obviously this particular series of scenes safely and wisely catered to the paying public's theological bent. While these scenes may have been presented in the form of static murals, O'Connor suggests they may instead have been projected via a magic lantern (e.g., "transparencies projected onto a white wall in a darkened room ... optical effects and illusions")[2] to accentuate the phantasmagoric qualities of the virtual time tour, and there may have even been accompanying sound effects, including the orchestral.[3] More sophisticated use of dissolving views where one scene faded into the next would have been possible.

Details of the mid–19th century's prehistorical phantasmogoric theatrical performances have been lost to time; however, Benjamin Waterhouse Hawkins' Victorian dinosaurs still haunting their "monster island" at Sydenham, England, serve as one lasting example. Many 19th century books and publications feature restored windows into the past featuring then poorly understood worlds recreated by noted paleoartists of the time. In a frenzy, talented writers dreamed up, first, literary re-creations offering speculative science, and then later, fictional stories illustrating what it would be like to confront in-the-flesh prehistoric beasts, in person. Accordingly, an accelerative decade-by-decade addition of new fossil monsters inflamed the public's longing for more. More what? Knowledge of weird prehistoric monsters coupled with thrills like never before. The movies came just in time!

Independently, Mark Berry, Donald F. Glut and Michael Klossner chronicled numerous films, of varying quality and accuracy, allied to the dinosaur or caveman movie genre.[4] But of all these, the two films that mattered most in the public's eye, in the sense of placing popular notions of where prehistoric monsters dwelled, were First National Picture's *The Lost World* (1925) and RKO's *King Kong* (1933). These films firmly rooted public perceptions of prehistoric monster islands, and it is time to explore their misty confines. Such movies are a profound interplay of imagetext; behind-the-scenes pre-production paleoart; sculptural restorations; with fictional tale theatrically transformed into spoken script — melded into a resultant phantasmagoria of sight and sound. Yet, at the outset, they had also had the impact of compressing the geological panorama of time which Buckland, Mantell and others had identified back into a single prehistoric age.

Aspects of Conan Doyle's work were outlined in chapters five and six, albeit without outlining how his novel transformed into Hollywood's now established lost world convention. Essentially, *The Lost World* broke one of Cuvier's rules, refined throughout the 19th century, namely that there were unique ages in geological time during which distinctive sets of organisms had prospered (and gone extinct). Mid-1920s Hollywood made no efforts to revert back to Cuvierian correctness. And, moreover, Hollywood only affirmed what people may have been already attuned to—that dinosaurs were far cooler than most of the other assortment of prehistoric beasts, such as fossil mammals and marine reptiles. So if Conan Doyle didn't reflect Cuvier's (and Mantell's) ideas of there having been distinct ages in prehistory, then Hollywood had no business doing so either. In any respect, far from bowing to geological correctness, Conan Doyle was simply heeding a "simple plan," stated in verse:

> I have wrought my simple plan
> If I give one hour of joy
> To the boy who's half a man,
> Or the man who's half a boy.[5]

One might (correctly) suggest that Conan Doyle's South American lost plateau isn't an oceanic island. However, although land-locked, the way in which his insulated plateau is utilized in the story would meet an ecologist's definition of island, that is, due to the relative isolation of its indigenous species.[6] In another sense, the large tabular movie jungle sets upon which Marcel Delgado's dinosaur models were animated and meticulously filmed represented in themselves miniature islands of imagineered fantasy.

In *The Lost World* (1925), garrulous and immensely self-assured Professor Challenger (played by Wallace Beery) launches an expedition to the Amazon, where Paula White's (played by Bessie Love) father, Maple White, was abandoned atop a lost and evidently prehistoric plateau hence dubbed Maple White Land. Maple White's journal was recovered though, and most amazingly therein are two drawings of in-the-flesh dinosaurs—showing both a living brontosaur, and the "most vicious pest of the ancient world"—*Allosaurus*. Despite this, during his Zoological Hall lecture a jocular crowd of British elite jeers Challenger, ridiculing his preposterous claims of living dinosaurs. Outside the lecture hall itself, skeletons of the American Museum's *Allosaurus* and a brontosaur are displayed; these happen to be the two genera noted by Maple White in his diary.

Few are persuaded by these ridiculous claims. Even London *Record Journal* reporter Ed Malone (played by Lloyd Hughes) realizes the fossil animals Challenger claims are still living in South America should rightfully have been "dead 10 million years." While Challenger calls for volunteers "to face death or worse in the name of science," an audience member taunts, "Bring on your mastodons. Bring on your mammoths." Caught up in the moment and desperate to prove his ruggedness to his love, the impulsive and fickle Gladys, Malone volunteers for the expedition, but is vehemently rejected by Challenger who hates all members of the abominable press. Sportsman John Roxton (played by Lewis S. Stone) volunteers as well, completing a love triangle that soon develops between himself, Malone and Paula. Besides subdued, rather comical Professor Summerlee, also along for the journey is "that mischievous beast," pet monkey Jocko, who, much later, in heroic Kong-like fashion climbs the steep rocky plateau wall in order to save Paula.

Following the public spectacle at Zoological Hall, later that night Challenger reluctantly invites Malone to join the expedition, provided he can enlist the support of his newspaper. Malone states, skeptically after paging through Maple White's journal, "You mean that you actually saw *living descendants* of these monsters that are supposed to have been dead for millions of years?" Assuredly, he says, and the expedition is soon underway. Rafting through the Amazon's 50,000 mile network of tributaries, following brief encounters with modern mammals (including a Brazilian tree-sloth, a modern evolutionary cousin of Great-Claw), they finally arrive at the base of the plateau's pinnacle. The very first prehistoric sign they spy from below is a hungry and gigantic winged *Pteranodon*, gripping prey with its talons. Here is proof that Maple White's outlandish claims are decidedly true! But the Challenger team is soon stranded atop the plateau when a felled tree facilitating their perilous venture over a chasm is rolled over the edge by a supposedly "completely harmless Brontosaurus."

Eight other dinosaurs soon make grand entrances; the staging becomes an homage to Charles R. Knight and the American Museum. The audience is treated to incredible dinosaurian splendor, the likes of which would have seemed most impressive to audiences unfamiliar with stop-motion technology. First, *Allosaurus* subdues Trachodon. The carnivore then attacks a *Triceratops* family; in a living embodiment of Knight's 1906 Rex versus Tops American Museum restoration, the mother of the Knightian tableau nudges its youngster aside to safety before dissuading the flesh-eater. Next, Cope's *Agathaumas* engages a tyrannosaur, which leaps onto the back of the horned dinosaur; the carnivore is mortally wounded by sharp nasal horn. But when *Agathaumas* fights another Rex that also leaps upon its back, this time the spiky-frilled ceratopsian becomes prey; its left foreleg crippled by cruel predatory jaws. Still unsatisfied with this fare, Rex leaps mightily — this time toward the skies, snaring a wayward pterodactyl out of midair.

A volcanic eruption triggers a dinosaurian stampede. (The prototypical volcanic eruption of the dino-monster genre was of course staged in Verne's *Journey*.) Many stop-motion animated dinosaurs are seen moving in the same scene. It must have been nerve-wracking to synchronize all these subtle movements for several dinosaurs at once, in frame-by-frame fashion or in effect simultaneously. Amidst the eruptive, Hadean turmoil, Malone observes running stegosaurs; carnivorous dinosaurs leap upon backs of fleeing brachiosaurs. Eventually the volcano simmers down and the team escapes from the prehistoric realm. Having descended the plateau, the team discovers that a brontosaur pushed over the cliff's edge by a tyrannosaur in an earlier scene has survived its fall; it is triumphantly rafted back to London aboard ship — where the real trouble begins.

In its colorful advertising, First National Pictures conflated dinosaurs on the rampage in London. Poster art shows a gigantic allosaur trampling a trolley car, and in other art crashing through Tower Bridge. Of course, theropodous carnivorous wrecking-crew dinosaurs didn't make it to London until 1960, when a pair of Gorgos made their way from Nara Island. But, in 1925, Brontosaurus was first to make militaristic landfall in civilized territory, as First National promoted the affair:

> If you should meet a ferocious prehistoric monster bigger than eleven elephants—? What if you should be strolling casually down the street and suddenly come face to face with a gigantic prehistoric monster — alive and with yawning jaws? Or, what if you were at home in your carpet slippers, comfortably perusing the sporting news, when suddenly the side of your home caved in and a terrifying giant of a beast, supposed to have been dead ten million years ago, came in upon you? Impossible you say? And yet you see in this picture Piccadilly, the heart of London, visited by one of these same prehistoric monsters. The huge beast, a brontosaurus, one hundred feet in length and bigger then eleven elephants ploughs down the busy thoroughfare, just at theatre time. His elephantine feet turn omnibuses and taxicabs into kindling wood. His long tail, terrific in strength, sweeps over monuments, iron lamp posts and corners of buildings. London bobbies with pistols and rifles fire at him. Their bullets are about as effective as a bean shooter. Nothing but a cannon or half a ton of dynamite could stop him. "Bronty" proves just as bewildered and frightened as London's populace. Rafted over the ocean by a party of explorers from his home in the Lost World, far up in the unexplored regions of the Amazon River, his one big idea is to get back there, as quickly as possible. Wheeling suddenly, he collides with a skyscraper. The building collapses under his weight. The monster continues in his mad race for liberty from civilization and comes finally to the famous Tower Bridge. The bridge collapses. The last seen of "Bronty," he is swimming down the Thames toward the Atlantic — and freedom.[7]

Yet, compared to the onslaught of military might unleashed upon dino-monster *daikaiju eiga* beginning three decades later, there is relatively little weaponry expended in this early dinosaur skirmish. Only two dinosaurs are fired upon with guns in *The Lost World*. The first is an allosaur confronting the Challenger team's flimsy jungle encampment atop the plateau. Secondly, when the enraged brontosaur goes on the rampage in London, a man who is about to be stepped upon fires his handgun at the descending foot, thereby sparing his life.

The miniature movie sets, or the real prehistoric monster islands inhabited by miniature sculptures, upon which O'Brien operated Delgado's dinosaur puppets before the stop-motion camera were outstanding for their time. As noted by Jeff Rovin:

> they were erected ... on an average of six feet long by three or four feet deep, and stood approximately three feet from the ground. Their height and size allowed O'Bie [i.e., Willis O'Brien] easy access to any corner of the set from cameraside ... every set piece was carefully secured; leaves and grasses were either heavily lacquered or cut from tin to prevent them from moving between exposures.... With the exception of one scene, the animation in *The Lost World* was done solely by O'Bie. The exception was a dinosaur stampede. Staged on a set that was 75 by 150 feet, the sheer spectacle of the sequence made it necessary for O'Bie to work with a handful of assistants. Working ten hours a day, O'Bie's daily output was 480 separate frames, thirty-five feet of film, which ran just over a half minute onscreen. At this rate, the stop motion photography took a total of fourteen months to complete.[8]

This film was extraordinary in nearly every sense of the word. As dinosaur film expert Mark Berry noted in 2002, "what it was like for 1925 audiences can only be imagined. Absolutely nothing like it had ever been seen. Not only was the world much bigger — big enough to make a 'lost world' seem a real possibility — but audiences were nowhere near as technology-savvy as they are now.... O'Bie's dinos in 1925 must have been something like 1993's audiences seeing *Jurassic Park* for the first time — only more so."[9] Stop-motion movie master Ray Harryhausen noted, "No one had ever seen a dinosaur on the screen, only in still paintings, in a museum, or skeletons ... and that, I believe, fascinated Willis O'Brien most — the idea of trying to put on the screen things that you *can't possibly* photograph."[10]

First National had scored a major hit for dinosaurs (rather than any other kind of prehistoric monster) in the movies, and underscored how dinosaurs' rightful place in (cryptozoological) nature was in a lost world, or prehistoric monster island settings. Accordingly, the prehistoric monster island concept became embedded in public mindset, traditionalized for over half a century on into the modern *Jurassic Park* era.

Conan Doyle influenced a century's worth of science fictional entertainment. The next film, albeit never completed, invoking a "lost world" theme, was RKO's planned *Creation*. O'Brien only shot test footage during (circa) 1929 to 1932. The plot, borrowing plotwise from Burroughs' *The Land That Time Forgot* story,[11] involved survivors of a shipwrecked submarine who encounter prehistoric monsters living on a volcanic South American island. The film was to have featured twin-horned Tertiary mammal *Arsinotherium*, pterodactyls, and a tyrannosaurid that haunts a ruined temple. Scant surviving footage reveals a gripping scene; an enraged *Triceratops* chasing a rifle-wielding man who has just cruelly shot its offspring in the eye.[12] But, perhaps predictably considering the declining status of fossil mammals then, *Arsinotherium* never made it to the silver screen. Mounting costs of production soon forced cancellation of ill-fated *Creation.*

As groundbreaking a film as *The Lost World* was in its day, *King Kong* set superlative standards for stop-motion dinosaur special effects and drama. As opposed to the 1932 *King Kong* novel, where Kong is referred to by Ann Darrow (played by Fay Wray) as belonging to a prehistoric age, in the movie, the "gigantic prehistoric gorilla"[13] is guilty by virtue of association with its prehistoric brethren, a choice selection of Jurassic and Cretaceous sauria. While there is a slightly suggested gradation from Jurassic to Cretaceous fauna as sequentially encountered by Carl Denham's (played by Robert Armstrong) rescue party behind the Great Wall, this temporally stacked ordering seems entirely incidental. First, a Knightian *Stegosaurus*, dating from the Jurassic Period, attacks the sailors and is killed. Next, the crew is trampled and brutalized by a marauding Brontosaurus, also from the Jurassic.

Next, three Cretaceous sauria are showcased, each in turn succumbing to mighty Kong's boxing and wrestling prowess. Kong spares Ann Darrow from a magnificent Rex's cruel jaws in a titanic struggle since unmatched on the silver screen.[14]

Dinosaur film expert Mark Berry recently published a list of "The Top 10 Dino-Movie Scenes of All Time."[15] He evidently prefers scenes involving rampaging Rexes. Four of the ten listed scenes features tyrannosaurids, while another stars Ray Harryhausen's "Gwangi" dinosaur (which resembles Charles Knight's 1906 Rex restoration). Two of these scenes appeared in *Jurassic Park* movies; "Tyrannosaur's San Diego charge" from *The Lost World: Jurassic Park* (1997) comes in at number ten, and *Jurassic Park*'s (1993) "Main Road Sequence," where the Rex escapes its pen in a downpour, is number five. The classic 1933 Kong/Rex battle scene vaulted to the number two spot, and Berry cannot disguise his enthusiasm when he writes:

> Often hailed as the greatest stop-motion sequence ever created, the extended fight between Kong and the *Tyrannosaurus rex* is remarkable for many reasons, not the least of which is the fact that it was made in the primitive, early days of movies yet is still so impressive today. The fight, pitting the "thinking" Kong against the brute force of the reptile, is a microcosm of the movie as a whole in its perfect rhythm of action — pause — action — pause, as the two combatants grapple, then regroup, then go at it again with another strategy, then back off.... Though you subconsciously know that Kong must win, you get the sense throughout the battle that all it would take would be one wrong move or misstep from Kong in order for the *T. rex* to turn the tide in his favor. And as thrilling as the animation is, the sequence has so much more going for it. The camera work and editing is highly effective, with a great variety of shots—long, medium, closeup, composite — and a matching variety of shot lengths between cuts. Then there's the evocative glass-painted jungle milieu in which it takes place, courtesy of Mario Larrinaga, and the groundbreaking,

Sculptor Bruce Horton's vision of the immortal "cosmic" battle between prehistoric "gods," Kong and Rex, is magnificently captured in this bronze sculpture.

naturalistic sound effects, courtesy of Murray Spivack. And there's Kong's administration of the coup de grace as he cracks the dinosaur's jaws into splinters, a scene for which Spivack provided a sound effect that still makes audiences squirm.... The famous Kong-Rex tilt is a triumph from start to finish, the undisputed highlight of King Kong's thrill-packed second act, and one of the highlights in the history of stop-motion.[16]

Next, atop his mountain lair Kong grapples with Cretaceous slippery and snaky-looking *Elasmosaurus*, obviously modeled after an 1897 Knight painting. Having subdued the aquatic reptile, Kong then alertly clobbers a gigantic, toothed, bat-winged *Pteranodon* before its talons can snatch Ann aloft. Battered pterosaur is readily dispatched and tossed over a steep cliff. Kong's next rounds are staged in Manhattan versus "monsters" from the present Holocene—an elevated train and four fighter biplanes. Yet, as Denham laments, it wasn't the planes, but "beauty killed the beast," a tragic victim of man's exploitation of nature.

In an under-budgeted sequel, RKO's *The Son of Kong* (1933), Carl Denham returns to Skull Island, this time seeking treasure. Along the way, however, Kong's righteous prehistoric, white-furred offspring repeatedly spares the adventurers from danger. This time, human interlopers sample another assortment of faunal inhabitants—spiky-horned Cretaceous *Styracosaurus*, a vicious Pleistocene cave bear, a sea-dragon, and *the* prototypical, quadrupedal, carnivorous sauriodinosauroid, vaguely resembling the much later screened Paleosaurus or Harryhausen's Rhedosaurus. Also, a Brontosaurus makes a cameo near the end of the film. Little Kong handily dispatches each of its worthy antagonists.

Like Conan Doyle's lost world plateau—as in the 1912 novel version especially—Skull Island, therefore, supports an odd admixture of prehistoric-aspect sauria and mammalia.[17] More so than any other, Skull Island became *the* isle affirming the undying pop-cultural concept of secluded prehistoric monster settings trapped in a pseudo-prehistoric age, where raw-willed Nature mysteriously conjured denizens from its majestic past.

While Kong's island genesis partly owes to Conan Doyle, significantly, further inspiration came from discoveries made on a real lost world island— Komodo, east of Java. In 1926, zoologist and American Museum of Natural History trustee W. Douglas Burden led an expedition to Komodo following documented sightings of its real indigenous monsters dating to 1912. Burden's objective was to bring back live specimens of the primitive-looking Komodo Dragon, creatures seemingly more suited to Mantell's dinosaur age rather than modernity; yet his adventure recalls Professor Challenger's. Arriving at Komodo, Burden described the island: "With its sharp serrated skyline, its gnarled mountains, its mellow sun-washed valleys and the giant pinnacles that bared themselves like fangs to the sky, [the island] looked as fantastic as the mountains of the moon.... We seemed to be entering a lost world.[18]

Burden trapped a live specimen that got away overnight, but eventually brought back a dozen trophies for the Museum, and a live pair which were donated to the Bronx Zoo. The live dragons proved a sensation, drawing thousands of visitors where they were seen by Merian C. Cooper, Kong's producer-director. The dragons didn't live long, however, in their new foreign surrounding. Meanwhile, Cooper befriended Burden, writing to him after the release of *King Kong*:

When you told me that the two Komodo Dragons you brought to the Bronx Zoo, where they drew great crowds, were eventually killed by civilization, I immediately thought of doing the same thing with my Giant Gorilla. I had already established him in my mind on a prehistoric island with prehistoric monsters, and I now thought of having him destroyed by the most sophisticated thing I could think of in civilization, and in the most fantastic way ... to place him on top of the Empire State Building and have him killed by airplanes.[19]

King Kong's breakneck action pace, Kong's charismatic appeal, Ruth Rose's adroit contributions to the tightly woven script, the wonderful miniature sets (more sophisticated than *The Lost World*'s), the improved camera effects, more accurate and robust (i.e., Knightian) animated puppets—coupled with thrilling drama—had synergistic impact, transforming into record setting box office returns. The movie earned $89,931 in its first four days following its March 2, 1933, release. Decades later, writers compared the relative appeal of 1925's *Lost World* versus 1933's *King Kong*.

The resulting near-ancestor of *King Kong* turned out spectacularly well, easily surpassing most other such films—especially the various attempts at explicit remakes. In other respects, *The Lost World* is far less effective than *King Kong*, which has the advantages of a strong central character capable of evoking both sympathy and terror, a more stronger love story amid the desperate adventure, vastly stronger dramatic emphasis and—most crucially—the breakneck pace.... *The Lost World* ... rambles annoyingly and too often introduces its dinosaurs in a painfully matter-of-fact way as they trudge and amble through the jungle.[20]

Who, let alone the ancients, would have guessed that our haunting eighth wonder of the world would be a miniature, 18-inch-tall, anthropoidal-shaped metal armature cloaked in rabbit fur![21] Few stories of the fantastic variety are more compelling and metaphorical for our age than Mary Shelley's *Frankenstein* and RKO's *King Kong*, both memorializing mankind's arrogance and folly.

Kong's Skull Island, located somewhere in the Indian Ocean near Sumatra, became the quintessential prehistoric monster isle. Together, Skull Island's Rex and Kong embody the idea of big, fierce, yet not quite so extinct in its most refined, pristine sense.

In my *Dinosaurs in Fantastic Fiction*, I refrained from digressing into what was considered to be the tangential (as well as plentiful) arena of children's fiction and comic books, in which prehistoric animals are mentioned. But there are no such restrictions herein, as two often neglected examples will be outlined here—one children's novel and one comic book, both derived from, as well as perpetuating, lost worlds mystique.

In 1941, Oskar Lebeck and Gaylord Dubois published a hardback for adolescents titled, *Hurricane Kids and the Lost Islands*.[22] William Ely's color front cover swiped a posing of Charles R. Knight's 1897 Laelaps leaping toward the protagonists. This book melds *King Kong* and *The Lost World* with a sprinkling of Burroughs' Tarzan for good measure. Certainly, by the late 1930s, the lost world–island concept had become the most popular fictional framework in which crypto-dinosaurs and other prehistoria could be merged with explorers from modernity. Since then, this ever-popular theme had been rehashed numerous times in fiction novels, short stories and in films. In this early genre example, Dave and Alan Burnham, nicknamed the Hurricane Kids—adventure-seeking survivalists in their early 20s or late teen years, who are adept at sailing—are shipwrecked in their 41-foot cruiser the *Southern Star*. (It's suggested they're somewhere near the Cape Verde Island group.) While the storm-tossed ship goes down, the kids drift toward nearby land, an island setting. They soon spot a herd of Charles Knight–inspired Trachodons moving through the jungle forest. More saurians follow. "Recalling their geology course in college, the twins were able to name a number of the huge vegetable eating lizards:—the *brontosaurus,* the *parasaurolophus* and the more delicately made ostrich dinosaur, called *ornithomimus* by the textbooks."

The kids find refuge in a cave, also the nesting grounds of a justifiably angered *Pteranodon*; the wing-finger is dispatched with a spear. Chapter four features a "Battle of Dinosaurs," a terrific tussle between *Tyrannosaurus* and *Triceratops*. The meat-eater savagely rips off a three-horn leg just before those sharp horns stab into Rex's belly. Both animals are left to their death throes as Dave and Alan scurry further into this incredible jungle. Observing smoke in the sky, the kids deduce the island is, of course, volcanic.

Fortunately they discover their supplies have washed ashore along with a smaller vessel, the still sea-worthy *Comet* which had been stowed away on board the *Southern Star*. Admittedly feeling rather like neo-cavemen, the kids rely on a friendly gigantic turtle, *Archelon*, to drag the *Comet* toward shore. Then they're soon menaced by a 100-foot-long sauropod guarding its offspring. Their rifle shots fail to penetrate its thick hide; so instead their safety is ensured by the *Comet*'s swift, powerful engine.

Next they encounter Stone Age (Caucasian) cavemen who are about to sacrifice a buxom-looking cave woman, Maya, to their bloodthirsty tribal god. But she is spared by her red-haired lover, Tarzan-like Omu. The two islanders scramble into the jungle and soon team with the Hurricane Kids. Then over several ensuing chapters, the group defeats a Knight-like sabertooth cat and—in a scene inspired by Jules Verne—vicious marine sauria. After discovering oil deposits, they survive a volcanic eruption, a tsunami, and a prehistoric stampede. The *Comet* carries them out to sea where they confront a band of pirates who make our heroes walk the plank. What a curious succession of predicaments; there's no rest for weary souls on these Lost Islands!

But the strangest, perhaps most predictable dilemma of all is saved for the climax (yes—with its reserved, Kong-inspired sexual innuendo and racial undertones). Maya is seized from their cave camp by an enormous, elephant-sized, gorilla. It seems that during an earthquake a pair of gorillas have migrated over a ravine after a huge log fell bridging the outer Stone Age human-inhabited portion of the island with the gorilla's usual habitat, in the interior. First, one gorilla is killed. Crossing over the ravine on the mossy, fallen log, the team catches up with the other gorilla — and *also* a band of non-caucasoid, Zulu-like spearmen (i.e., referred to as "black savages") who reside in the island's interior with the huge gorillas species. After the Hurricane Kids rescue Maya from the gorilla's Kong-like clutches, it seems a *One Million B.C.*–type war is about to erupt between the two primitive societies. But rifle prowess and muscle power send the tree bridge careening into the chasm with a band of warriors aboard, including their monarch. Of course, the chief of the victorious caucasoid Stone Age tribe is so grateful at this outcome that he spares Maya from the sacrificial rite.

Ultimately, the Hurricane Kids sail off into the sunset toward a timely rescue ship — with a final plug thrown in advertising their next thrilling adventure, another prehistoric-themed title (possibly never published), "The Hurricane Kids in the Canyon of Cliff Dwellers."[23]

In a second example, the "War That Time Forgot"[24] was more than just another of those military conflicts lacking an exit strategy, because it involved the most massive terrorists of all — gigantic dinosaurs, theropods chief among them — battling with U.S. forces on places with nifty sounding names like Dinosaur Island, Mystery Island, and Fireworks Island. It all began in the pages of DC Comics' *Star Spangled War Stories* no. 90 in the spring of 1960 with its featured story, "Island of Armored Giants." With a forever war declared against bloodthirsty creatures from the dinosaur age, the main assault continued amid unyielding machine-gun fire — dinosaurs of every persuasion toppling tanks, buzzing planes from skies above and sinking subs — through the next 8 years (issue no. 137), with skirmishes lingering on down through the 1980s. It was like Conan Doyle gone stark raving mad! But as the series title goes, most of us have *forgotten* about a war that was, well, forgotten through decades of time.[25]

In *War That Time Forgot*, DC dispensed with the singular Tarzanian sort of hero-primitive, instead making heroes of just regular Joes— U.S. GI s who had enlisted to fight our country's enemies, but got far more than they bargained for.

Arguably, DC's *The War That Time Forgot* wielded more science fictionally oriented overtones and plots than did most predecessor or contemporary

Opposite: The essence of DC Comics' "War That Time Forgot" is conveyed in this stylistic rendering by Jack Arata.

books featuring dinosaurs, mirroring the period in which *War* was conceived. In post–Korean War 1950s, repercussions of atomic gloom and a pervading Cold War reverberated globally. American television programs and movies of the time shadowed atomic war angst and aura, precipitating unsettling TV shows like *The Twilight Zone,* movies such as *Fail Safe, Dr. Strangelove,* and *On the Beach* and of course those giant bug and colossal man movies. Until the Soviets blinked, America became embroiled in a short yet nerve-wracking Cuban Missile Crisis. Fantasy dinosaurs of doom like Godzilla — a symbolic embodiment of living radiation — underscored horrors of exposure to nuclear fallout.

Accordingly, as in those big bug films, movieland dinosaurs — like Gorgo, Rodan, Anguirus, the Rhedosaurus and Paleosaurus — were also *big* and, to our dismay, threatened shores of civilization! Science fiction had already veered thematically in dangerous directions — pointing out mankind's folly. Look what mad scientists had wrought from Pandora's box — out of Dr. Frankenstein's tainted crucible! Now mankind must pay the price. And so, we waged our metaphorical war with dinosaurs-of-doom in the movies, and *also* in the comics — in the war that time forgot.[26] Well, perhaps DC's *War* could only be forgotten because it was fought solely on distant shores, in a way presaging Vietnam's atrocities.

In part, DC's *War* was *daikaiju eiga*–inspired. Consider the monstrous fauna — dinosaurs and other prehistoric aquatic and flying reptiles and enormous fish. They're all *huge*, like the dinosaurs in *King Kong*, or like Godzilla. While most of DC's illustrated monsters are in the 50-foot-tall range, some are over well over 100 feet long (or tall). Mighty big fauna for a limited *insular* habitat! And like self-respecting filmic *daikaiju* they largely resisted our bullets, grenades and rocket shells. But they *weren't* invulnerable; unlike monsters like Gorgo and Godzilla it *was* possible to kill them with regularity.

In DC's mystic realm even herbivorous dinosaurs can destroy battleships or overturn tanks. In "The Island of Thunder" (issue no. 98), for example, three-horned *Triceratops* stabs and *lifts* a Sherman tank with its horns! In "The War on Dinosaur Island" (issue no. 105) tank-lifting honors go to *Stegosaurus*. A few of the dino-monsters faintly resembled the fin-backed *Dimetrodons* from *Journey to the Center of the Earth* (1959), which played to matinee audiences in the early 1960s, reptilian alligator-monsters featured in Irwin Allen's 1960 production *The Lost World,* or even stranger contemporary movieland dino-monsters like Gamera and Gappa. There are also giant crustaceans, probably inspired by Ray Harryhausen's stop-motion cinematic successes.

In some cases plots or individual story scenes were clearly inspired by then newly released movies such as *Gorgo,* or *King Kong vs. Godzilla.* However, DC's unique brand of monsters from the dinosaur age wouldn't qualify as *daikaiju* themselves (per J. D. Lees' definition) because in the *War* they behaved too much like real dinosaurs (or animals) known to science. Assuredly DC's dino-monsters were mysterious in nature and origin but, inherently, not sufficiently so as to qualify as *daikaiju*.

Story plots do seem formulaic and repetitive. Typically, a Second World War expeditionary strike force of some specialized nature is tasked on a daring rescue mission, behind enemy lines, a mission impossible. The oft-anonymous enemy is usually suggested to be the Japanese, especially since most of the stories are staged in the South Pacific. Thus there are distinctive associations between DC's *War* and the World War II Godzillasaurus flashback Lagos Island segment from Toho's *Godzilla vs. King Ghidorah* (1991). Next, the specially assigned military team (scuba frogmen, parachutists, experimental weapons team, etc.) is suddenly averted off course by odd atmospheric phenomena (clouds, fog, haze, wind storms, etc.) onto island settings inhabited by vicious, oversized dinosaurs. Dinosaurs erupt right out of the ground from earthquakes, or infest the local waters, while Rodan-sized winged horrors (screaming "Skreee!") dive-bomb from the heavens. Around every steamy jungle bend and turn GIs confront new dino-monsters, prompting volleys of grenades and machine gun bursts. Then, following by-the-skin-of-your-teeth rescue from cruel snapping jaws of doom, the few survivors decide that it's best to keep this all under wraps, lest they be judged Section 8 insane by their commanding officer.

As for the science behind the fiction, well there isn't too much, but after all, it's a comic book. In one tale, a GI muses, "That strange cloud we passed through!— It must be a warp in time — through which we must pass to return to our own time again."[27] That cloud cover warp pseudo-explanation suffices for most but not all of what happens on these uncharted reptile-ravaged islands. There may be some degree of temporal stasis at large here too, preserving the millions of years old fauna to modernity. And there's clearly more than one Pacific island, possibly in the same chain, infested with dinosaurs because early on in the adventures one is destroyed volcanically, and another by a tidal wave.[28]

Frozen dinosaurs discovered by military forces in issue no. 106 (possibly inspired by Universal's *Dinosaurus!* (1960) are linked loosely to theories that perhaps the dinosaurs had gone extinct (everywhere else) during an Ice Age. The ice theme is dealt with more prominently beginning in issue no. 112, "Dinosaur Sub-Catcher," as a submarine venturing under the polar ice encounters prehistoric creatures, suggesting "War That Time Forgot" was also influenced by Edgar Rice Burroughs who had by 1918 already envisioned *The Land That Time Forgot* (1918). However, according to Don Markstein:

It wasn't until the 1980s that an explanation, of sorts, was offered for the characteristic fauna of Dinosaur Island (as the locale has become known). There seems to have been a connection, through hundreds of miles of solid rock, between it and Skartaris, the setting of DC's *Warlord* series, an inner–Earth land loosely based on the Pellucidar books by Edgar Rice Burroughs. Dinosaurs are, of course, common in such inner–Earth lands, as everyone familiar with the clichés of fantasy fiction knows.[29]

Obviously *Star Spangled War Stories* series' creators had a lot of fun concocting these imaginative tales. And, golly, what an assortment of tales![30]

"Awful Changes" on Monster Island

Curiously, if the 19th century's most celebrated, influential geologist, Charles Lyell,[31] had his way, prehistoric monsters would have been denied their most distinctive natural history characteristic: *uniqueness* in geological time. Lyell objected to a planetary history viewed by his contemporaries as directional and progressive through time. Instead, Lyell's Planet Earth operated in a non-catastrophic[32] steady state tempo and mode, essentially unchanging macroscopically since the beginning; a pseudo-cyclic world destined to resurrect denizens from the fossilized past quite similar or generic in form, yet upon closer inspection, different from prehistoric extinct species. Lyell's paleontological ideals became widely circulated through three volumes of his lauded and immensely popular *Principles of Geology* (1830–1833, and subsequent editions).[33] However, Lyell's contemporaries objected to the most lasting and erroneous tenet of his geotheory, an overbearing reliance on uniformitarianism caricatured through De la Beche's 1830 science fictional "Awful changes. Man found in a fossil state — Reappearance of Ichthyosaur! 'A change came o'er the spirit of my dream.' Byron." Here, De la Beche lampooned a futuristic, Lyellian period following man's demise, or a return to *another* reptilian age when:

> those genera of animals return, of which the memorials are preserved in the ancient rocks of our continents. The huge *Iguanodon* might reappear in the woods, and the ichthyosaur in the sea, while the pterodactyle might flit again through umbrageous groves of tree-ferns. Coral reefs might be prolonged beyond the arctic circle, where the whale and the narwhal now abound. Turtles might deposit their eggs in the sand of the sea beach, where the walrus sleeps, and where the seal is drifted on the ice-floe.[34]

In other words, according to Lyell, biological speciation was wired, operating in a fashion unfamiliar to current (Darwinian) comprehension. To Lyell, ceaselessly changing Planet Earth could mysteriously conjure biological forms out of its fossil past into modernity, or geological periods of the future. However, Lyell's foreign sense of non-directional, a historical geological time conflicted with a growing body of evidence and consensus opinion suggesting that distinctive, never-to-be-repeated ages of increasingly complex organisms had indeed succeeded one another through a perceived ladder

De la Beche's science fiction cartoon, "Awful Changes," poked fun at Charles Lyell's inflexible notion that the Earth's geological "history" and future had no apparent (evolutionary) direction. Here, "Professor Ichthyosaurus"— generically analogous, that is, anatomically, to Jurassic ichthyosaurs — is lecturing to a reptilian audience in some futuristic geological age about the paleontological implications of prehistoric (i.e., modern) man (Amguedda Genedlaethol Cymru, National Museum of Wales).

of progressive organic development. Life's history had apparently progressed from a sterile, oldest azoic origin through an "Age of Marine Invertebrates," succeeded by an "Age of Fishes," followed by the "Age of Reptiles," then an "Age of [placental] Mammals," culminating in man.[35] Evidently, there were also former interspersed "ages of coal" characterized by the ascendance of luxuriant land plants; the most impressive and economically significant of these followed the Age of Fishes. Geologists found another vast coal measure in early Tertiary deposits; indicative of tropical conditions formerly prevailing in the now temperate northern hemisphere.[36]

Although founded on the strength of his authoritative *Principles*, Lyell's position was, at best, tenuous and forced. So William Buckland countered

Lyellian geo-mechanics with an alternative perspective, exemplified in a chart, "Ideal section of a portion of the earth's crust," published in his popular *Geology and Mineralogy* (1836). Perhaps for the first time, visually, this diagram placed a succession of prehistoric animals including *Iguanodon*, pterodactyls, marine reptiles and other extinct varieties within their respective ages.[37] In other words, fossil organisms had lived at *certain* times and would not be expected to reappear individually or as anachronistic groupings of species in *any* age. To Buckland's more correct reasoning, there was no indication or reason to believe that such animals, or even their generic forms, should rise again from the ashes of prehistory. Prehistoric, extinct species were *unique* in time; it would be improbable for extinct fauna and flora to simply reappear. Prehistory could not rerun: no do-overs allowed. Buckland's rather tepid position became conceptually affirmed decades later by Louis Dollo who formulated an evolutionary "Law of Irreversibility," that "once evolution had taken a particular path, there was no retracing it."[38] Planetary climate itself evidently had cooled considerably since the earliest periods; gradually cooling conditions consequently drove corresponding organismal changes as recorded in the fossil record.

If consequences of Lyell's "Awful Changes" were true, Planet Earth's past and future history would demonstrate little overall *uniqueness* or apparent direction.[39] Dino-monsters of the past *could* conceivably be resurrected during any age; that is, when shifting climatic and hence geographical conditions fostered their return. Furthermore, given the proper blend of ecological conditions, prehistoric-looking creatures considered generically similar to genuine fossil varieties could become reconstituted de novo in the future, cohabitating with others resembling forms from any other geological age. So, according to Lyell and his witty detractors, creatures similar in form to Jurassic animals could someday exist alongside creatures similar in body outlines to Late Cretaceous monsters.

While this hadn't yet occurred, Lyell thought it *could happen* in *futuristic* settings pregnant with orgasmic possibilities. Moreover, was it plausible that such circumstances already exist in unexplored and unexploited regions of modern day Earth? This is more or less the impression influentially projected a century later through *The Lost World* and *King Kong*, perhaps *the* films shaping popular Lyellian notions as to where prehistoric monsters live and which ones might be found, cryptozoologically, living together on primitive monster islands. Thus, beyond Lyell's archetypical speculations, we owe the modern rendition of the monster island concept to the imaginations of *Lost World* author Conan Doyle, and paleoartists Willis O'Brien (1886–1962) and Marcel Delgado (1901–1976).

Whereas *Lost World* (1925) and *King Kong* were founded on a pseudo–Lyellian premise (a commingling of beasts apparently derived from separate geological periods), preceding fantastic literature penned by Verne, Boitard,

A section of Thomas Webster's "Ideal section of a portion of the earth's crust," published in William Buckland's *Geology and Mineralogy* (1837 ed.). Note that the figures of animals seen above the cross section are grouped per their relative, distinctive position in geological time. This illustrates how, in contrast to Lyell, Buckland (and others) thought geological time was directional, and life's history was progressive.

and Edgar Rice Burroughs ("ERB" as in his Caspak trilogy) featured contrasting and more accurate life through time concepts. In their respective novels, *Journey* and *Paris Before Man*, French writers Verne and Boitard each emphasized how organic life changed (one would say evolved today) throughout geological time. Then, over half a century later, beginning with his *The Land That Time Forgot* (1918), Burroughs introduced fearsome products of paleontological, evolutionary progress, in the process exposing readers to a thrilling, imaginative time tour.

Edgar Rice Burroughs's insular Caspak or Caprona of a trilogy beginning with *Land That Time Forgot* was certainly a prehistoric insular habitat; although, there, evolution was vibrant.[40] Conversely, today's popular conventionalized notion of prehistoric monster islands is far less complex than Burroughs imagined during the 1910s. More significantly, prehistoric monster islands as popularly regarded today are not ERB–ian, but instead Lyellian. The latter are usually imagined as mysterious lost world places where prehistoric animals ostensibly representing different geological ages coexist; spawned out of time through awful changes—to be discussed next, mechanistically.

Unlike conditions in ERB–ian prehistoric monster places, in Lyellian lost world settings, due to limited geographical range the few relic survivors—last races or members of their kind—are barely hanging on to survival. Another often explored theme suggests that preserved prehistorical creatures found by the human interlopers seem indistinguishable from original fossil genera living in remote geological times. Regardless, they're living together, in cahoots with primitive humans no less! Such circumstances would confound one's ability to assign fossil species to distinct geological periods, past or present.[41]

But are Lyellian prehistoric monster places truly non-evolutionary in nature: places where time truly stands still or has only shifted imperceptibly since long, long ago? To what extent are prehistoric monster islands, well, Lyellian? Cryptozoological denizens of Maple White Land's and Skull Island's fauna would appear Lyellian at first glance because such prehistoria are an anachronistic blend; animals apparently from *different geological ages living together*. Theoretically, several circumstances may have permitted prehistoria populations to arise on monster islands. Such fauna may have been *preserved* in original form since, say, Mesozoic times: same species. Or their lineages may have *persisted* since the Mesozoic, although with varying degrees of evolutionary transformation resulting either in nearly identical, related *species* (or subspecies), or *entirely new genera* generally resembling fossil ancestors. Or, finally, they could have resulted through the mysterious operation of awful changes.

So how would Professor Challenger or Carl Denham know with certainty whether prehistoria they encountered on these islands, respectively, are genuine relic survivors from the past (e.g., same dinosaur species as those known from the fossil record), or perhaps if they were less or completely unrelated to authentic Mesozoic dinosaurs?

However similar Kong's 1933 living tyrannosaurid-like foe may seem to the American Museum's 5037 *T. rex* skeletal composite, it remains virtually impossible to determine whether they're the same Rex species. You don't have fossil DNA or soft tissues to compare to Skull Island's famous alleged living fossil. And how exceedingly difficult to prove live in-the-flesh specimens are true living fossils; that is, identical to possible fossil ancestors known only from partially complete skeletons! Without an absolutely complete fossil record showing the persistence of lineages of dinosaur species from the Mesozoic surviving through Pleistocene ice ages into modernity, one could never prove that observed, living prehistoric fauna are genuinely related to fossil ancestors. (And we'd also require explanations for how a western North American *T. rex* ancestor migrated far away to a geologically recent island in the

Indian Ocean.) Superficially, comparative bone structure analysis might suggest genetic connections, but it's hard to ignore the overwhelming fact that Kong's 1933 Rex and AMNH-5027 are separated temporally by a gulf of *sixty-five million* years. Live species might appear generically similar to fossil specimens; for all practical purposes, however, the exact nature of their relationships and occurrences defy explanation.

Unless time has stood steadfastly still through the ages, places such as Maple White Land (e.g., *Lost World*) and Kong's Skull Island are unlikely to have preserved dinosaur species identical to those that lived during the Mesozoic Era. Hence, in a compelling, lengthy and thrilling CGI-aided scene featured in Universal's *King Kong* (2005), the pack of tyrannosaurs menacing Ann Darrow and Kong (mimicking and surpassing in intensity the impact of the original 1933 Rex/Kong battle) are not *Tyrannosaurus rex* species, but instead as thoughtful Ann Darrow decides in Christopher Golden's 2005 *King Kong*, a new genus: *V. rex*, that is, V for voracious.[42]

During his younger years, Lyell wasn't an evolutionist; he demurred on how new species came into existence. Lyell's awful changes seem mysterious to us now because he initially disclaimed biological evolution. Pragmatically, however, hypothetical awful changes must represent evolutionary dynamics at play; most likely, an alternate or *convergent evolutionary* process generating living lost world dinosaurs from *non-dinosaurian* ancestors instead. So Lyell's view is not inconsistent with Dollo's Law of Irreversibility in that, under awful (i.e., *evolutionary*) changes, biological organisms would not follow the same unlikely path regenerating, say, DNA–exact *Tyrannosaurus rexes* from ancestral chicken stock in some future age. For as paleobiologist George Gaylord Simpson (1902–1984) noted, "evolution is irrevocable. When a series of complex changes has been encoded in the DNA, it is not likely that these will be undone and the original plan restored. However, there may be modifications that appear to 'reverse' earlier traits"; that is, only at a generic level of scrutiny.[43] Nevertheless, because the absolute ancestry of hypothetical living dinosaurs may be indecipherable, the usual mixed bag paleontological assortment of prehistoria discovered in monster places such as Maple White Land and Skull Island—whether preserved as original species, or generated genetically with evolutionary modifications into similar generic descendants—can today be considered a fantastic reflection of pre–Darwinian, Lyellian awful changes.

Other Prehistoric Monster Worlds

The Lyellian prehistoric monster island idea never faltered since its inception a century ago, and thrives today despite competition from rival habitats where live prehistoria may likely be encountered by mankind either in printed fiction or film. Following *Son of Kong*, Hollywood soon followed thematically with a pair of dinosaur-packed cheapies—*Unknown Island* (1948) and *The Lost Continent* (1951).[44] During the early atomic age self-respecting geographers should have been humiliated to have somehow missed an entire continent on the globe; still, bearing Komodo's genuine example in mind, the prospect of encountering a previously uncharted island bearing unknown fauna and flora then seemed more likely than the possibility of finding dinosaurs living inside the planet. So while it may be easier to dismiss *The Lost Continent* merely on its unlikely titular premise, an oddity enacted in *Unknown Island* is worth mentioning.

By the mid–20th century, thanks to Verne's *Journey*, *King Kong* and several other examples, it had become conventional to imagine prehistoric animals as ceaselessly fighting Darwinian, survival-of-the-fittest battles within their misty prehistoric confines, or increasingly via *daikauju eiga* examples, in populated cities. Correspondingly, producers and writers sought to emulate Kong's classic 1933 Rex battle with surrogate tyrannosaurs: Gorosaurus, Godzilla, V-Rex, or Gaw.[45] Peripherally, that's just what seems to be happening in *Unknown Island*, where prehistoria ranging from the Permian through the Pleistocene cohabitate. Curiously, however, the ape featured in this picture isn't a gorilla, but a suitmation (Cuverian, if you will) example of Jefferson's 18th century Great-Claw! This creature's adversary in *Unknown Island*'s climactic battle scene is a stumbling suitmation *Ceratosaurus* (moonlighting from its usual engagement with *Stegosaurus*). And so we must proclaim, "Long live Jefferson," whose prehistoric Megalonyx brainchild inspired cinematic exertions later mimicked by titular monsters featured in Toho's *King Kong vs. Godzilla* (1963)!

Prehistoric monster islands are inhabited by relic populations, vulnerable to modernity and representatives from the civilized world who would study or, however unwittingly, destroy surviving prehistoria for personal gain. Inhabitants of such worlds are primitive, symbolizing Europeans' gradational conquest of the Americas or the African interior. That such throwback fauna has been spared the fate of extinction owes greatly to their seclusion on uncharted islands or locations. The idea of horrifying, mysterious or otherwise enchanting lost worlds coupled with the incredible journeys needed to reach such places has been out there in public consciousness for centuries. For instance, a short list of favorite examples would include Daniel Defoe's *Robinson Crusoe* (1719), Jonathan Swift's *Gulliver's Travels* (1726), H. Rider Haggard's *She* (1887), H. G. Wells' *The Island of Dr. Moreau* (1896), L. Frank Baum's *The Wonderful Wizard of Oz* (1900), James Hilton's *The Lost Horizon* (1933) and William Golding's *Lord of the Flies* (1954). Not all of these would qualify as hard science fictional works, per se, although as Brian W. Aldiss noted, "the link between science fiction and travel is probably as close in the [nineteen] eighties as the link between science fiction and science."[46] Unlike *Journey* and *Lost World*, these classics don't rely upon or embody prehistoric elements, contrivances and themes.[47]

Verne's *Journey* predates Conan Doyle's *Lost World* by nearly half a century. Verne simply assimilated the hollow earth idea. By the 1860s, hollow science had been circulating in pop-culture for nearly two centuries. However, Verne's sublime setting of a cavernous Earth became the most famous fictional example of a honeycombed planetary interior that was also prehistoric. While Verne's (non-evolutionary) prehistoric tableau was not Lyellian due to its life through time tendencies, later writers inspired by *Journey* such as Burroughs and Lin Carter (1930–1988)—who in turn emulated Burroughs—both fabricated prehistoric, Lyellian hollow Earth environments named, respectively, Pellucidar and Zanthodon, populated by representatives from several geological horizons. Thematically, hollow earth paleo-environments may be distinguished from the nature of prehistoric monster islands in that the former are characterized as zoologically thriving, *utopian* settings where all manner of heroes abound; this is unlike latter settings were relic primitives on the verge of extinction are threatened by maniacal societal castoffs.

Burroughs' Pellucidar is often discussed, while Carter's fascinating inner world of Zanthodon, viewed as derivative to Burroughs' classics, has not received similar attention — so let's indulge. The adventures of American thrill-seeker Eric Carstairs in legendary Zanthodon commence in Linwood ("Lin") V. Carter's (1930–1988) *Journey to the Underground World* (1979).[48] The real author (i.e., Carter) claims to have known Carstairs, although he won't reveal exactly how he obtained his account, which seems more fictional than true despite its absolute veracity. The fabled underground world, many miles under the surface — Zanthodon — is clearly derived from Edgar Rice Burroughs's Pellucidar, borrowing elements and scenes from classics such as *At the Earth's Core* (1914), *Tarzan the Terrible* (1921) and *Back to the Stone Age* (1936, 1937). Desperately seeking employment in Egypt, helicopter pilot Carstairs encounters Professor Potter, a polymath/polyglot — who fortunately and most conveniently, considering where they're headed off to, happens to know quite a bit about paleontology. Potter, who has been translating ancient texts, has learned of the entrance to a legendary underground world which (as in Jules Verne *Journey to the Center of the Earth* fashion) will lead to the center of the world. The entrance is to be found through a hollow mountain core in North Africa's Ahaggar Mountains. The adventure begins.

Zanthodon's geological origin is, well, fairly original — given the long history of underground and sometimes prehistoric settings encountered in fantastic literature. Professor Potter's theory is that Zanthodon was formed 150 million years ago when a relatively small meteorite made of antimatter with a blast force of "*dozens* of hydrogen bombs" penetrated the earth's crust, annihilating solid rock, forming a huge underground chamber. Over time since the Jurassic Period, animals now extinct on the earth's outer surface accidentally fell into or migrated into the vast underground cavern where their kind was preserved to the present day. Their plan is to descend into and through the 200-foot-diameter hole at the lip of Mount Zanthodon's crater in Carstairs's helicopter, well stocked with supplies for their incredible journey.

The walls of the shaft begin to glow phosphorescently a few dozen miles down into the crust, although fortunately this isn't due to residual radioactivity; huge mushrooms are spotted growing far below the surface (as in Verne's "Journey"). Greatly fatigued, Carstairs fatefully hands over control of the aircraft to Potter, who not long thereafter accidentally causes the helicopter to crash in a strange jungle below the volcanic orifice. The last recorded reading within the shaft prior to Carstairs' retirement was at 70 miles' depth. (Later, Carstairs estimates Zanthodon's depth inside the Earth to a couple of hundred miles.) As they come to their senses, a Triassic croc — *Protosuchus*—crawls toward them. Of this apparition, Potter muses how "it will evolve into your true and genuine *Crocodylus* ... unless, of course, evolution and its forces have been suspended here, as I more than half suspect to be the case."[49] But Potter never rationalizes how the presence of a living Triassic croc may falsify his theory of origin for Zanthodon's immense subterranean, phosphorescently lit cavern (which as he claimed formed later in the subsequent *Jurassic* age).

Shortly after, Potter and Carstairs flee to the branches of a tall tree as they're menaced from below by a *Triceratops*, the presence of which even the professor is hard-pressed to explain. But, even stranger, he's at a loss to account for its (*gryf*-like) *carnivorous* habit as it devours a pterodactyl! From their vantage point above, they watch a titanic battle between Tops and a Woolly Mammoth which rushes in suddenly, causing Potter to exclaim, "the two monsters come from ages nearly one hundred and fifty million years apart.... This is utterly fantastic!"[50] Next an armored Scolosaurus trudges by before they stumble into a Devonian forest; then they're captured by a band of Neanderthal men who have enslaved several Cro-Magnon, tethered to a rope. Now Potter surmises that "both early man and mammoth must have fled down from the advancing glaciers when the Ice Age came down from Europe."

But all this theorizing must soon yield to basic survival instincts as they plot their escape. What happens next forges much of what the rest of Carter's Zanthodon series of novels is all about. Carstairs has a handgun, but with only a few bullets left which he'd prefer not to waste. During their weary march he learns the universal native Zanthodonian language from the beautiful Cro-Magnon, Darya, a chieftain's daughter. As their ability to converse (rapidly) improves, however, beautiful blond Darya is left feeling insulted, believing Carstairs is making a fool of her, lying to her when he tries to explain the wonders of New York City with its tall buildings, subway systems and jet airplanes. Eventually, the slaves make a break for freedom, with Carstairs musing, "life in the savage jungles of Zanthodon *is* cruel and unfair; in this primitive realm beneath the earth's crust, survival does not always go to the best, but often to the luckiest." And, yes, reflecting the male utopian vision, he's also enchanted by the nakedness of its most fetching, buxom women![51]

By the time Conan Doyle wrote his 1912 *Lost World* classic concerning the forbidden land of Curupuri, interest and scientific integrity of the possibility of inner earth worlds had declined due to the exploration of both north and south polar regions. Because the hollow idea was tied to the concept of polar holes, and intrepid explorers crossing tundra and ice floes had not discovered holes at the poles, writers of fantastic fiction could no longer readily suspend disbelief using a hollow earth contrivance. However, besides Carter's dramatic Zanthodon series, Nature's defiant lack of cooperation with fantasy didn't prevent three of the better holes at poles genre entries from being written long after conquering flags had been planted on icy poles— Vladimir A. Obruchev's *Plutonia* (1924); Burroughs' *Tarzan at the Earth's Core* (1929); and James Rollins' *Subterranean* (1999).[52]

Besides prehistoric monster island and hollow earth refugia, mankind also 'encountered' prehistoric monsters in other favored paleo-environments: for example, undersea, marine or aquatic settings. Preying on our innate fear of water, such monsters have included assorted creatures such as Nessie, Karel Capek's war-like Newts, the Black Lagoon's Gillman, demonic Jurassic sharks such as Steve Alten's *Meg* (1997), and even the great sauropods, that is when they were considered aquatic — as opposed to terrestrial — animals. Like Verne's 1864 prototypical *Ichthyosaurus* versus *Plesiosaurus* pairing, marine dino-monsters represent Nature's fury.[53] Thematically, these paleo-settings and their adapted prehistoric monsters contrast with terrestrial, monster island habitats. For the sea represents Nature's vengeful wrath. Therefore, its unique monsters are the epitome of big, fierce, yet again, not quite so extinct. Oceanic prehistoric monsters are the most vibrant, energetic and ominous of all the cryptozoological inventions. They are amphibious engines of might and invincibility, hellbent on punishing mankind — Godzilla, Gorgo, Ray Bradbury's literary "Foghorn" monster from the deeps, and Cthadron, for example. The sea is vibrantly alive. What was spawned within cannot be nullified. Tucked away under undulating waves for eons, these prehistoric agents of destruction have survived the longest and therefore are most invulnerable to nature's paroxysms. To them, mankind and his self-destructive tendencies are but mere afterthought.

In the Epilogue to *The Time Machine*, H. G. Wells's narrator suggests that the time traveler protagonist may, on another exploration of time may have been swept back into the past, and fell among the blood-drinking, hairy savages of the Age of Unpolished Stone; into the abysses of the Cretaceous Sea; or among the grotesque saurians, the huge reptilian brutes of the Jurassic times. He may even now — if I may use the phrase — be wandering on some plesiosaurus-haunted Oolitic coral reef or beside the lonely saline lakes of the Triassic Age.[54]

While Wells never penned such a sequel, accordingly, another traditional setting for encountering prehistoric monsters— especially so given that pale-ontologists are in a sense genuine time travelers— is of course in their own stomping grounds, prehistory. Time travel is perhaps the most sophisticated of all the contrivances used by fantastic story writers for bringing mankind face to face with tyrannosaurs, horned dinosaurs and pterosaurs. While, arguably, early examples of a pseudo-travel through geological time to sample life's history included John Mill's *The Fossil Spirit: A Boy's Dream of Geology* (c. 1855),[55] Boitard's *Paris Before Man*, and Verne's *Journey*, reliance on time travel machines, per H. G. Wells' influence, first appears through various contrivances in the early 1950s. But, relative to monster island and hollow earth scenarios, the resultant is more accurately *non*–Lyellian, as the time destinations are to discrete geological periods such as the Jurassic and most commonly the Cretaceous Periods.[56]

The farthest lost worlds imaginable are prehistoric monster *planets*. These are outer space worlds inhabited by creatures that, relative to terrestrial life, seem primitive. Because exobiology and genetic ties to terrestrial organisms are often uncertain in such stories, however, and because it is too tricky ascertaining whether another fictional planet's geological conditions are prehistoric relative to Earth's, alien species conflate proper categorization — even if they outwardly appear dinosaurian.[57]

What would appear to be the most recent categorical entry —*bioengineered* prehistoric monsters, as in Crichton's *Jurassic Park*[58] — is really a tumultuous throwback to the prehistoric monster island theme. Not only are Crichton's bioengineered dino-monsters for the most part confined to two lonely Pacific islands, but they support a Lyellian admixture of creatures dating from the Early Jurassic through the Late Cretaceous. Furthermore, just as in Conan Doyle's *Lost World* plateau and like *King Kong*'s lost island, *Jurassic Park*'s Isla Sorna and Isla Nublar are infested with leaping raptor dinosaurs dignifying Edward Cope's and Charles Knight's perceptions of the vibrant, bounding Laelaps. We see this display as well in a cheap *Jurassic Park* knock-off, *Raptor Island*, first aired on the Science Fiction Channel on August 21, 2004. Even in the earliest example of (science fictional) dinosaur bioengineering this writer is aware of — John Taine's 1929 novel, *The Greatest Adventure*— the enormous dino-monsters are discovered in what could only be regarded as a veritable lost world of the time — Antarctica! There is one crucial difference, however. While Lyellian prehistoric monster islands often seem on the verge of extinction when discovered by human intruders, hi-tech bioengineered dino-monsters threaten world domination. Arguably, Godzilla — transformed by the bomb and stemming from a marine habitat near Odo Island — may be viewed as a Lyellian, lost world denizen.

Mad Scientist Explainers and Conquerors

Why does it always seem that scientists and other explorers discover and penetrate into the forbidden recesses of prehistoric monster islands just about when that nearby menacing volcano erupts? It's as if Nature doesn't want trespassers from modernity mucking up the place — hence, the fiery, self-destructive display. In fact, dime store dino-monster novels and B films are rife with those essential climactic exploding volcanoes. It happened on Skull Island (at the end of *Son of Kong*), and it happened in the land of Curupuri — Conan Doyle's lost world plateau — in the 1925 movie production. The exploding volcano contrivance owes its convention to Jules Verne's *Journey*. Verne's classic, prototypical volcanic eruption, in turn, owes its genesis to science fiction's mad scientist intellectual progenitor — Sir Humphry Davy (1778–1829), who made a spectacle of the first *miniature* model volcanic eruption, since repeated *ad infinitum* by grade schoolers at so many science fairs. Except Davy's theoretical premise for the *cause* of volcanic eruptions was more involved than simple acid and base (i.e., baking soda and vinegar) chemistry; in fact, his flawed geochemistry was essential — one might say central — to Verne (and inspirational to Mary Shelley at the time she wrote *Frankenstein*) shortly after Mt. Tambora's 1816 eruption.

No, Davy wasn't mad, per se, but rather a precocious chemist who made many important scientific discoveries. That his good name appears in both Shelley's as well as Verne's signal novels, however, indicates how influential was this highly competent gentleman scientist whose science, became instilled in two mad scientist archetypes of fantastic literature (and hence much later the movies), Shelley's Dr. Victor Frankenstein, and Verne's Professor Otto Lidenbrock, who we shall now briefly consider as the prototypical lost world explainer and conqueror.

Davy isn't by any means the only visionary scientist upon whose sturdy shoulders visual paleoartists, film makers, and writers of fantastic fiction have been carried, in the popular elucidation of prehistoric monster worlds. Men such as Cuvier, Buckland, Mantell, Owen, Lyell, both Erasmus *and* Charles Darwin, Barnum Brown, Osborn, Bakker and, lately, Paul Sereno all have become prominent sources to artist contemporaries, for example. Yet how curious that both Verne and Shelley turned to a common source — Davy (besides other influential scientists upon whom they relied) in the crafting of two seminal works leaving such lasting impact upon legions of prehistoric

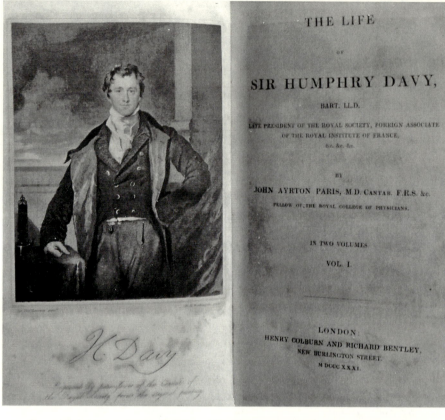

In an odd sense, chemist/geochemist Humphry Davy was the original "mad scientist" when considered as an inspirational source for (1) Mary Shelley, who created the Frankenstein legend, retaining such import on today's prehistoric fiction, and (2) Jules Verne, who wrote the most enduring, prototypical tale concerning scientists confronting fossil animals in sublime, prehistoric settings (collection of Allen G. Debus, photographed by the author).

monster aficionados. And so, with Davy as muse, we shall now briefly consider a handful of mad scientist discoverers of lost worlds, eventual conquerors of their prehistoric monstrosities.[59]

Accordingly, a short digression into the fictional explorers and conquerors of lost worlds and those obligatory mad scientists who comprehend the nature of fossil beasts paradoxically dwelling therein offers further perspective on the dino-monsters' mysterious realm.

If you are presented with evidence for a prehistoric monster island, or in modern parlance a *lost* world, this implies that it's been found by someone. And who could best explain the nature of the inhabitants of such forbidden places, but *paleontologists* who—transformed by the romantic and rugged nature of their profession—by the early 20th century were sometimes regarded as mad. Eventually, scientist-interpreters of past environments resurrecting the dead (from fossils) were also increasingly characterized as powerless to stave off apocalyptical prehistoric invasions. Ironically, civilized modernity—symbolized through Kong's New York City, *Gojira*'s Tokyo, or *Tyrannosaurus*' San Diego—could become lost world settings too. Reflecting mankind's plight and fragility of civilization, the most significant prehistoric monster islands of the dinosaur movies or fantasy literature became those from which prehistoria invaded, sometimes aided by Man's importation.

Beyond the monsters themselves, let's consider the humans who "decipher" or interpret the encountered prehistoric creatures. These are prototypically scientist-brainiac types (usually men) providing the essence and meaning of the prehistoric monsters—elucidating the monster's origin and why it's (or they're) on the rampage against mankind. These sometimes indispensable men also enunciate laws of the beast, that is, what we can expect from the prehistoric monster regarding its behavior and formidable powers, and even an inkling of how man may destroy it. We find that most such later filmic scientists are stereotypes of original story characters, although it is also noted that many such films and stories have dispensed with the scientist-interpreter altogether because we've grown so accustomed to those quasi-scientific explanations that we can automatically graft theories expressed in another predecessor film already in public consciousness onto the latest one. So, here, the theme of the mad scientist role in fictionalized tales of prehistory will be tackled in two stages.

Let's examine, first, how originators of the genre (Jules Verne and Arthur Conan Doyle) dealt with such curious characters in a pairing of influential novels—*The Beast from 20,000 Fathoms* and *The Giant Behemoth*—and, second, how the concept of paleontologist-as-scientific-monster-interpreter materialized in pivotal movies, omitting further consideration of Japanese *daikaiju eiga* (addressed in Chapter Nine of this book). The premise of most latter day prehistoric monster films is founded in those fabulous 1950s movies featuring characters curiously derived from Jules Verne's original scientific persona projecting paleontological ambience in *Journey to the Center of the Earth*—Professor Lidenbrock and his young geologist protégé, Axel. Among filmic examples, *Beast*'s Rhedosaurus is purportedly a genuine relic preserved in a lost world setting—the Arctic, locked in ice since prehistory, not unlike *Jurassic Park*'s DNA–preserving insects sealed in amber—while *Behemoth*'s Paleosaurus is one of those undersea doomsday draconians.[60]

Of course, Verne wasn't the first to meld paleontology with fiction. Although mostly forgotten today and less revered for writing in this particular genre, there were predecessors. Three decades before Verne's *Journey to the Center of the Earth* (e.g., 1864, 1867 eds.), Rodolphe Toepffer published a peculiar geological tale, *Journeys and Adventures of Doctor Festus* (1833, with its mastodons and mammoths living in polar regions). In chapter four, we've already noted Cornelius Mathews, who conjured an 1839 tale about a mastodon attacking Native American Indians, *Behemoth: A Legend of the Mound-Builders*, which (according to Perry Miller) may have inspired Herman Melville's *Moby-Dick*. Next, Melville incorporated paleontological themes in *Mardi* (1849, with a chapter staged on an Isle of Fossils) and *Moby-Dick* (1851), the latter replete with an entire chapter devoted to "Fossil Whales." And then a few years prior to publication of Verne's *Journey*, Pierre Boitard's illustrated book *Paris Before Men* (1861) was released posthumously. Here, Asmodeé the demon, masquerading as a paleontologist, explains monstrous creatures of prehistory as they're revealed to a reluctant time traveler. And too, George Sand's *Laura: A Journey into the Crystal* (1864) is laced with geological/paleontological undertones.[61]

In Verne's *Journey to the Center of the Earth*, the first truly influential book of its kind, the prototypical paleo-mad scientist not only comprehends but also at times dramatically "unveils" prehistoric elements confronted by an expedition party as they penetrate the underworld on a quest following 16th century alchemist Arnie Saknussemm's murky trail. Certainly, having knowledge of geology and life's history proves invaluable as young geologist-in-training Axel must depend on his uncle, Professor Otto Lidenbrock's, understanding of prehistory for his very survival. Soon, the expedition finds itself deep within the bowels of the Earth. A geologically ordered succession of paleontological marvels, fossils and fleshed-out specimens is encountered (though they sometimes live only in Axel's fevered imagination), threatening safe passage.

Lidenbrock is a bad-tempered sort, bordering on megalomania, wedded to theoretical geological opinions even when Axel notes apparent observational discrepancies. How did the prehistoric monsters get into the underground cavern (e.g., central sea)? Lidenbrock proclaims such fossil antediluvia fell through fissures caused by episodic contraction of overlying crust, allowing seawater to spill into the Earth's subterranean realm. In facing prehistorical titans, Lidenbrock comments on the morphology of two reptiles observed through his telescope, the 100-foot-long *Ichthyosaurus* as it battles a similarly sized *Plesiosaurus*. Meanwhile, Axel, who narrates Verne's novel, provides considerable commentary during the reptiles' awesome struggle to the death. Of course, it's safe to presume that Axel's knowledge of the marine monsters has been acquired through Lidenbrock's teachings. Much later, after Lidenbrock discovers a fossil anthropoid, the great scientist expounds pompously at length on the nature of human prehistory. Thus, armed with (Verne's summary of) contemporary scientific knowledge, an 1860s dissertation on the conceptual nature and origin of humanoid fossils, Axel and party are soon confronted by a massive living representative, a gigantic anthropoid.

Whereas despite his overall decent intentions Lidenbrock seems rather demented at times, Conan Doyle's Professor Challenger from *The Lost World* novel (1912) is even more mad, although usually in the sense that he's so often vehement and angry — an affliction of his passion for doing science accurately. To him, the exalted science of paleontology should not be practiced by nincompoop colleagues. To a certain degree Conan Doyle's paleontologist Challenger's persona seems derived from Verne's Lidenbrock. However, rather than Lidenbrock, Conan Doyle experts contend Challenger's character was apparently patterned after Conan Doyle's anatomy instructor at Edinburgh University, William Rutherford, who had an "Assyrian beard ... prodigious voice ... enormous chest and ... singular manner." Two others whose personalities may have been interwoven into Challenger's character were Conan Doyle's acquaintances, boisterous physician George T. Budd and evolutionist E. Ray Lankester. Melded from these forceful and influential figures, fictional Professor Challenger positively *wields* science as a force to be reckoned with! For exalted science, and presumably paleontology which ironically would be viewed obsolete by the mid–20th century, was considered optimistically beneficial to humanity, "making the world a safer place to live via the improvement of knowledge."[62]

And yet what a bizarre, insolent character is Professor Challenger of the 1912 novel! Bickering with Professor Summerlee (who eventually is converted by the veracity of Challenger's claims) nearly every pace of their journey toward and along the lost South American plateau where prehistoria dwell, Challenger deduces that the extant yet prehistoric life must have reached the plateau summit ages ago after which they became isolated from the rest of the world. Except, more recently, there were several episodes where geologically younger species fortuitously accessed the plateau's summit as well, thus creating the unusual faunal mélange. For whatever reason, environmental conditions permitting their survival throughout prehistory just haven't changed considerably since. More or less like Verne's Lidenbrock, Challenger's supreme knowledge of paleontology and evolution permits him to take command of the lost world frontier. At first he uses the plateau's anachronistically mixed fauna as an evolutionary laboratory substantiating his rather outlandish theories. Then quite literally he leads a bloody charge in a war waged between a modern Indian tribe (the good guys) and hostile ape-men, whom ragged-looking Challenger has not only begun to resemble physically, but who, because they're primitive, are also simply destined to be extinguished. Such scientific arrogance and benighted irresponsibility wouldn't be so easily tolerated today.

Science fiction author Robert Silverberg has stated that scientists appearing in science fiction tales prior to 1941 (i.e., when writer Robert Heinlein, 1907–1988, entered the field) are typically "stereotypes of the mad-scientist era."[63] Well, for the most part they may be, but we may also perhaps conclude that in Verne's seminal *Journey* (novel) scientists are also fearful instructors (and reconstructors) of the Earth's past. Half a century later, Conan Doyle's *Lost World* paleo-scientists evolved into overconfident optimists, meddling with natural history just to suit their evolutionary ideals. Then, by the early 1950s, world circumstances had changed considerably through the threat of atomic power, and consequently so did the idea of the mad scientist persona, especially as reflected in Toho's classic *Gojira*. Increasingly, paleontologists take a back seat to the sinister doings and overruling impact of physicists, nuclear scientists and chemists.

The paleo-scientists in 1950s giant paleo-monster films are correspondingly reduced to uncertain, fearful caricatures of their former (literary) selves. Rather than optimists, while they do still instruct about the past and the encountered prehistoria, they're warily projecting certain doom, angst concerning mankind's future. Death is when the giant monsters come! And we also increasingly see the paleo-scientists of such movies pitted against mad-scientists trained in other disciplines, usually atomic physics or chemistry, wielding powers beyond their control. Symbolically, the latter have opened Pandora's box, spawning monsters of metaphoric doom upon mankind, menaces which not even a Challenger or a Lidenbrock (or prototypical Dr. Frankenstein) could repel from our shores.

Perhaps a key difference between prototypical, grandiose figures from *literature*, like Lidenbrock, Challenger and their 1950s counterparts—*filmic* paleontologist giant monster explainers, is that the latter fulfilled their roles at a time when individuals had reason to look askance at science, which by then had brought us the means of total annihilation: atomic and hydrogen bomb weaponry. Thus, for their own good, paleontologists appearing in films such as *Gojira* and *The Beast from 20,000 Fathoms* act more subdued as they soberly, solemnly and resignedly predict the fate of our world, extrapolated from the very presence of these horrible monsters wielding such awful powers.

Geologists and paleo-scientists don't appear in all the early prehistoric monster films—take *King Kong* (1933), for example, where Carl Denham is as close to a mission scientist as it comes. In the 1932 novel Ann Darrow, not a scientist, uses her intuition to declare that Skull Mountain's [*sic*] fauna may be prehistoric. Yet in such examples there's usually enough banter and dialog among the menaced (non-scientist) characters to convey the message (as if we couldn't already tell!) that we're dealing with a *prehistoric* monster. But when the monsters become ever more threatening, then it's time for mankind to send in our most potent weapon—the metaphorical, esteemed professional scientist, who so often is mad to boot! Whereas awe and majesty of science was projected through Verne's and Conan Doyle's triumphant revelation of prehistoric monsters, by the 1950s, occurrences of giant and more powerful prehistoria were much more menacing, rife with ominous implications.

In 1953's *The Beast from 20,000 Fathoms* (predating Toho's *Gojira*, 1954), the mix of scientists becomes a bit more complicated and arguably more interesting than we've so far seen with the two prototypical novels (where *two* men of scientific background joined each of Verne's and Conan Doyle's expeditions into prehistory). In *Beast* we enjoy interactions between physicist Professor Thomas Nesbitt (played by Paul Christian), the world's foremost paleontologist — Dr. Thurgood Elson (played by Cecil Kallaway), and his pretty assistant Lee Hunter (played by Paula Raymond). First, Nesbitt, a government scientist, is involved with experimental testing of a nuclear weapon unleashing enormous Rhedosaurus from its Arctic confines and long slumber. So after he spies the huge saurian growling and strutting around the icy terrain in the midst of a blizzard, he's judged mad by his colleagues. A psychologist declares Nesbitt victim of a traumatic hallucination, even though Nesbitt is certain of his observation.

So Nesbitt seeks out kindly old Dr. Elson. Nesbitt suggests that at the end of the Mesozoic era when the planet's seas turned icy cold (which was truly once regarded as a possible cause for the Late Cretaceous extinctions), animals like the monster he's seen would have been trapped in the ice. Then the atomic bomb explosion could have freed the monster. But realizing large dinosaurian creatures expired one hundred million years ago wouldn't have survived such a prolonged hibernation, Elson pooh poohs Nesbitt's silly and outlandish theory — that is, even after his own assistant, Lee Hunter, comes rushing to Nesbitt's defense with a story of frozen mastodon cadavers recently found in the Arctic. Recalling 1925's *Lost World* allosaur and sauropod skeletons mounted in Zoological Hall, this dialog takes place in front of two dinosaurian skeletons. A *T. rex* skeleton stands behind Professor Nesbitt while he's speaking, and in more panoramic shots we see a curious brachiosaur-like looking quadrupedal fossil being assembled. The latter is later described as being an evolutionary ancestor of the animal that will soon terrorize New York City. Simply morph the two fossils together and you've got the skeletal essentials melded into a Rhedosaurus. Relying on the principle that seeing is believing, such images lend a sense of authenticity to their discussion. Well, a physicist can perhaps be forgiven for theorizing outside the realm of his own expertise, but Dr. Elson won't permit such stuff and nonsense.

Now the importance of dinosaur *images* becomes vital to the plot. For, perhaps because Nesbitt is handsome, yet ostensibly because she has a "deep abiding faith in the work of scientists," Lee Hunter invites him over to her apartment to look through a pile of pictures of "all the known prehistoric animals," a questionable undertaking which wouldn't have taken quite as long then as it would today. After all, though, as Hunter suggests even if the animal he's seen is (more likely) an unknown one, "Just think of what it would mean if you were right." Well, as Dr. Elson muses quixotically in Lyellian terms, the "future is a reflection of the past." Strangely, yet predictably,

In *The Beast from 20,000 Fathoms*, scientists identify the reawakened, soon to be plundering Rhedosaurus from an artist's drawing. This filmic practice of identifying or otherwise illustrating offending prehistoria on the basis of "paleoart" began with *The Lost World* (1925), becoming cliché by the time *Gorgo* (1961) was released. But this on-screen, "disbelief-suspending" ploy became most pivotal in 1953's *Beast*.

Nesbitt actually identifies the prehistoric animal in a stack of pictures, prompting Hunter to rationalize that if another observer of the monster could *independently* verify this illustration, that might be proof enough of its existence. Fortunately, there *is* another observer, a sailor deemed insane whose ship and mates were destroyed allegedly by a "sea serpent," who later selects the same picture from Hunter's assorted images in Dr. Elson's office, corroborating Nesbitt's story.

Quaint Dr. Elson, oblivious to the fact that he could lose his reputation supporting nonsensical ideas, completely unravels from practicing paleontology rationally, drifting toward the fringe of reckless mad science. Elson is convinced beyond any shadow of doubt after he notes an inexorable chronological pattern to the observations of sea serpents and marine catastrophes (including destruction of a lighthouse) since the time of Nesbitt's initial observation. Furthermore, the sea monster seems headed toward the Hudson submarine canyons, where its only known fossils have been found! No longer a mere fuddy duddy, Elson fatefully enters the diving bell in the name of science, proclaiming that he's "leaving a world of untold tomorrows for a world of countless yesterdays." Does mankind's fate rest in the likes of such men? Don't we need a fitter knight to slay the dragon?

Moments before being crushed within the Rhedosaurus' jaws, Dr. Elson deduces that the monster is a "paleolithic survival" and then lapses into paleo-techno-blather about how the dorsal is singular not bilateral, simply to remind the audience that, he really knows his stuff. Yet, he continues, the "most astonishing thing about it is"— then silence; sigh, we'll never know. Our chief paleontologist has died and now only two scientists remain to battle the monster. And it is Nesbitt — wielder of radiation and presumably chemistry for better living — who in a quantum leap pronounces how it may be killed, using a radioactive isotope to destroy plague-releasing diseased tissue and hence the creature itself. If only destroying Godzilla were that easy! Following *Beast*, paleontologists fulfill important, prescient yet secondary roles in deference to scientists having other expertise and whose knowledge factors more forcefully into destruction of the monsters.

Later with the release of *The Giant Behemoth* (1959), we find scientists' anti-nuclear messages flourishing although conspicuously at the expense of paleontological intrigue. Here, Dr. Steve Karnes (played by Gene Evans) warns of harrowing radiological bio-magnification circumstances. This is caused by fallout from testing of ecologically adverse nuclear weapons, ultimately manifesting itself into a radiation-spewing paleo-monstrosity known as Paleosaurus. But while Karnes and British colleague Dr. James Bickford (played by André Morell) approach the problem soberly and somberly, offering false hope through laboratory scientific methods, this time, natural history museum Curator and paleontologist Dr. Sampson's (played by Jack MacGowran) quirky, cornball character suggests how hopelessly out-of-control the giant paleo-monster situation has become. Whereas in their prior fictional roles, Professors Yamane of Godzilla fame and Rhedosaurus explainer Elson at least had the decency to sound authoritative and sincere, instead Sampson seems completely out of touch with dire circumstances threatening modernity. Sampson is no Lidenbrock.

Karnes has command of startling facts and ominous figures, suggesting that seabirds consuming fish tainted with radioactivity bio-magnify radiological concentrations in their tissues by a factor of five hundred thousand times relative to seawater. Further, while Karnes presages a "biological chain reaction — a geometrical progression of deadly menace," poor Dr. Sampson only gibbers about how his colleagues would laugh at him if they knew his cryptozoological leanings. While Karnes soberly lectures on the 143 atom bomb testings conducted so far and warns of the dangers of dumping nuclear wastes in the ocean, benighted Sampson can't tell whether a photo of a huge footprint is fresh versus fossilized. Then Sampson ridiculously intones that animals of the old Paleosaurus family are electric like an eel and that they instinctively migrated to the "shallows which gave them their birth." How can such a kook possibly know such things? Ultimately, lunatic Sampson fulfills his obligatory movie function by producing a picture of the Paleosaurus for everyone to look at, which predictably turns out to be the same animal foreshadowed by Karnes' ecological concerns.

Even after the first Paleosaurus is killed, we have no one to turn to and no place to hide from the monster, apparently, because now a visit on the opposite side of the Atlantic from another member of the old Paleosaurus family has the Americans whipped into a frenzy. While the paleontologists are evidently hapless to stem the tide against giant paleo-monsters, now not even a coterie of atomic scientists, however penitent for their colleagues' misdeeds, can save us now from the invasion.

Tsutsui states in his *Godzilla on My Mind* (2004) that in the Godzilla series of films, "the answer to that fundamental concern of science fiction — can one really trust those people with specialist knowledge and with broad institutional responsibility to be competent, honest, humble and cautious? — is a resounding maybe. Go ahead and put your faith in the men in lab coats, business suits and uniforms, but don't ever turn your back on them."[64] But can we trust the emboldened, maniacal enthusiasm of paleontologists Lidenbrock, Challenger, or the prophetic resignedness of Yamane, Sampson and the rest who warn of ecological dangers wreaked by lab-coated men and who also inform us of monsters spawned by man's sinister knowledge? To a greater degree — yes, although since the advent of prehistoric monster story-telling their role has degraded from pompous know-it-alls into, at best, helpless and sometimes ridiculous personae. As we learned in the previous chapter, Dr. Serizawa died for our sins, remember? Death is when the monsters come — but *who* are the monsters?

It's been over 80 years since audiences thrilled to the sight of O'Brien's and Delgado's silver screen *Lost World* phantasgamoria. Since then, the United States has suffered several major wars, and issued an atomic assault on Japan. The mighty, symbolic prehistoric monsters, once horrific, now seem rather tame in comparison to risks we now face in daily reality, and accordingly, the movie industry offers bloodier thrills—for those who dare. As stated in *Spawn of Skull Island*, "we as a people have grown increasingly difficult to shock or thrill in the decaying generations since *The Lost World* (1925)" and 1933's *King Kong/Son of Kong* combo.[65] Furthermore, "the monsters of paleontological discovery and heroic imagination had become irrelevant to a society coarsened by the more pressing and inescapable horrors of Vietnam, Charles Manson and the Son of Sam and Jeffrey Dahmer, and even the confrontational quasi-escapism of such picture-show anti-personalities as ... Jason Voorhies and the spectral pervert Freddy Krueger." Or as *Alley Oop* cartoonist Vince T. Hamlin (1901–1993) stated in 1969–1970, "Nobody's interested in prehistory or dinosaurs or imaginative science anymore."[66]

But there's no need to grieve the loss of our most beloved prehistoria — Great-Claw, Mastodon, "Bronty," Rex and the others— because in the *Jurassic Park* era of dino-monster moviedom, Hamlin's statement must be refuted. For within a handful of years following Hamlin's lament, America entered its Dinosaur Renaissance scientific period, flowering throughout the 1980s, perhaps culminating at a pop-cultural level during the 1990s and thereafter with products such as Michael Fredericks' *Prehistoric Times* magazine for the masses devoted to paleoart; new, well-illustrated paleontological books and documentaries by the score; Michael Crichton's and Steven Spielberg's sensational *Jurassic Park* industry; and Universal's tremendous *King Kong* (2005) remake with its incredible CGI special effects, arguably conveying the central beauty and the beast theme more effectively than in the original. Dino-monsters and their prehistoric brethren are definitely in today like never before.

The old school time stands still theory relied upon by Conan Doyle in 1912, suspending disbelief through purportedly explaining why primitive organisms still exist in remote monstrous places, has been tweaked for a generation attuned to late 20th century Darwinian schoolroom biology founded on modernistic survival-of-the-fittest culled by natural selection concepts. Now, as Tony Phillips suggests in his 2006 *Prehistoric Times* article, "Modern Lost Worlds,"[67] prehistoria crafted by modern fantastic authors are resultant of *continually* operating evolutionary processes. So while dinosaurs and other prehistoric animals of Kong's new millennial (i.e., as perceived post–2004) Skull Island environment may in outline *resemble* fossilized creatures dug up by paleontologists, they're not necessarily the *same* species.

Furthermore, we find that during the 1990s and in the new millennium, movie makers and writers resorted to their old bag of tricks, time-honored devices, while incorporating a sounder science framework for the '90s and beyond with which to suspend disbelief in the existence of dino-monsters alive in modernity. The new films are a tasteful, wedded mix of something old, something borrowed and something new. As previously noted, in homage to Charles Knight's (and Cope's) famous 1897 painting of leaping Laelaps, theropods *Allosaurus* and *Tyrannosaurus* infiltrated their kangaroo-attack style of locomotion in *The Lost World* (1925). Perpetuating this tradition, half a century later, for instance, we see Toho's nondescript theropodian, Gorosaurus, leaping several times at Kong's chest in *King Kong Escapes* (1967). Even colossal Godzilla, ensconced upon his Monster Island resort, has occasionally and incongruously performed incredible feats of leaping toward enemies, as in *Godzilla vs. Megalon* (1976).[68]

And likewise, today, *Jurassic Park's* forlorn island pair are infested with leaping Raptors (i.e., *Velociraptors*), as is the case with an updated rendition of Kong's Skull Island. For what might passingly appear to be a pack of hungry, vicious Laelaps menaces and kills members of Carl Denham's exploration team in *King Kong* (2005). Bounding raptor-like dinosaurs also prove significant inhabitants of Skull Island as perceived by Joe DeVito, Brad Strickland and John Michlig, authors of *Kong: King of Skull Island* (2004).[69] Symbolically, in *Skull Island*'s triumphant climactic battle between the sentient Queen Raptor dino-monster named Gaw, Kong stabs a Tops horn into through Gaw's neck. A derivative Knight-inspired, if not downright draconian, scene!

In another Kong-related publication, *Kong: A Natural History of Skull Island* (2005) — one of several *King Kong* (2005) spinoffs — the ornately horned variety of dinosaur inhabiting Skull Island is named Ferrucutus (although it resembles the extinct *Styracosaurus* that appeared in 1933's *Son of Kong*). Much of Skull Island's curious, hypothetical fauna is illustrated and explained in *Kong: A Natural History*.[70] While in this exquisitely illustrated book, the dinosaurs are the primary, dominant fauna, other (non-dinosaurian) prehistoria are conjured into existence by artists and authors: truly a distorted natural history. Yes, leaping raptor-like dinosaurs dwell here as well.[71]

Not unlike the influential, symbolic Knightian Rex versus Tops battle, the dinosaur leaping motif (first observed fictionally on prehistoric monster islands), the dinosaur stampede, the oft-encountered pair of fighting filmic dinosaurs falling over a cliff, the dinosaur on the rampage in modern civilization, and the lost prehistoric monster island with its climactic erupting volcano have all become traditional clichés of the dino-monster movie industry, handed down from one generation to the next since the 1920s, when dinosaurs gained their first wave of heightened popularity.

Tyrannosaurs, giant (prehistoric) apes, those ubiquitous raptors (especially feathered varieties fortifying the avian-dinosaur evolutionary link), and a host of other prehistoria predictably terrorize humans stranded on prehistoric monster islands! Where would the annals of science fiction be without prehistoric monster worlds and their most ancient fauna, remarkably preserved for mankind's symbolic destruction, cruelly exploited by those incorrigible, meddling mad scientist-geologists?

Thanks to Davy's archetypal mad scientist persona and Charles Lyell's awful — no, *wonderful*—changes, we may embrace them all!

Appendix:
The Most Famous
Prehistoric Monsters
in Popular Culture

So many and varied kinds of monsters judged prehistoric have appeared in books, pulp stories, movies, and even science popularizations through the decades! Rather than attempting to list every filmic and literary circumstance — so difficult to keep up with in current markets!— here is a listing of, arguably, yet simply, the most historically *famous* prehistoria in popular culture. I am omitting alien dinosaurs and other exo-prehistoria inhabiting lost worlds in outer space because of their generally uncertain origins and exotic natural history. However, readers can enjoy chapter six ("Dino-Trek") of my *Dinosaurs in Fantastic Fiction* for more on those ever-popular alien prehistoria. I am also excluding creatures representing modern species mutated through exposures to radiation or chemical wastes into larger, more grotesque forms that outwardly appear prehistoric. For brevity, each creature's most popular or influential pop-cultural roles and appearances are indicated. I simply can't list each case or provide further details about each occurrence here. If you consult Mark Berry's fantastic *Dinosaur Filmography* (2002), however, you can read more about your favorite dinosaurs in film. Consult works by J. D. Lees and Steve Ryfle listed in the Bibliography for more on the Godzilla movies. Jeff Rovin's *The Encyclopedia of Monsters* (1989) is another valuable source for other odd species. (Because I've striven to keep this presentation as concise as possible, my descriptive 2006 *Scary Monsters* fanzine articles, listed in the Bibliography, might also be of service. Further information on many of these entries may also be gleaned from my *Dinosaurs in Fantastic Fiction*.) The object here is, beyond alphabetized entries, to direct readers to the best or most representative movies and books in which these prehistoria made their monstrous appearances; their grandest or most definitive exploits! In deciding whether to include particular dino-monsters and prehistoria deemed more obscure, I've strayed conservatively.

Allosaurus: *One Million Years B.C.*, 1966 (Hammer Films).

Ammonite: *Mysterious Island* (Columbian/Ameran, 1961); that gigantic squid-like, marine beast with the coiled shell.

Anguirus: *Gigantis the Fire Monster* (Toho, 1959); Quadrupedal Angilas (aka Angurus, Angilas, or Anzilla) is about 200 feet high with a length from 328 to 500 feet. See Chapter Nine for more on this horned ankylosaur.

Antarctic Giants: Colossal dinosaurians found by scientists surviving on the Antarctic continent, appearing in John Taine's sensational 1929 novel, *The Greatest Adventure*.

Ape Gigans: A 14-foot-high antediluvian gorilla added to 1870s English translations of Jules Verne's *Journey to the Center of the Earth*.

Ape-men: See Chapter Five.

Arctic Giant: Godzillean tyrant lizard on the loose in Superman's Metropolis in a 1943 Paramount cartoon directed by Dave Fleischer.

Baragon: *Frankenstein Conquers the World* (Toho, 1966); a 150-foot long, large-eared (like "Gorgo") creature with armor carapace.

Belshazzar: Huge meat-eating dinosaur featured in John Taine's 1934 novel, *Before the Dawn*.

Bernie: Ingenious spoof on Barney the purple daytime TV dinosaur — a raptor dinosaur appearing in Ian McDowell's 1994 story.

Brontosaurus: See chapters Six and Ten.

Carcharodon: An enormous 60+-foot long Miocene shark often featured in stories such as Steve Alten's 1997 novel *Meg*.

Carnotaurus: *Dinosaur* (Walt Disney, 2000), *Dinosaur Valley Girls* (Frontline Entertainment, 1996); a real twin-horned South American carnivorous brute also appearing in Michael Crichton's 1995 novel, *The Lost World*.

Cave Bear: *Son of Kong* (RKO, 1933); an evolved, blind variant of this creature appears in Arthur Conan Doyle's 1910 story, "The Terror of Blue John Gap."

Cave Man: See Chapter Five.

Ceratosaurus: *One Million Years B.C.* (Hammer/20th Century–Fox, 1966), *Journey to the Beginning of Time* (Czech, 1955); see chapters Six and Ten.

Chasmosaurus: *When Dinosaurs Ruled the Earth* (Hammer, 1970).

Compys: *The Lost World: Jurassic Park* (Universal/Amblin, 1997); venomous, crow-sized, bipedal meat-eating dinosaurs (*Procomgsognathus*).

Crater Lake Monster: *The Crater Lake Monster* (Crown International Pictures, 1977); a 25-foot-long, aquatic Plesiosarus.

Creature: *The Creature from the Black Lagoon* (Universal, 1954); also known as the Gillman, and America's most revered suitmation monster. Evidently a Devonian, living fossil anthropoid. See Chapter Five.

Deadly Mantis: *The Deadly Mantis* (Universal, 1957); a 150-foot-long, "million-year-old" giant species of Praying Mantis unleashed from an iceberg following a volcanic eruption.

Deinonychus: *Carnosaur* (Concorde/New Horizons, 1993); a 9-foot-long bipedal Early Cretaceous raptor dinosaur, first vivified in the annals of science fiction in David Gerrold's 1978 novel, *Deathbeast*.

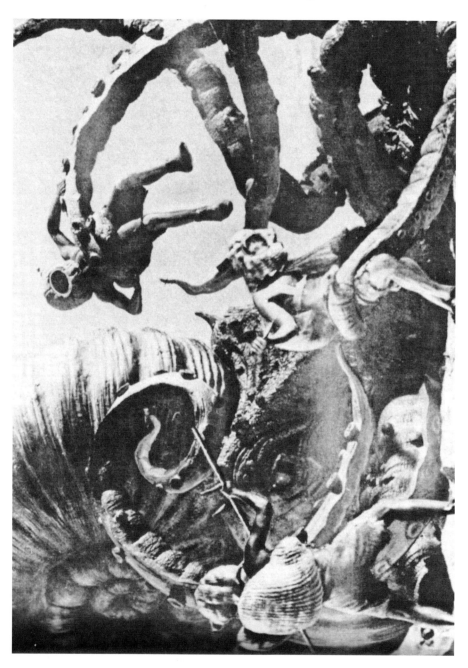

Deep sea divers are menaced by Ray Harryhausen's stop-motion animated ammonite in *Mysterious Island* (1961). Ammonites lived predominantly during the Mesozoic Era, although they never grew to sizes such as shown in this movie still.

Deinotherium: *Journey to the Beginning of Time* (Czech, 1955); a large elephantine creature with down-turned tusks (unlike Mastodon). Fossilized bones found by ancient Greeks belonging to this genus were mistaken for the remains of mythological giants.

Diatryma: *The Ghost of Slumber Mountain* (World Cinema Distributing Company, 1919); a large, flightless Tertiary bird. More recently, this monster also appeared in Stephen Baxter's 1995 novel, *The Time Ships*, as well as in Penelope Banka Kreps' 1993 thriller, *Carnivores*.

Dilophosaurus: *Jurassic Park* (USA–Universal/Amblin, 1993); Michael Crichton's (*Jurassic Park*, 1990) famous spitter, has a voracious appetite.

Dimetrodon: *Journey to the Center of the Earth* (20th Century–Fox, 1959); see Chapter Six.

Dragon Fly: *Monster on the Campus* (Universal, 1958), *Godzilla vs. Megaguirus* (Toho, 2000); giant dragon flies assaulted a Cro-Magnon hunter in Charles G. D. Roberts' *In the Morning of Time* (1919).

Dragons: *Dragons: A Fantasy Made Real* (Animal Planet, 2005); rather outside the scope of this volume. However, see Chapter One for historical context.

Elasmosaurus: *King Kong* (RKO, 1933), *The Land Unknown* (Universal-International, 1957); Among its many appearances, this aquatic monstrosity also terrorized victims in Harry Adam Knight's 1984 novel *Carnosaur*. Also see Chapter Six.

Fire Monster: *The Lost World* (20th Century–Fox, 1960); strangely tagged a tyrannosaur by Professor Challenger (played by Claude Rains) in Irwin Allen's production, this creature enjoys splashing in lava. This animal was re-dubbed an "Irwinosaur" by Mark Berry.

Fog Horn Creature: The amphibious, titular dino-monster, which destroys a lighthouse in Ray Bradbury's famous 1951 short story, "The Foghorn" originally titled "The Beast from 20,000 Fathoms." In pop-culture, ancestral to Ray Harryhausen's Rhedosaurus.

Gamera: *Gammera the Invincible* (Daiei Motion Picture Co., 1965), *Gamera — The Guardian of the Universe* (Daiei, 1995); a 200-foot long flying, fire-breathing turtle with upwardly jutting tusks released from Arctic ice.

Gappa: *Gappa — The Triphibian Monster* (Nikkatus Corp., 1967); a 300-foot-tall, beaked and winged reptilian gains its name from its ability to fly, walk bipedally on land and even swim in the sea. A retelling of Rodan's and Gorgo's stories.

Gargantosaurus: See Chapter Nine.

Gaw: See Chapter Ten.

G-Fantis: A usually quadrupedal, godzilla-sized dinosauroid created and edited by J. D. Lees for his magazine, *G-Fan* (obviously named to honor the many prehistorical *daikaiju* and their avid fans).

Giant Claw: *The Giant Claw* (Clover Productions/Columbia, 1957); a mangy, giant bird, possibly resulting from a terrestrial ice age 17 million years ago (yet also possibly derived from an antimatter galaxy).

Gigantis: *Gigantis the Fire Monster* (Toho, 1959); the second Godzilla. See Chapter Nine.

Glyptodont: *One Million B.C.* (United Artists, 1940); large, armored mammalian herbivores, often with spiky war clubs at the ends of their tails. A glyptodont also appears in Penelope Banks Kreps's pulp 1993 thriller, *Carnivores*.

Godzilla: *Gojira* (Toho, 1954); see Chapter Nine.

Godzillasaurus: *Godzilla vs. King Ghidorah* (Toho, 1991); in Godzilla's second origin myth, Godzilla is the result of a 1950s hydrogen bomb detonation causing mutation of a 40-foot-tall tyrannosaur,

John Ryder's restoration of a Pleistocene glyptodont, *Panochthus*, described by Richard Owen in 1847. Note the rigid tail weaponry; other glyptodont genera such as *Doedicurus* possessed spiky "ankylosaurian" war clubs (from *Popular Science Monthly*, June 1878).

which has survived on Lagos Island in the Pacific.

Gorgo: *Gorgo* (King Brothers/MGM, 1960); an iconic 200 foot tall *daikaiju* mother dino-monster, a manifestation of primordial evil who comes after her son following its capture in the North Sea near Nara Island. The idea was remade, conceptually, by Nikkatsu in *Gappa*.

Gorosaurus: *King Kong Escapes* (Toho, 1967); a nondescript, bipedal theropod dinosaur killed by Kong on Mondo Island. See Chapter Ten.

Great-Claw: See chapters Two and Ten.

Gwangi: *The Valley of Gwangi* (Warner Bros.–Seven Arts, 1969); a 15-foot-tall crypto-species of *Allosaurus* captured in Mexico's desert wilderness, brought to life through the talents of Ray Harryhausen.

Gyaos: *Gamera vs. Gaos* (Daiei, 1967); although featured as a nocturnal, 200-foot tall blood-drinking bird, its pterodactyl features seem more prominent. Daiei's equivalent to Toho's Rodan.

Ichthyosaurus: See Chapter Six.

Iguanodon: See chapters Six and Ten.

Irwinosaurus: A term apparently coined by Mark Berry referring to several movie dinosaurs created as special effects by adding horns, sails and other appendages to live lizards and reptilians for Irwin Allen's *The Lost World* (20th Century–Fox, 1960).

Kong: See chapters Six, Eight, Nine and Ten.

Jurassic Punk: *Dinosaur Valley Girls* (Frontline Entertainment, 1996); a trademark, customized allosaur menacing prehistoric cave-babes and cavemen.

Laelaps: (Scientifically now known as *Dryptosaurus*.) See chapters Six, Eight and Ten.

Loch Ness Monster: (aka "Nessie"); *Loch Ness* (Polygram, 1996); as many tourists have remarked, "It's much more fun to believe there is a plesiosaur here than that there isn't." See Chapter Ten.

Mahars: Edgar Rice Burroughs' monstrous 8-foot-tall, winged pterodactyls

A drawing of one of Irwin Allen's nondescript pseudo-dinosaurs contrived from the addition of horns and a dorsal fin glued-on to a live caiman alligator that appeared in *The Lost World* (1960).

that inhabit the Earth's inner surface introduced in his delightful 1914 novel, *At the Earth's Core*. Mahars are absolute rulers of the Saurozoic world of a "million years ago" known as Pellucidar.

Mammoth: See Chapter One.

Mastodon: See chapters Three, Four and Ten.

Megalania: Imagine a *giant* form of Komodo Dragon which grew to 26 feet in length and may have preyed on aborigines in prehistoric Australia two million years ago. *Megalania* appeared in L. Sprague de Camp's 1993 time travel story, "The Honeymoon Dragon."

Megalosaurus: See Chapter Six.

Megatherium: See Chapter Two.

Mosasaurus: See Chapter Six.

Newts: Prehistoric "living fossil" descendants of a Miocene marine anthropoid/

amphibian race soon conquer the world in Karel Capek's 1936 novel, *War with the Newts*. After the newts defeat the human race, they turn their arms against one another, repeating mankind's mistakes.

Nightstalkers: A vicious breed of raptor dinosaurs infesting the late Cretaceous Period in Will Hubbell's 2002 novel, *Cretaceous Sea*. These cunningly intelligent, feathered dino-monsters stalk a pair of human time-travelers who have survived a world-altering asteroid impact.

Nitolans: In Barry Longyear's 1989 novel, *The Homecoming*, a fleet of dinosaurs has returned to 20th century Earth ("Nitola") after 70 million years to reclaim their planet. Nitolans decide to let humanity self-extinguish over the course of a few centuries, rather than destroying us with their insurmountable weaponry.

Ogra: The 200-foot-tall mother *daikaiju*

dino-monster, a manifestation of primordial evil, which tramples London, mocking British military forces in its relentless pursuit of her 60-foot-tall son (aka billed as "Gorgo"), stolen from the waters off Nara Island. Ogra was the name author Carson Bingham assigned to the lead monster in his 1960 pulp novel, *Gorgo*, based on the movie script for the later released film, *Gorgo* (King Brothers/MGM, 1961).

Paleosaurus: *The Giant Behemoth* (David Diamond Productions/Allied Artists, 1959); see chapters Nine and Ten.

Plesiosaurus: See Chapter Six.

Pteranodon: *King Kong* (RKO, 1933; *Rodan* (Toho, 1956); Perhaps the most frequent flyer out of Mesozoic skies into pop-cultural filmic and literary prehistoric adventures.

Pterodactyl: *One Million Years B.C.* (Hammer/20th Century–Fox, 1966); see Chapter Six.

Pteronychus: In Thomas Hopp's novels, *Dinosaur Wars* (2000) and *Dinosaur Wars: Counterattack* (2002), an intelligent, space-faring feathered dinosaur species, which hightailed it for the Moon 65 million years ago to avoid the asteroid impact extinctions, return to conquer Earth.

Quetzalcoatlus: *Q* (United Films, 1982).

Raptors: This term, referring to DNA-bioengineered *Velociraptors*, vaulted into dino-lore following release of the influential *Jurassic Park* (1993). Springing on *JP*'s lively heels, soon other raptors permeated through the filmic and literary annals of dinosaurabilia. See Chapter Ten.

Reptar: Reptar is the cartoon television monster watched in terror by *The Rugrats* kids in episodes of a popular 1990s cartoon. Reptar resembles a towering tyrannosaur, but destroys cities much like the famed *daikaiju* monsters of filmdom.

Reptilicus: *Reptilicus* (Saga Studio, 1962); the story is more somberly portrayed in Dean Owen's 1961 novel from the script, *Reptilicus*.

Rhedosaurus: See Chapter Ten.

Rodan: See Chapter Nine.

Sagoths: These are eight-foot-tall, primitive, gorilla-like humanoids that dwell in Pellucidar in Burroughs' *At the Earth's Core* (1914), and sequels. Sagoths are enslaved by the Mahars with whom they communicate using a strange sixth sense transmitted through the "fourth dimension."

Shadow: In H. P. Lovecraft's *The Shadow Over Innsmouth* (1936), these are a race of scaly, bipedal anthropoids ("blasphemous fish-frogs") that have survived since "Palaeogean" times (at least since 80,000 years ago) inhabiting coastal waters along the northeast Atlantic. When their piscine features become most pronounced, they resemble the "Gillman," anticipating Universal's *The Creature from the Black Lagoon* (1954).

Shark-crocodile: Dating from an 1870s English edited translation of Jules Verne's *Journey to the Center of the Earth*, Verne didn't author passages inserted into the novel involving this mosasaurian beast. Shark-crocodile is pitted against Ape-gigans in a short dream sequence.

She-creature: *The She Creature* (American International, 1956); a woman becomes a reincarnation of a prehistoric spirit, which lived in an amphibious humanoid one million years ago. Under hypnosis, her She Creature persona walks the Earth again.

Silurians: These are seven-foot-tall reptilian humanoids starring in 1970s *Dr. Who* (BBC) episodes that hid inside the Earth since the Cretaceous Period, when the Moon was (supposedly) torn from Earth's Pacific basin. In Malcolm Hulke's 1983 novel, *Dr. Who and the Cave Monsters*, the Silurians—the first intelligent life form to evolve on Earth — are lured to the surface after humans conduct in-ground electrical resistivity experiments in England.

Sleeping Dragons: In a 1988 episode of *Monsters*, written by Michael Reeves, intelligent dinosauroids survive to the present day through suspended animation in a time capsule programmed by radioactivity to open in 65 million years.

Smilodon: *Sinbad and the Eye of the Tiger* (Columbia, 1977); The most famous prehistoric feline of all. Jeff Rovin's 2000 novel, *Fatalis*, featured a sabertooth.

Spinosaurus: *Jurassic Park III* (Universal/Amblin, 2001).

Stegosaurus: See chapters Six and Ten.

Styracosaurus: *Son of Kong* (RKO, 1933).

Terror Bird: *Mysterious Island* (Columbian/Ameran, 1961); "Terror Birds" lived principally in South America some 15 million years ago, with some genera living into the Pleistocene. The most popular example, *Phororachos*, a six-foot-tall dino-avian debuted in Conan Doyle's 1912 novel, *The Lost World*. Another genus, *Titanis*, was featured in James Robert Smith's 2006 novel, *The Flock*.

Thakdol: Lin Carter introduced this giant variety of pterodactyl in his 1979 novel *Journey to the Underground World*, which inhabits the inner–Earth realm known as Zanthodon. A winged Thakdol carries off a heroine as in Ray Harryhausen's scenes for *One Million Years B.C.*

Thipdars: These 40-foot long pterodactyls — referred to in Burroughs' *Tarzan at the Earth's Core* as pteranodons — inhabit the inner–Earth world of Pellucidar.

Titanosaurus: *Terror of Godzilla* (Toho, 1975).

Triceratops: See chapters Six and Eight.

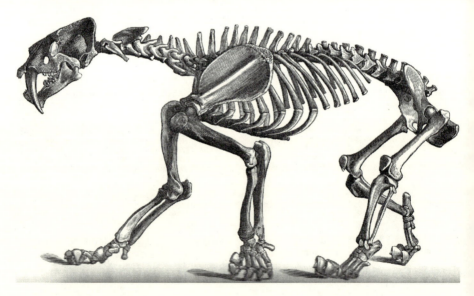

Smilodon, is one of the most famous prehistoric mammalian vertebrates associated with the Pleistocene age (from Joseph Le Conte's *Elements of Geology: A Text-Book for Colleges and for the General Reader.* New York: D. Appleton and Company, 1903; author's collection).

Alice B. Woodward's restoration of the *Stegosaurus,* published in Henry R. Knipe's *Evolution in the Past* (New York: J. B. Lippincott Company, 1912; photograph from the volume by the author).

Trodon: A 15-foot-tall, pterodactyl-like, winged reptilian marsupial (a "winged kangaroo") that first appeared in Burroughs' 1937 novel, *Before the Stone Age.*

Trog: *Trog* (Warner Brothers, 1970); the titular monster is a primitive human living in a cave in the British countryside.

Troodon: Similar to *Jurassic Park*'s raptors, a sentient *Troodon* pack attacked a USS *Enterprise* landing party in Diane Carey's and Jim Kirkland's 1995 novel, *Star Trek: First Frontier.*

Tyrannosaurus: See chapters Six and Eight.

Varan: *Varan the Unbelievable* (Toho, 1961).

Vastatosaurus: (aka "V. Rex")

King Kong (Universal, 2005); see Chapter Ten.

Terror bird, *Andalgalornis,* sculpted by the author, 2001.

Velociraptor: *Jurassic Park* (Universal/ Amblin, 1993); Michael Crichton, *Jurassic Park* (1990).

Vermithrax: *Dragonslayer* (Paramount Pictures & Walt Disney Productions, 1981); a pterodactyl-like winged dragon.

Xinli: Predatory dinosaurs of a lost world realm relied upon by lizard men for sport in Abraham Merritt's 1931 novel, *The Face in the Abyss.*

Yilane: Dastardly descendants of mosasaurs which share the planet with primitive humans in Harry Harrison's trilogy of novels, beginning with his *West of Eden* (1984). A premise of Harrison's parallel universe setting is that the dinosaurs never were extinguished 65 million years ago.

Yongary: *Yongary, Monster from the Deep* (Toei/American International Television, 1967).

Yunga Beast: In Willis Knapp Jones' 1927 short story, "Beast of the Yungas," an aquatic *Diplodocus* still roaming the Amazon literally frightens a man to death.

Chapter Notes

Introduction

1. F.W. Holiday, *The Great Orm of Loch Ness* (New York: Avon Books, 1969), p. 20.

2. Ralph O'Connor, *The Earth on Show: Fossils and the Poetics of Popular Science, 1802–1856* (Chicago: University of Chicago Press, 2007), p. 329.

3. Ibid., pp. 328–329.

4. W. Douglas Burden, *Dragon Lizards of Komodo: An Expedition to the Lost World of the Dutch East Indies* (New York: Putnam's, 1927), p. 7; A.A. Debus, *G-Fan* no. 88, pp. 26–30, 2009.

5. Stephen J. Gould, *Dinosaur in a Haystack: Reflections in Natural History* (New York: Harmony Books, 1996), p. 223.

6. Adrienne Mayor, *The First Fossil Hunters: Paleontology in Greek and Roman Times* (Princeton: Princeton University Press, 2000); Charles Gould, *Mythical Monsters* (New York: Crescent Books, 1989; originally published 1886).

7. Allen A. Debus, "From *Incognitum* to Odo: Prehistoric Roar," *G-Fan*, no. 76 (Summer 2006), pp. 38–41.

8. O'Connor, *The Earth on Show*, p. 231.

9. O'Connor states that "Poetry was still a language of authority and geology's status could be buttressed by borrowing its voice." Ibid., p. 345; A. Debus, *Dinosaurs in Fantastic Fiction: A Thematic Survey* (Jefferson, NC: McFarland, 2006), pp. 187 (n. 4), 188 (n. 4).

10. O'Connor, *The Earth on Show*, pp. 195, 280–281, 337.

11. Ibid., p. 197.

12. Ibid., p. 439; here O'Connor stated, "How did geology rise to this position of authority? On one level, simply because it worked.... But the science's power to make the past spectacularly visible had been constructed by its popularizers, above all in literature. During the first half of the nineteenth century, writers associated with the Geological Society's redefinition of geology fashioned new ways of narrating earth history from the ruins of ex-ploded cosmologies. To begin with, these stories were told at the peripheries of the science, in privately-circulated comic verses; but, as the conflict between rival theories became more apparent, writers of both literalist and non-literalist persuasions increasingly courted the burgeoning middle-class reading public. Using the sites and genres that this public was already familiar with, they developed a rhetoric of spectacular display which dovetailed the theatrical tropes of eighteenth-century theories of the earth with the new vogue for virtual tourism and apocalyptic drama, mediated by the culturally authoritative voice of the poet. Geology's truth-claims were build on an uneasy yet highly effective patchwork of image and text, prose and poetry, history and prophecy, fact and fiction. The reader's imagination became a stage, equipped with all the latest technological innovations, on which the invisible was made visible."

13. Ibid., pp. 276–279, 300.

14. Ibid., p. 187.

15. Martin J.S. Rudwick, *Worlds Before Adam: The Reconstruction of Geohistory in the Age of Reform.* (Chicago: University of Chicago Press, 2008), p. 466. Here Rudwick refers to paleoartist "(John) Martin's darkly Gothick interpretation of Mantell's extinct monsters."

16. Ibid., pp. 420–421.

17. Crichton, Michael, *Jurassic Park* (New York: Ballantine Books, 1990), p. 115.

Chapter One

1. Guy Endore, *The Werewolf of Paris*, 1933 (Citadel Press–First Carol Publishing Group ed., 1992), p. 62.

2. Allen G. Debus, *Man and Nature in the Renaissance* (Cambridge: Cambridge University Press, 1978), p. 35.

3. Adrienne Mayor, *The First Fossil Hunters: Paleontology in Greek and Roman Times* (Princeton: Princeton University Press, 2000).

4. Athanasius Kircher, *Mundus Subterraneus* (Rome, 1665 ed.).

5. Mayor, *The First Fossil Hunters*, p. 190.

6. Ibid., p. 141.

7. *Deinotherium* was a fossil elephant. One of the first documented attempts, made by the artisan Pulcher, to restore a fossil vertebrate from a one-foot-long fossilized tooth resulted in a human giant's head, sculpted in clay. See Mayor, *The First Fossil Hunters*, pp. 144–146.

8. Edith Hamilton, *Mythology* (New York: New York American Library, 1942 ed.), pp. 64–65.

9. Note there are also horrible monsters mentioned in biblical Revelations as well as in *The Epic of Gilgamesh*, although to date there is no scholarship supporting such creatures, which may have been inspired on the basis of real fossil bones.

10. Mayor, *The First Fossil Hunters*, p. 163.

11. Ibid., p. 149.

12. Hamilton, *Mythology*, p. 67.

13. Perhaps not unlike lightning emanating from Toho creations' King Ghidorah's three electrified heads.

14. A. Mayor, *Fossil Legends of the First Americans* (Princeton: Princeton University Press, 2005). Mayor's definition of a "fossil legend" is stated on p. xxix, "as a story or belief that relates extraordinary creatures of myth and legend to observations of the mineralized remains of extinct animals, or otherwise attempts to explain fossil traces of prehistoric species, including marine and plant fossils, and the bones, teeth, claws, burrows, nests, eggs, and footprints of extinct animals." Also see Naomi Oreskes' article, "The Humanistic and Religious Foundations of Deep Time," *Science*, Vol. 314 (October 27, 2006), p. 597.

15. Mayor, *Fossil Legends of the First Americans*, pp. 40–45.

16. Ibid., p. 221.

17. Charles Gould, *Mythical Monsters* (originally published 1886, reprinted by New York: Crescent Books, 1989), p. 2.

18. Ibid.

19. For more on dragons, see Richard Carrington's *Mermaids and Mastodons: A Book of Natural and Unnatural History* (London: Chatto and Windus, 1957), pp. 64–77, and an entry for "Dragon" in the *Encyclopædia Britannica* (London: William Benton, 1964 ed.), 7:623–624. William J. T. Mitchell considers the dragon as "the cultural ancestor of the dinosaur. It might be described as the ruling reptile of premodern social systems, associated with kings and emperors, with buried treasure (cp. fossil fuels), and with the fall of dynasties. If the dragon is the object of a knightly quest in which it is killed in order to save the community, the dinosaur is the object of a heroic paleontological quest in which the monster is brought back to life and displayed as a trophy of modern political and economic systems." Mitchell, *The Last Dinosaur Book: The Life and Times of a Cultural Icon* (Chicago: University of Chicago Press, 1998), p. 86.

20. Gould, *Mythical Monsters*, pp. 196–198.

21. Ibid., p. 199.

22. Ibid., p. 233.

23. Ibid., p. 208. Regardless, in most cultures dragons have most commonly represented evil, rather than good.

24. Although beyond the scope of this book, another important early pioneer of the biological sciences was Aristotle 384 B.C.E.–322 B.C.E. For more on Aristotle's biological writings, see Isaac Asimov's *Biographical Encyclopedia of Science and Technology* (Garden City, NY: Doubleday, 1964), pp. 16–18.

25. Gould, *op. cit.*, p. 201; For further perspective on geographical origins of dragons, naturalist W. Douglas Burden (*Dragon Lizards of Komodo: An Expedition to the Lost World of the Dutch East Indies*, New York: Putnam's, 1927, p. 92) noted several decades later that, "...it is just possible the Komodo lizards are the very ones which have given rise to all the dragon mythology."

26. Gould, *Mythical Monsters*, pp. 234–235.

27. Allen A. Debus and Steve McCarthy, "A Scene from American Deep Time: New York's Palaeozoic Museum — Revisited," *The Mosasaur — The Journal of the Delaware Valley Paleontological Society*, Vol. 6 (May 1999), pp. 105–115.

28. Claudine Cohen, *The Fate of the Mammoth: Fossils, Myth and History* (Editions du Seuil, 1994; Chicago: University of Chicago Press, translated by William Rodarmor, 2002 ed.), Chapter 3, pp. 42–43.

29. Martin J.S. Rudwick, *The Meaning of Fossils: Episodes in the History of Palaeontology* (Chicago: University of Chicago Press, 2nd ed., 1985).

30. R. Hooke, *Micrographia, or Some Physiological Descriptions of Minute Bodies Made by Magnifying Glasses* (London: Martyn and Allestry, 1665; Dover Press, 1961). Hooke's *Micrographia* was published a few years before Nicolas Steno's *Prodomus* (1669); in fact Steno, who had read *Micrographia*, shared Hooke's precocious geological perspectives on the origins of fossils and geological strata.

31. Inwood, Stephen. *The Forgotten Genius: The Biography of Robert Hooke, 1635–1703* (MacAdam/Cage, 2002), pp. 118–119.

32. W.N. Edwards, *The Early History of Palaeontology* (London: Trustees of the British Museum, Natural History, 1967), p. 32.

33. Cohen, *The Fate of the Mammoth*, pp. 45–54.

34. Ibid., pp. 51–55; Rudwick, *The Meaning of Fossils*, pp. 30–32.

35. Claude C. Albritton, *The Abyss of Time: Changing Conceptions of the Earth's Antiquity after the Sixteenth Century* (New York: Jeremy P. Tarcher, 1986), chapters Two and Three. Tongue-stones were sometimes the size and shape of human tongues.

36. Cohen, *The Fate of the Mammoth*, p. 45.

37. Ibid., p. 57. Also see Herbert Wendt's *Before the Deluge* (Garden City, NY: Doubleday, 1968), Plate 11 and p. 24.

38. Leibniz had formerly believed in a mystical theories for origin of fossils, but later changed his mind to a more modernistic view. See Cohen, *The Fate of the Mammoth*, pp. 53–54.

39. Ibid., p. 60.

40. Without further discussion here, as they pertain to fossils, "diluvialist" ideas of William Whiston (1667–1752), John Woodward (1665–1728) and Johann Jakob Scheuchzer (1673–1733) would seem rather odd today. See John C. Greene's *The Death of Adam* (Ames: The Iowa State University, 1959), and Cohen, *The Fate of the Mammoth*, for more on these individuals. Another excellent reference is Martin J.S. Rudwick's *Bursting the Limits of Time: The Reconstruction of Geohistory in the Age of Revolution* (Chicago: University of Chicago Press, 2005).

41. Semonin, Paul. *American Monster: How the Nation's First Prehistoric Creature Became a Symbol of National Identity* (New York: New York University Press, 2000), pp. 19–21.

42. Ibid., p. 65.

43. Ibid., p. 66.

44. Ibid., pp. 47–78.

45. Ibid., p. 69. Hiob Ludolphus is further discussed on p. 68.

46. Ibid., pp. 76–77.

47. Ibid., pp. 78–83.

48. Lopez Pinero, Jose. "Juan Bautista Bru (1740–1799) and the description of the genus *Megatherium*," *Journal of the History of Biology*, Vol. 21 (1988), pp. 146–163.

49. These specimens inspired Merian C. Cooper's creation of King Kong. See "King Kong" entry, pp. 252–254 in Richard Milner's *The Encyclopedia of Evolution: Humanity's Search for Its Origins* (New York: Facts on File, 1990); A.A. Debus, *G-Fan* no. 88, pp. 26–30, 2009.

50. Jose Luis Sanz, *Starring T. Rex!: Dinosaur Mythology and Popular Culture* (Bloomington: Indiana University Press, 2002), pp. 109, 115–127.

51. Details of this intriguing story, encompassing a scope far beyond my present purpose, are absorbingly outlined in Martin J.S. Rudwick's illuminating *Bursting the Limits of Time*.

Chapter Two

1. In her 2008 book, *A History of Paleontology Illustration* (Bloomington & Indianapolis: Indiana University Press), art historian Jane P. Davidson discusses print-making processes for early books that contained fossil illustrations or paleoart.

2. Allen A. Debus and Diane E. Debus, *Paleoimagery: The Evolution of Dinosaurs in Art* (Jefferson, NC: McFarland, 2002).

3. Ibid., p. 242; Allen A. Debus, "Sorting Fossil Vertebrate Iconography in Paleoart," *South Texas Geological Society Bulletin* (San Antonio), Vol. 44, no. 1 (Sept. 2003), pp. 5, 10–24.

4. George Gaylord Simpson, "The Beginnings of Vertebrate Paleontology in North America," *Proceedings of the American Philosophical Society*, Vol. 86, no. 1 (Sept. 1942), pp. 130–188. (See p. 131 for six periods.)

5. Buffon quoted in Stephen J. Gould, *The Lying Stones of Marrakech: Penultimate Reflections in Natural History* (New York: Harmony Books, 2000), p. 79.

6. Ellis S. Yochelson. "Peale's 1799 Theory of the Earth," *Earth Sciences History*, Vol. 10, no. 1 (1991), pp. 51–55.

7. Benjamin Franklin. "Conjectures Concerning the Formation of the Earth," *Am. Philos. Soc. Trans.*, Vol. 3 (1793), pp. 1–5.

8. Not much has been written on this important paleoartist, Bru; however, the following references may be of interest. Jose M. Lopez Pinero, "Juan Bautista Bru (1740–1799) and the Description of the Genus *Megatherium*," *Journal of the History of Biology*, Vol. 21 (1988), pp. 147–163; Jose M. Lopez Pinero, *Juan Bautista Bru de Ramon: El Atlas zoologoico, el megaterio y las tecnicas de pesca valencianas 1742–1799* (Valencia: Ayuntamiento de Valencia, 1996); Julian P. Boyd, "The Megalonyx, the Megatherium, and Thomas Jefferson's Lapse of Memory," *Proceedings of the American Philosophical Society*, Vol. 102, no. 5 (October, 1958), pp. 420–435.

9. George Gaylord Simpson, *Discoverers of the Lost World* (New Haven: Yale University Press, 1984), pp. 8–10.

10. Martin J.S. Rudwick. *Georges Cuvier, Fossil Bones, and Geological Catastrophes* (Chicago: University Press, 1997), p. 26.

11. Simpson, *Discoverers of the Lost World* (1984), p. 10.

12. Boyd, "The Megalonyx, the Megatherium, and Thomas Jefferson's Lapse of Memory," pp. 425–429.

13. Rudwick, *Georges Cuvier, Fossil Bones, and Geological Catastrophes* (1997), pp. 25–32.

14. Ibid., p. 32.

15. Ibid., p. 34.

16. Ibid., p. 198.

17. "Interesting papers" quote is excerpted from Boyd, "The Megalonyx, the Megatherium, and Thomas Jefferson's Lapse of Memory," p. 434; Thomas Horrocks, "Thomas Jefferson and the Great Claw," *Virginia Calvacade*, Vol. 35, no. 2 (Fall 1985), pp. 70–79; I. Bernard Cohen, *Science and the Founding Fathers* (New York: W.W. Norton, 1995), pp. 290–292.

18. Cohen, *The Fate of the Mammoth*, p. 292.

19. Ibid.

20. Thomas Jefferson. "A Memoir on the Discovery of Certain Bones of a Quadruped of the Clawed Kind in the Western Parts of Virginia," reprinted from *Am. Philos. Soc. Trans.*, Vol. 4 (1799), pp. 246–260, pp. 246–258, in *Benchmark Papers in Geology/151 North American Geology: Early Writings*, Robert M. Hazen, ed. (Stroudsburg, PA: Dowden, Hutchington & Ross, 1979), pp. 74–86. In a *Postscript*, printed on pp. 259–260 of the *Am. Philos. Soc. Trans.* paper, Jefferson stated, "The megatherium is not of the cat form, as are the lion, tyger, and panther, but is said to have striking relations in all parts of its body with the bradypus, dasypus, pangolin, etc. According to analogy then, it probably was not carnivorous."

21. Ibid.

22. Ibid., p. 248.

23. Ibid., p. 251.

24. Ibid., p. 252.

25. Boyd, "The Megalonyx, the Megatherium, and Thomas Jefferson's Lapse of Memory," pp. 421–422.

26. Simpson, *Discoverers of the Lost World* (1942), pp. 168–173.

27. Jefferson, "A Memoir on the Discovery of Certain Bones of a Quadruped of the Clawed Kind in the Western Parts of Virginia" (1799), pp. 255–256.

28. H.F. Osborn, "Thomas Jefferson as a Paleontologist," *Science*, Vol. 82 (1935), pp. 533–538. (Quote excerpted from p. 537.) See Boyd, "The Megalonyx, the Megatherium, and Thomas Jefferson's Lapse of Memory," for details concerning the relationship between Jefferson and Stuart.

29. Thomas Jefferson, *Notes on the State of Virginia* (Philadelphia, 1782).

30. The relationship between Jefferson and Buffon is well outlined in Paul Semonin's *American Monster* (New York: New York University, 2000).

31. Boyd, "The Megalonyx, the Megatherium, and Thomas Jefferson's Lapse of Memory."

32. John C. Greene, *American Science in the Age of Jefferson* (Ames: Iowa State University Press, 1984), p. 44.

33. Boyd, "The Megalonyx, the Megatherium, and Thomas Jefferson's Lapse of Memory."

34. Osborn, "Thomas Jefferson as a Paleon-

tologist," p. 537; Joseph Leidy, "A Memoir on the Extinct Sloth Tribe of North America," *Smithsonian Contributions to Knowledge*, Vol. 7 (New York: Putnam, 1855), pp. 3–68, quote excerpted from p. 3.

35. Simpson, *Discoverers of the Lost World* (1942), pp. 155, 157.

36. Ibid., p. 153.

37. Ibid., p. 154.

38. Herbert Wendt, *Before the Deluge* (Garden City, NY: Doubleday, 1968), pp. 187–193; Charles Darwin, *The Origin of Species: By Means of Natural Selection or the Preservation of Favoured Races in the Struggle for Life* (New York: Collier Books, 1962 ed. Reprinted from 1872 6th ed.); Charles Darwin. *The Voyage of the Beagle* (New York: Mentor, 1988 ed.).

39. William Buckland, *Geology and Mineralogy Considered with Reference to Natural Theology* (London: William Pickering, 1837 ed.).

40. Richard Darwin Keynes. *Fossils, Finches and Fuegians* (New York: Oxford University Press, 2003), pp. 107–111, 380.

41. William B. Hodgson. *Memoir on the Megatherium and Other Extinct Gigantic Quadrupeds of the Coast of Georgia with Observations on Its Geologic Features* (New York: Bartlett & Welford, 1846).

42. Charles Lyell. *A Second Visit to the United States of North America*, Vol. 1 of 2 (New York: Harper & Brothers, 1849), pp. 234–260.

43. Hodgson, *Memoir on the Megatherium and Other Extinct Gigantic Quadrupeds of the Coast of Georgia with Observations on Its Geologic Features*, p. 40.

44. Darwin, *The Origin of Species* (1872); Darwin, *The Voyage of the Beagle* (1988).

45. Darwin, *The Voyage of the Beagle* (1988), p. 149.

46. Ibid., p. 150.

47. Joseph Leidy, "Notices of Remains of Extinct Reptiles and Fishes, Discovered by Dr. F.V. Hayden in the Bad Lands of the Judith River, Nebraska Territory" (originally published in *Proceedings of the Academy of Natural Sciences of Philadelphia*, Vol. 8, March 1856, pp. 72–73), reprinted in *Dinosaur Papers (1676–1906)*, ed. David B. Weishampel and Nadine M. White (Washington, DC: Smithsonian Institution, 2003), pp. 259–261.

48. Ibid, pp. 7–8; E-mail correspondence from Ronald L. Richards, Curator of Paleobiology and Chief Curator of Natural History at the Indiana State Museum in Indianapolis communicated that "Much of a *Megalonyx jeffersoni* skeleton from Henderson Kentucky had been secured by David Dale Owen. The skeleton was eventually mounted at Indiana University, Bloomington, but later destroyed in a fire. The Indiana State Museum eventually recovered what remained of the skeleton: mandible, par-

tial (one tooth); one lumbar vertebra; one medial phalanx (pes); one terminal phalanx (pes); one terminal phalanx (manus). The Drake County, Ohio specimen (Orton Geological Museum, Ohio) is the best current specimen in existence of *M. jeffersoni."* A posted article titled "David Dale Owen's *Megalonyx"* provides a photo of the skeletal mount, and states the 9,400-year-old carbon date established in 1958 for wood samples collected from the stratum of interest. Further, "The age is accepted, with some reservation, as the terminal record for *Megalonyx."*

49. Castor Cartelle and Gerardo De Iuliis, "*Eremotherium laurillardi*: The Panamerican Late Pleistocene Megatheriid Sloth," *Journal of Vertebrate Paleontology*, Vol. 15, no. 4 (December 1995), pp. 830–841.

50. Buckland, *Geology and Mineralogy Considered with Reference to Natural Theology,* pp. 159–162, Plate 5.

51. A popular outline of the relation between Richard Owen and Charles Darwin, in context of South America's Pleistocene faunal extinctions may be read in Wendt, *op. cit.*

52. More on Owen's musings on the *Megatherium* can be read in Steve McCarthy's and Mick Gilbert's *The Crystal Palace Dinosaurs: The Story of the World's First Prehistoric Sculptures* (Croydon: Crystal Palace Foundation, 1994).

53. Leidy, "A Memoir on the Extinct Sloth Tribe of North America," pp. 24–25; also see Wendt, *Before the Deluge,* pp. 191–192; Henry A. Ward, *Catalogue of Casts of Fossils from the Principal Museums of Europe and America, with Short Descriptions and Illustrations* (Rochester, NY: Benton & Andrews, Printers, 1866), p. 13.

54. McCarthy and Gilbert, *The Crystal Palace Dinosaurs.*

55. William A.S. Sarjeant. "Crystal Palace," *Encyclopedia of Dinosaurs,* Philip J. Currie and Kevin Padian, eds. (San Diego: Academic Press, 1997), pp. 161–164.

56. Allen A. Debus and Steve McCarthy, "A Scene from American Deep Time: New York's Palaeozoic Museum — Revisited," *The Mosasaur — The Journal of the Delaware Valley Paleontological Society,* Vol. 6 (May 1999), pp. 105– 155.

57. Martina Kolbl-Ebert, "Female British Geologists in the Early Nineteenth Century," *Earth Sciences History,* Vol. 21, no. 1 (2002), p. 21. Quoting from another 1894 source of an 1834 (?) lecture given by Buckland, Kolbl-Ebert cites, "Still more remarkable was the interest excited by his lecture upon the megatherium, which was delivered on the last day of the meeting. The occasion was the first on which a fossil monster had been described to an unscientific audience of ladies and gentlemen"; William Clift, *Transactions of the Geological Society,* Vol. 3 (1835),

pp. 437–450. Note that while Buckland had already popularized "antediluvian" fossilized hyenas, rhinos and cave bears before the time of his 1830s lectures— and given there were erroneous reports of human "fossils" found in cave deposits— monstrously sized *Megatherium* represented a clear-cut case of a fossil, prehistoric mammal that had no living counterpart. Also see Michael Shortland, "Darkness Visible: Underground Culture in the Golden Age of Geology," *History of Science,* Vol. 32, Part 1, no. 95 (March 1994), p. 34, who refers to an 1832 lecture given by Buckland.

58. Todd Womack. "Plentifully Charged with Fossils: The 1822 Discovery of the *Eremotherium* at Skidaway," *Fossil News: Journal of Avocational Paleontology,* Vol. 6, no.7 (July 2000), pp. 14–16.

59. Ward, *Catalogue of Casts of Fossils from the Principal Museums of Europe and America, with Short Descriptions and Illustrations.*

60. For example; H.G. Wells, *Mr. Blettsworthy on Rampole Island* (Garden City, NY: Doubleday, Doran, 1928); Charles G.D. Roberts, *In the Morning of Time* (London: Hutchinson, 1919); William D. Matthew, "Scourge of the Santa Monica Mountains," *The American Museum Journal,* Vol. 16, no. 7 (1916), pp. 469–472.

61. Charles R. Knight's restorations were the most famous of all. See Donald F. Glut and Sylvia M. Czerkas, *Dinosaurs, Mammoths and Cavemen: The Art of Charles R. Knight* (New York: E.P. Dutton, 1982).

62. For more on the relation between giant sloths and prehistoric humans, see Charles Lyell, *Charles Lyell on North American Geology* (reprinted from *Am. Journal of Science,* Vol. 2, no. 3, 1847; New York: Arno Press, 1978), pp. 36–37, 267–269, 322–323; Alfred Russell Wallace, "The Antiquity of Man in North America," printed in the 1887 issue of *The Nineteenth Century,* Nov. 1887, online at http://216.239. 51.104/search?q=cache:0QHPd2t654EJ:www. wku.edu/~smithch/wallace/S4.

63. Gavin Menzies, *1421: The Year That China Discovered America* (New York: 1st Perennial ed., 2004), pp. 145–155, 209, 220–223, 272–273; cryptozoological reports that giant ground sloths persisted remain abundant, underscoring how ever-popular Jefferson's "Great-Claw" remains through the present day.

64. R. A. Farina and R. E. Blanco, "*Megatherium,* the Stabber," *Proceedings of the Royal Society of London,* B, Vol. 263 (1996), pp. 1725– 1729.

Chapter Three

1. The reason why it became known as the Ohio Animal is given in G.G. Simpson, "The Beginnings of Vertebrate Paleontology in North

America," *Proceedings of the North American Philosophical Society*, Vol. 86, no. 1 (Sept. 1942), p. 150.

2. Ibid., 151.

3. Paul Semonin, *American Monster: How the Nation's First Prehistoric Creature Became a Symbol of National Identity* (New York: New York University Press, 2000), pp. 62–63.

4. Simpson, "The Beginnings of Vertebrate Paleontology in North America," p. 150.

5. Ibid., pp. 67–68.

6. Claudine Cohen, *The Fate of the Mammoth: Fossils, Myth and History* (Chicago: University of Chicago Press, 2002), pp. 63–66.

7. See Simpson, "The Beginnings of Vertebrate Paleontology in North America," p. 150, and H.F. Osborn, "Thomas Jefferson as a Paleontologist," *Science*, Vol. 82, no. 2136 (Dec. 6, 1935), p. 534. Kerr applied the name "Elephas americanum" to the American Mastodon in 1792, "without otherwise contributing knowledge to it"; Simpson, "The Beginnings of Vertebrate Paleontology in North America," p. 150.

8. Simpson, "The Beginnings of Vertebrate Paleontology in North America," p. 150. Rembrandt Peale attempted to justify the name "Mammoth" for the Mastodon.

9. Cohen, *The Fate of the Mammoth: Fossils, Myth and History*, pp. 186–188.

10. Semonin, *American Monster*, p. 181.

11. Jefferson quoted in Osborn, "Thomas Jefferson as a Paleontologist," p. 534.

12. Ibid., 534.

13. Cohen, *The Fate of the Mammoth*, p. 96.

14. Simpson, "The Beginnings of Vertebrate Paleontology in North America," p. 130.

15. Semonin, *American Monster*, p. 181.

16. Ibid., p. 183.

17. Simpson, "The Beginnings of Vertebrate Paleontology in North America," p. 142.

18. Ibid.

19. Ibid., p. 146.

20. Peter Collinson, "Of Some Very Large Fossil Teeth Found in North America, and Described by Peter Collinson," *Philosophical Transactions* (Anno 1767, London), p. 477.

21. Ibid., p. 478.

22. Semonin, *American Monster*, pp. 146–147.

23. William Hunter, "Observations on the Bones, Commonly Supposed to Be Elephant Bones, Which Have Been Placed Near the River Ohio in America," in *Philosophical Transactions* (Anno 1768, London), p. 505.

24. Semonin, *American Monster*, p. 147.

25. Hunter, "Observations on the Bones, Commonly Supposed to Be Elephant Bones, Which Have Been Placed Near the River Ohio in America," p. 507.

26. Ibid., p. 506.

27. Simpson, "The Beginnings of Vertebrate Paleontology in North America," pp. 148–149.

28. Semonin, *American Monster*, p. 159.

29. Ibid., p. 160.

30. Ibid., p. 307.

31. Ibid., p. 308.

32. Ibid., p. 309.

33. Ibid.

34. Ibid., p. 310.

35. Ibid., pp. 169–170, 185. Dissenters' Cresswell and Gen. Rogers Clark still did not believe that the mastodon was carnivorous.

36. My italics. Osborn, "Thomas Jefferson as a Paleontologist," p. 36.

37. Adrienne Mayor, *Fossil Legends of the First Americans* (Princeton: Princeton University Press, 2005).

38. See Martin J.S. Rudwick, *Bursting the Limits of Time* (Chicago: University of Chicago Press, 2005), pp. 270–271. Michaelis' paper was titled "Thiergeschelt der Urwelt."

39. Robert West Howard, *The Dawnseekers: The First History of American Paleontology* (New York: Harcourt, Brace, Jovanovich, 1975).

40. Rembrandt Peale, *Historical Disquisition on the Mammoth or, Great American Incognitum, an Extinct, Immense, Carnivorous Animal Whose Fossil Remains Have Been Found in North America* (London, 1803), reprinted in *Natural Sciences in America: Selected Works in Nineteenth Century North American Paleontology* (New York: Arno Press, 1974), introduction by Keir B. Sterling.

41. Ibid., pp. 38–39.

42. Note the symbolism! Rather facetiously, Franklin had suggested the wild turkey should be recognized as America's symbol.

43. Semonin, *American Monster*, p. 357.

44. R. Peale, *Historical Disquisition on the Mammoth*, p. 24.

45. Semonin, *American Monster*, p. 340.

46. Readers familiar with Martin J.S. Rudwick's *Scenes from Deep Time* (Chicago: University of Chicago Press, 1992), will note similarities between Peale's "Exhumation" and illustrations found as Figures 30 and 31 in Rudwick's volume describing the exhumation of the Tertiary elephant *Dinotherium*, conducted by August Klippstein and Johann Kaup in 1836. *Dinotherium* was thought to be allied to modern Tapirs.

47. Semonin, *American Monster*, p. 336.

48. Howard, *The Dawnseekers*, Plate opposite p. 158; another early painting (circa 1806) of a Mastodon with tusks shown *both* raised and recurved downward was recently printed on page 401 of Martin J.S. Rudwick's *Bursting the Limits of Time* (Chicago: University of Chicago Press, 2005). Rudwick wrote, "This drawing was sent by the London anatomist Everard Home to Cuvier in Paris; it highlighted the missing bones for which Rembrandt had carved replacements in wood.... The drawing showed the upward

curve of the tusks that Home thought more likely to be correct than the downward curve (in dotted outline) of the Peales' reconstruction."

49. See Sterling's Introduction to *Natural Sciences in America: Selected Works*.

50. Peale, *Historical Disquisition on the Mammoth*, pp. 62–73.

51. For considerably more on Bewick and Anderson, see William Martin Smallwood's *Natural History and the American Mind*, in collaboration with Mabel Sarah Coon Smallwood (New York: Columbia University Press, 1941).

52. Simpson, "The Beginnings of Vertebrate Paleontology in North America," p. 160.

53. Semonin, *American Monster*, pp. 335–337.

54. Rembrandt Peale's 1802 46-page publication in which a figure showing mastodon's down-turned tusks appeared is *Account of the Skeleton of the Mammoth, a Non-descript Carnivorous Animal of Immense Size, Found in America* (London: E. Lawrence, 1802).

55. Ibid., p. 26, quoted in Semonin, "The Beginnings of Vertebrate Paleontology in North America," p. 334.

56. R. Peale, "A Short Account of the Mammoth," *Philosophical Magazine*, Vol. 14 (1803), p. 166, as cited in Semonin, "The Beginnings of Vertebrate Paleontology in North America," p. 335. Rembrandt's reconstruction of a mammoth with down-turned tusks appeared in this article. See p. 336.

57. The walrus paradigm has seen limited use in monster fare, however two examples are Ray Harryhausen's stop-motion animated puppet of the "Walrus giganticus" featured in *Sinbad and the Eye of the Tiger* (1977), as well as Toho Production's giant walrus "Magma" appearing in *Gorath* (1962).

58. R. Peale, "A Short Account of the Mammoth," in *Historical Disquisition on the Mammoth*, pp. 42, 80.

59. Ibid., pp. 51–53.

60. Ibid., pp. 75–76.

61. Ibid., p. 80.

62. Semonin, *American Monster*, p. 364.

63. Ibid., p. 370.

64. Ibid., p. 379.

65. For more, see Martin J.S. Rudwick's translations of Cuvier's papers in *Georges Cuvier, Fossil Bones, and Geological Catastrophes: New Translations and Interpretations of the Primary Texts* (Chicago: University of Chicago Press, 1997).

66. Cuvier was aided in his elucidation of Mastodon's diet through correspondence with Benjamin Smith Barton, of which more may be found in referring to notes 68 and 69, below.

67. Louis Figuier, *Earth Before the Deluge* (London: Cassell, 1867), translated by H.W. Bristow, p. 346.

68. Barton's letter was published as "Facts, Observations, and Conjectures, Relative to the Elephantine Bones (of a Different Species), That Are Found in Various Parts of North America," *Phila. Med. Phys. Jour.*, 1st suppl. (1806), pp. 22–35, and also reprinted in *Benchmark Papers in Geology 51: North American Geology — Early Writings* (Stroudsburg, PA: Douden, Hutchinson & Ross, 1979), Robert M. Hazen, ed., pp. 88–89.

69. Barton, "Facts, Observations, and Conjectures, Relative to the Elephantine Bones (of a Different Species)," pp. 94–95.

70. Mayor, *The First Fossil Hunters*, p. 94.

71. Simpson, "The Beginnings of Vertebrate Paleontology in North America," p. 174.

72. Ibid., p. 174. Also see Francis W. Pennell's "Benjamin Smith Barton as Naturalist," *Proceedings of the American Philosophical Society*, Vol. 86, no. 1 (Sept. 1942), pp. 108–122. Here we note, p. 120, that in April 1815, Barton visited Cuvier in Paris.

73. Semonin, *American Monster*, p. 403.

74. Howard, *op. cit.*, 79.

75. Semonin, *American Monster*, pp. 175–178.

76. Peale, "A Short Account of the Mammoth," pp. 32–33.

77. Yochelson, "Peale's 1799 Theory of the Earth," *Earth Sciences History*, Vol. 10, no. 1 (1991), p. 55.

78. Semonin, *op. cit.*, 328–329.

79. José Luis Sanz, *Starring T. Rex*, p. 13; Also, Lord Byron's (1788–1824) "verse-drama" *Cain: A Mystery* (1821) contained references of denizens of Hades including "extinct pre–Adamites, mammoths, and leviathans … and into the realm of the geological past." For more on Byron, see Ralph O'Connor's *The Earth on Show: Fossils and the Poetics of Popular Science, 1802–1856* (Chicago: University of Chicago Press, 2006), pp. 102–104.

Chapter Four

1. Albert C. Koch, *Description of the Missourium, or Missouri Leviathan; Together with Its Supposed Habits and Indian Traditions Concerning the Location from Whence It Was Exhumed, Also Comparison of the Whale, Crocodile and Missourium with the Leviathan as Described in 41st Chapter of the Book of Job* (2nd Edition, Enlarged, 1841, published by Prentice and Weissinger of Louisville, Kentucky). Many quotations attributed to Koch in this chapter are derived from this short work.

2. In Aurele La Rocque's "Contributions to the History of Geology," 3 — Biographic Index (Columbus, Ohio, 1964), Koch's death is given as 1867.

3. Anon., *The Presbyterian*, Jan. 12, 1839.

4. A. C. Koch, *Journey through a Part of the United States of North America in the Years 1844 to 1846*, translated by Ernst A. Stadler in 1972 (Carbondale: Southern Illinois University Press, 1972).

5. Carl Zimmer, *At the Water's Edge: Macroevolution and the Transformation of Life* (New York: Free Press, 1998).

6. Paul Semonin, *American Monster: How the Nation's First Prehistoric Creature Became a Symbol of National Identity* (New York: New York University Press, 2000), p. 383.

7. MacDonald stated his belief that "Leviathan" was a *Tyrannosaurus*. For critique on his thesis, see chapter 16, titled "Dinosaurs and Divinity," in Allen Debus and Diane Debus, *Dinosaur Memories: Dino-Trekking for Beasts of Thunder, Fantastic Saurians, "Paleo-people," "Dinosaurabilia, and other "Prehistoria"* (Lincoln, NE: Author's Choice Press, 2002).

8. Semonin, *American Monster*, p. 382.

9. While most accounts have it that Missourium was sold before Koch's return to America, Pat Middleton's recent book, *America's Great River Road* (Heritage Press, 1998) cites an 1844 Centennial report. "An account, by James A. Michener, relates a visit by Levi and Elly Zendt to 'see the Gigantic Elephant Discovered in These Regions by Dr. Albert Koch...' when they passed through St. Louis on their way west in 1844. The Zendts, like most other immigrants in the western expansion, had never seen an animal so fearsome as the skeleton suggested. The phrase 'seeing the elephant' as in 'I saw the elephant' symbolized the amazing new world of America's west. It also referred to the great awe, fear, terror the pioneers faced when making the trip west. 'When a man saw the elephant, ... rising out of the darkness with those beady, flaming eyes, he must heed its warning' to go further. Many of those who passed Koch's museum on Market Street 'saw the elephant' both literally and figuratively!"

10. Herbert Wendt, *Before the Deluge*, New York: Doubleday, 1968, p. 264.

11. Herman Melville, *Moby-Dick* (1851; New York: Signet Classics, 1961 ed.).

12. Dennis Dean. "The Influence of Geology on American Thought," in *Two Hundred Years of Geology in America: Proceedings of the New Hampshire Bicentennial Conference on the History of Geology* (Cecil J. Schneer, ed., published for the University of New Hampshire by the University Press of New England: Hanover, New Hampshire, 1979).

13. Perry Miller, *The Raven and the Whale: Poe, Melville and the New York Literary Scene* (Baltimore: Johns Hopkins University Press, 1997).

14. Cornelius Mathews, *Behemoth: A Legend of the Mound-Builders* (1839; reprinted, Whitefish, MT: Kessinger Publishing, 2007).

15. Herman Melville, *Mardi: And a Voyage Thither* (1849; 1970 ed., Northwestern University Press and the Newberry Library, Evanston and Chicago).

16. Elizabeth S. Foster, "Melville and Geology," *American Literature*, Vol. 17, no. 1 (March 1945), pp. 50–65.

17. Ibid.

18. Robert Silverberg, *Scientists and Scoundrels: A Book of Hoaxes* (New York: Thomas Y. Crowell, 1965).

Chapter Five

1. William Mullen, "Darwin: Many Attacks Like Those of 19th Century," *Chicago Tribune*. June 13, 2007, Section 1, pp. 1, 22.

2. Martin J.S. Rudwick, *Bursting the Limits of Time: The Reconstruction of Geohistory in the Age of Revolution* (Chicago: University of Chicago Press, 2005), pp. 280–281, 592, 602, 620.

3. Charles Lyell, *The Geological Evidence of the Antiquity of Man* (originally published as *The Geological Evidences of the Antiquity of Man: With Remarks on Theories of the Origin of Species by Variation* by J. Murray, 1863; New York: Dover, 2004), p. 161.

4. For more on Koch, see Chapter 4 in this volume.

5. Barbara Franco, *The Cardiff Giant: A Hundred Year Old Hoax* (Cooperstown, NY: New York State Historical Association, 1990).

6. Regarding the Cardiff giant, according to Richard Milner, "Fossil evidence for human evolution was still so scant in the 1870s that a bewildered public paid a fortune to see the most brazen scientific hoax in history.... It is not clear whether the hoax was originally planned as a swindle or if ... the giant [was] built to ridicule clergymen who insisted on the literal truth of every word in Genesis including 'there were giants in the earth in those days.'" (Milner, *The Encyclopedia of Evolution*, New York: Facts on File, 1990, p. 70.) For Piltdown Man, see entry in Milner, as well as note 50 below.

7. Lyell, *The Geological Evidence of the Antiquity of Man*, pp. 46–57.

8. Charles Darwin, *The Origin of Species by Means of Natural Selection or the Preservation of Favoured Races in the Struggle for Life* (originally published as the 6th ed., 1872; New York: Collier Books, 1962), p. 483.

9. Erik Trinkaus and Pat Shipman, *The Neandertals: Changing the Image of Mankind* (New York: Alfred A. Knopf, 1993), p. 190.

10. Charles De Paolo, *Human Prehistory in Fiction* (Jefferson, NC: McFarland, 2002), p. 78.

11. Stephanie Moser, *Ancestral Images: The Iconography of Human Origins* (Ithaca, NY: Cornell University Press, 1998), p. 29.

12. Ibid., chapter two, "Mythological Visions of Human Creation," pp. 21–38.

13. Ibid., p. 49.

14. Ibid., p. 18.

15. Ibid., p. 7.

16. Ibid., p. 65.

17. Ibid., pp. 44–45.

18. Ibid., p. 90.

19. Ibid., pp. 90–96.

20. Lyell, *The Geological Evidence of the Antiquity of Man.*

21. Moser, *Ancestral Images*, pp. 108–109.

22. Ibid., pp. 125–131.

23. Rupke, quoted in Ibid., p. 125.

24. Ibid., p. 136; Boitard's novel was written during the 1830s.

25. Martin J.S. Rudwick, *Scenes from Deep Time: Early Representations of the Prehistoric World* (Chicago: University of Chicago Press, 1992), p. 168.

26. Moser, *Ancestral Images*, pp. 137–138.

27. Ibid., pp. 16–217; *Pithecanthropus* is now known as *Homo erectus.* Also see Donald Johanson's and Maitland Edey's *Lucy: The Beginnings of Humankind* (New York: Warner Books, 1982) p. 37.

28. Moser, *Ancestral Images*, pp. 151–152; Allen A. Debus, *Dinosaurs in Fantastic Fiction: A Thematic Survey* (Jefferson, NC: McFarland, 2006), pp. 51–53.

29. Moser, *Ancestral Images*, pp. 138–139.

30. Ibid., pp. 139–141.

31. Trinkaus and Shipman, *The Neandertals*, p. 399.

32. Charles Knight, J.H. McGregor and Henry F. Osborn were on the forefront of the American Museum's contingent. See Moser, *Ancestral Images*, pp. 156–160. Also see Osborn's *The Hall of the Age of Man*, revised by William K. Gregory and George Pinkley (New York: American Museum of Natural History, 1938), and Henry Field's *Prehistoric Man — Hall of the Stone Age of the Old World*, Leaflet 31 (Chicago: Field Museum of Natural History, 1933), the latter featuring the art of sculptor Frederick Blaschke.

33. From Clive Gamble's Foreword to Moser, *Ancestral Images*, p. xxii.

34. Ibid. As noted by Gamble, in 1911, Sir Arthur Keith commissioned a restoration of a Neandertal cave figure appearing much like a modernistic human. "Boldly titled 'Not in the Gorilla stage,' Keith's illustration was clearly flinging down the gauntlet to the Frenchman (Boule) and even gilding the lily somewhat by showing the man wearing a necklace for which there is no evidence at all. The move, artistically, from Mugger-waiting-behind-a-rock to Toolmaker-by-the-fire-of-progress could not be more dramatic…. Which image prevailed? The answer was emphatically Boule's…. More-

over, I think the events of the First World War which shortly followed were also set against Keith's image. The carnage of war stripped away the veneer of civilization and revealed an older moral ancestry that people wanted to expel, ignore or blame on someone else. It was only after the Second World War that Boule's anatomical conclusions turned out to be wrong and a complete rehabilitation of Neanderthals took place. But not before the damage to their public image was irrevocable." (Quote from Gamble, p. xxii.)

35. Trinkaus and Shipman, *The Neandertals*, p. 402.

36. De Paolo, *Human Prehistory in Fiction*, p. 144.

37. Moser, *Ancestral Images*, p. 145.

38. Ibid., p. 158; Moser discusses Knight, Burian, Wilson and Gurche.

39. De Paolo, *Human Prehistory in Fiction*, p. 35.

40. Ibid., p. 145.

41. Marc Angenot and Nadia Khouri, "An International Bibliography of Prehistoric Fiction," *Science Fiction Studies* Vol. 8, no. 1 (1981), pp. 38–53.

42. Debus, *Man and Nature in the Renaissance*, Chapter One, "Verne's Subterranean 'Museum,'" pp. 17–35, 51; A.A. Debus, "Reframing the Science in Jules Verne's *Journey to the Center of the Earth*," *Science Fiction Studies*, Vol. 33, no. 100, part 3 (November 2006), pp. 405–420; A.A. Debus, "Two Mysterious Monsters of Science Fiction Literature," *Scary Monsters*, no. 57 (Jan. 2005), pp. 46–49.

43. Brian W. Aldiss, with David Wingrove, *Trillion Year Spree: The History of Science Fiction* (New York: Avon Books, 1986), p. 110.

44. Ibid.

45. De Paolo, *The Encyclopedia of Evolution*, discusses these titles extensively in chapters One and Five, respectively, pp. 9–17 and 46–58. For more on *Dr. Moreau*, see Frank McConnell's *The Science Fiction of H.G. Wells* (New York: Oxford University Press, 1981), pp. 88–106.

46. De Paolo, *Human Prehistory in Fiction*, p. 10.

47. H.G. Wells, *The Island of Dr. Moreau*, in *The Complete Science Fiction Treasury of H.G. Wells, with a Preface by the Author* (New York: Avenel Books, 1978), p. 156.

48. De Paolo, *Human Prehistory in Fiction*, p. 46.

49. Ibid., p. 47. For more on the evidence for cannibalism in prehistoric men, see Milner, *The Encyclopedia of Evolution*, pp. 69–70.

50. John Evangelist Walsh, *Unraveling Piltdown: The Science Fraud of the Century and Its Solution* (New York: Random House, 1996), p. 121.

51. Coincidentally, Doyle's novel *Strand* serialization overlapped Charles D. Roberts' sim-

ilar (Wellsian) entry, *In the Morning of Time*, appearing in issues of a rival publication, *The London Magazine* (1912). Here, an early human family withstands attacks from a less evolved, rapacious pack of anthropoids. Charles G.D. Roberts, *In the Morning of Time* (London: Hutchinson, 1919).

52. De Paolo, *Human Prehistory in Fiction*, pp. 37–45.

53. John Taine, *The Iron Star* (1930. Reprint, Westport, CT: Hyperion Press, 1976).

54. For more on these titles, see chapters two and three in A.A. Debus' *Dinosaurs in Fantastic Fiction: A Thematic Survey*.

55. De Paolo, *Human Prehistory in Fiction*, pp. 18–24, 79–93; Pierre Boulle, *The Planet of the Apes*, trans. Xan Fielding (New York: Gramercy Books, 1963); Clarke, Arthur C. *2001: A Space Odyssey* (New York: Signet Books, 1968), pp. 13–37.

56. Petru Popescu, *Almost Adam* (New York: Avon Books, 1996); John Darnton, *Neanderthal* (New York: St. Martin's, 1996); Stephen Baxter, *Evolution* (New York: Ballantine Books, 2000); Brad Strickland with John Michlig, *Kong: King of Skull Island*, created and illustrated by Joe De-Vito (Milwaukie, OR: DH Press, 2004); Jonathan Green, *Pax Brittania: Unnatural History* (Oxford, UK: Abaddon Books, 2007).

57. William D. Allmon, "The Pre-Modern History of the Post-Modern Dinosaur: Phases and Causes in Post-Darwinian Dinosaur Art," *Earth Sciences History*, Vol. 25, no. 1 (2006), pp. 30–31.

58. In a reprint of the *Lost World* movie pressbook (A First National Picture), reissued with the Image Entertainment DVD, the off-camera relationship between the chimp named Mary and Bull Montana in his apeman costume was summarized. Evidently, Mary accepted the apeman as a compatriot ape, even pulling a flea from his ape suit.

59. Bradbury's classic tale is discussed thoroughly in A.A. Debus, *Dinosaurs in Fantastic Fiction*, pp. 85–102; Richard Marsten, *Danger Dinosaurs!* (Philadelphia: John C. Winston, 1953).

60. Quotes are from Trinkaus and Shipman, *The Neandertals*, pp. 405–406. For more on cavemen in film, read Michael Klossner's *Prehistoric Humans in Film and Television: 581 Dramas, Comedies and Documentaries, 1905–2004* (Jefferson, NC: McFarland, 2006). Added to the sexuality element was fear of miscegenation.

61. See Donald F. Glut's *Classic Movie Monsters* (Metuchen, NJ: Scarecrow, 1978), p. 259; Vincent Di Fate, "Milicent and the Monster: The Strange Saga of the Mystery Woman Who Designed the Universal Gill Man!" *FilmFax*, no. 100 (2003), p. 56; Also see Debus and Debus, *Dinosaur Memories*, chapter fifteen, "The Truth

IS Out There: On the Trail of 'Living' Dinosaurs," where some of the South American legends of allegedly prehistoric cryptozoological species are debunked.

62. Samantha Weinberg, *A Fish Caught in Time: The Search for the Coelacanth* (New York: HarperCollins, 2000). See page 89, where Weinberg proposes that, "Hollywood latched on: *"The Creature from the Black Lagoon*, featuring a finned monster that emerged from the sea, apparently was inspired by the discovery.'"

63. See Allen A. Debus' *Dinosaurs in Fantastic Fiction*, pp. 65–68, 147, for more on antecedent varieties of marine "prehistoric" crypto-anthropoidal monsters of fantastic literature.

64. Baxter quoted from *Science Fiction in the Cinema* (New York: A.S. Barnes, 1970, p. 121), as cited in Glut, *Classic Movie Monsters*, p. 264.

65. Glut, *Classic Movie Monsters*, p. 265; Allen A. Debus, "The Creature Lives around Us—Evolution of the Beast," *Mad Scientist*, no. 19, Spring 2009, pp. 8–14.

66. Ibid., p. 269; "A distinctly *dinosaurian* Gillman variant slashed its way through 'Carl Dreadstone's'" (an unidentified writer, not Ramsey Campbell), *Creature from the Black Lagoon* (New York: Berkley Publishing, 1977), a gripping novel *inspired by* the classic film (based on a story by Maurice Zimm). The novel doesn't follow the movie script by any means, although the Amazonian setting may seem familiar. After spying the gigantic creature, which is 30 feet tall and has a powerful tail, scientists onboard the *Rita* debate whether the animal is prehistorical (e.g.—descendant from plesiosaur or diplodocine stock), or an atomic mutant. Does it represent our future or the past? Although they name the monster "Mutantus giganticus," its reptilian-amphibian characters seem predominant. Plus—it's highly intelligent and in a Kong-like fashion holds strange reverence for the expedition's sole female, Kay. In the end, after the immensely strong creature has been subdued by a torpedo, Kay stares at its decapitated head lying on the ship. "Looking down at the eye, Kay screamed repeatedly, for the eye was accusing her, reproving her. The eye was the eye of a betrayed friend." Nature betrayed?

67. Donald Johanson, and Maitland Edey, *Lucy: The Beginnings of Mankind* (New York: Warner Books, 1981).

Chapter Six

1. Dean Hannotte, a collector of many dinosaur things, including rare memorabilia, may have coined the term "dinosaurabilia" during the 1980s. There are many other collectors who own vast and valuable items. Another term often encountered is "dinosaurologist," referring to individuals who conduct dinosaur sci-

ence professionally, serious-minded amateur collectors who support dinosaur science, or those who are actively engaged in producing dinosaur art.

2. Allen A. Debus and Diane E. Debus, *Dinosaur Memories: Dino-Trekking for Beasts of Thunder, Fantastic Saurians, "Paleo-People," "Dinosaurabilia," and other "Prehistoria"* (Lincoln, NE: Author's Choice Press, 2002), pp. 85–110; Allen A. Debus, "Why Not the Mammals?" (Parts 1 and 2), *Fossil News: Journal of Avocational Paleontology*, Vol. 8, nos. 8–9 (August-Sept. 2002), pp. 4–9; pp. 8–11, 14–16.

3. Debus, and Debus, *Dinosaur Memories*, p. 97.

4. Ibid., p. 103.

5. Benjamin Waterhouse Hawkins, "On Visual Education as Applied to Geology" (originally published in *Journal of the Society of Arts*, Vol. 2, 1853–54, pp. 444–449), reprinted in *The Dinosaur Papers (1676–1906)*, ed. David B. Weishampel and Nadine M. White (Washington D. C.: Smithsonian Institution, 2003), p. 220.

6. For more on Crystal Palace's dinosaurs, see *Encyclopedia of Dinosaurs*, ed. Philip J. Currie and Kevin Padian (New York: Academic Press, 1997), entry by William A.S. Sarjeant, "Crystal Palace," pp. 161–164.

7. Jane P. Davidson, *A History of Paleontology Illustration* (Bloomington & Indianapolis: Indiana University Press, 2008), p. 183.

8. Martin J.S. Rudwick, *Worlds Before Adam: The Reconstruction of Geohistory in the Age of Reform* (Chicago: University of Chicago Press, 2008), pp. 158–159.

9. L.B. Halstead and W.A.S. Sarjeant, "*Scrotum Humanun* Brookes— The Earliest Name for a Dinosaur," *Modern Geology*, Vol. 18 (1993), pp. 221–224. In their Abstract, these authors wrote, "The earliest dinosaur bone to be illustrated was by Plot (1677). Brookes (1763) applied it to the binomen *Scrotum humanum*; the specimen appears to have been a condyle of *Megalosaurus*. Brookes's binomen, though senior, has been forgotten and is now formally set aside, being considered 'not an available name.'"

10. Stukeley's paper is discussed in Davidson, *A History of Paleontology Illustration*, pp. 35–36. The plesiosaur may rather be a related form, a nothosaur.

11. Adrian J. Desmond, *The Hot-Blooded Dinosaurs: A Revolution in Palaeontology* (New York: Warner Books, 1975), p. 9.

12. Ralph O'Connor, *The Earth on Show: Fossils and the Poetics of Popular Science, 1802–1856* (Chicago: University of Chicago Press, 2007), p. 97. O'Connor reproduces a picture of such a dragon-pterodactyl on p. 97 of his book, painted by George Howman in 1829. Gideon Mantell is often credited for distinguishing the

"Age of Reptiles." (See, for example, Weishampel and White, *Dinosaur Papers*, p. 161, and Rudwick, *Worlds Before Adam*, 2008, p. 441.) For more on the earliest pterosaur discoverers and their discoveries, see Peter Wellnhofer's *The Illustrated Encyclopedia of Pterosaurs: An Illustrated Natural History of the Flying Reptiles of the Mesozoic Era* (New York: Crescent Books, 1991), pp. 20–31; Martin J.S. Rudwick, *Bursting the Limits of Time: The Reconstruction of Geohistory in the Age of Revolution* (Chicago: University of Chicago Press, 2005), p. 501.

13. Of course, not all the prehistoric mammals known to early 19th century science were of large beasts. Rudwick, *Bursting the Limits of Time*, p. 23.

14. Ibid., p. 32.

15. W.D. Conybeare, "On the Discovery of an Almost Perfect Skeleton of the *Plesiosaurus*," (originally published in *Transactions of the Geological Society of London*, 2nd series, Vol. 1, 1824, pp. 381–390), reprinted in Weishampel and White, *Dinosaur Papers*, ed. Weishampel and White, p. 61.

16. Many such specimens had been collected by Mary Anning (1799–1847) at Lyme Regis, England. Several of her fossil vertebrate discoveries had major impact on Britain's emerging geological sciences. For more on Anning, see Christopher McGowan, *The Dragon Seekers: How an Extraordinary Circle of Fossilists Discovered the Dinosaurs and Paved the Way for Darwin* (Cambridge, MA: Perseus Books, 2001).

17. The ammonites are shown floating on the ocean surface with peculiar wing sails attached. The related, extinct fossil belemnites were also incorporated into this restoration.

18. Martin J.S. Rudwick, *Scenes from Deep Time: Early Pictorial Representations of the Prehistoric World* (Chicago: University of Chicago Press, 1992), p. 47. The original painted version of *Durior antiquior* appears on the front of the book jacket. As noted by Rudwick, lithographs were made for sale.

19. Ibid., pp. 82–83; O'Connor, *The Earth on Show*, p. 231.

20. McGowan, *The Dragon Seekers*, p. 74; paleontologist Robert Bakker referred to ecological changes of the Triassic Period as "titanic ecological struggles," or "the clash of mighty empires." Robert T. Bakker, *The Dinosaur Heresies* (New York: William Morrow, 1986), p. 416.

21. This will be more evident in chapter eight of the present volume. Also, see chapters one and three of my *Dinosaurs in Fantastic Fiction*.

22. Allen A. Debus, "Reframing the Science in Jules Verne's *Journey to the Center of the Earth*," *Science Fiction Studies*, Vol. 33, no. 100 (Nov. 2006), pp. 405–420; Louis Figuier, *The World Before the Deluge* (London: Cassell, 1867 ed.), see Plate XV facing p. 230. Famed pale-

oartist Zdenek Burian's (1905–1981) paintings of battling sea monsters were published in books written by Joseph Augusta, such as *Prehistoric Animals* (London: Spring Books, 1956), see Plate 20, and *Prehistoric Sea Monsters* (London: Paul Hamlyn, 1964); see Burian's tylosaur versus elasmosaur in a frothy sea battle completed in 1963.

23. Allen A. Debus and Steve McCarthy, "A Scene from American Deep Time, New York's Palaeozoic Museum — Revisited," *The Mosasaur: The Journal of the Delaware Valley Paleontological Society*, Vol. 6 (May 1999), pp. 105–115.

24. Steve McCarthy and Mick Gilbert, *The Crystal Palace Dinosaurs: The Story of the World's First Prehistoric Sculptures* (London: Crystal Palace Foundation, 1994), p. 77.

25. E.D. Cope, "The Fossil Reptiles of New Jersey," in *Dinosaur Papers*, ed. Weishampel and White, pp. 328–329; Archibald M. Willard's (1836–1918) 1872 painting of prehistory incorporates similarly posed, sparring marine reptiles and a pterosaur as well. See Debus and Debus, *Paleoimagery*, pp. 84–85.

26. The theme is lasting. A life-sized sculptural display of two battling sea reptiles (*Elasmosaurus* and the ichthyosaur *Ophthalmosaurus*) may be witnessed at Ogden, Utah's, George S. Eccles Dinosaur Park. This outdoor park opened in the early 1990s. Also, at the South Dakota School of Mines and Technology in Rapid City, skeletal reconstructions of two large marine sauria are posed as if in mortal combat — a 29-foot-long *Mosasaurus*, and long-necked plesiosaur *Alzadasaurus*.

27. Dennis R. Dean, *Gideon Mantell and the Discovery of Dinosaurs* (Cambridge: Cambridge University Press, 1999).

28. Rudwick, *Scenes from Deep Time*, pp. 52–54.

29. An exceptional modern example is Luis Rey's restoration showing three megalosaurs chasing after a *Polacanthus* (an armored form related to *Hylaeosaurus*). This was published on the front cover of The Dinosaur Society UK's *Quarterly*, Vol. 1, no. 4 (Winter 1995/96).

30. O'Connor, *The Earth on Show*, p. 125, stated "the *Iguanodon* was a monster according to the strict usage of the term, seemingly composed from parts of known animals. (The only other extinct land saurian then known, the *Megalosaurus*, was less extraordinary: conceived as a carnivorous lizard, it was pictured as a kind of crocodile and was also thought to be smaller than the *Iguanodon*, which counted for a great deal.)"

31. Dean, *Gideon Mantell and the Discovery of Dinosaurs*, book jacket and p. 123; as noted by O'Connor, *The Earth on Show*, p. 276, "this was evidently a 'sketch for a picture 3 yards long' which may have hung in the Mantellian mu-

seum." In (circa) 1834, De la Beche also drew a silly cartoon showing a similarly poised *Iguanodon* "celebrating Mantell's 1834 acquisition of the 'Maidstone Iguanodon' the best preserved specimen of the time." (O'Connor, *The Earth on Show*, p. 429.)

32. Rudwick, *Scenes from Deep Time*, pp. 76–77; O'Connor, *The Earth on Show*, pp. 275–277.

33. O'Connor, *The Earth on Show*, p. 78.

34. Ibid., p. 80; Desmond, *The Hot-Blooded Dinosaurs*, p. 19.

35. Gideon Mantell, "An Excerpt from *The Geology of the South-East of England*," (London, 1833, pp. 260–333 and Plates II–V), in *Dinosaur Papers*, ed. Weishampel and White, p. 103.

36. Richard Owen, "An Excerpt from the Report on British Fossil Reptiles" (originally published in *Report of the Eleventh Meeting of the British Association for the Advancement of Science*, Plymouth, England, July 1841, John Murray, publisher, London, 1842, pp. 60–204), in *Dinosaur Papers*, ed. Weishampel and White, pp. 173–174, 199. The term "dinosaur" is documented to have been used for the first time in Owen's printed report.

37. Gideon Mantell, "On the Structure of the Jaws and Teeth of the *Iguanodon*" (originally published in *Philosophical Transactions of the Royal Society of London*, Vol. 138, part 1, 1848, pp. 183–202), reprinted in *Dinosaur Papers*, ed. Weishampel and White, p. 208; O'Connor, *The Earth on Show*, p. 117.

38. *Hylaeosaurus* did turn out to be quadrupedal after all, indicating, as Owen claimed, an "approach to the mammalian type," although, like Scharf two decades before, Owen and Hawkins got its body armor positioned incorrectly on the Crystal Palace statue. Quote is from *Dinosaur Papers*, ed. Weishampel and White, p. 179.

39. Edward Hitchcock, "Description of the Foot Marks of Birds (Ornithichnites) on New Red Sandstone in Massachusetts" (originally published in *American Journal of Science and the Fine Arts*, Vol. 29, 1836, pp. 307–340 and Plates 1–3 with 21 figures), reprinted in *Dinosaur Papers*, ed. Weishampel and White, p. 151.

40. The entire poem is reprinted in Jordan D. Marche II, "Edward Hitchcock's Poem: The Sandstone Bird (1836)," *Earth Sciences History*, Vol. 10, no.1 (1991), pp. 5–8.

41. Joseph Leidy, "Notices of Remains of Extinct Reptiles and Fishes, Discovered by Dr. F.V. Hayden in the Bad Lands of the Judith River, Nebraska Territory" (originally published in *Proceedings of the Academy of Natural Sciences of Philadelphia*, Vol. 8, March 1856, pp. 72–73), reprinted in *Dinosaur Papers*, ed. Weishampel and White, pp. 259–261.

42. Joseph Leidy, "Remarks Concerning *Had-*

rosaurus" (originally published in *Proceedings of the Academy of Natural Sciences of Philadelphia*, Vol. 10, 1858, pp. 215–218), reprinted in *Dinosaur Papers*, ed. Weishampel and White, p. 265; Joseph Leidy, "An Excerpt from Cretaceous Reptiles of the United States" (originally printed in *Smithsonian Contributions to Knowledge*, Vol. 14, article VI, 1865, pp. 111, 1–4, 76–102, with Plates II, VIII, XII–XVII, and XX), in *Dinosaur Papers*, ed. Weishampel and White, p. 299.

43. Warren D. Allmon, "The Pre-Modern History of the Post-Modern Dinosaur: Phases and Causes in Post-Darwinian Dinosaur Art," *Earth Sciences History*, Vol. 25, no. 1 (2006), p. 13.

44. Ibid., pp. 14–18; Allen A. Debus and Diane E. Debus, *Paleoimagery: The Evolution of Dinosaurs in Art* (Jefferson, NC: McFarland, 2002), pp. 120–125.

45. Edward D. Cope, "The Fossil Reptiles of New Jersey" (Part 1), *The American Naturalist*, Vol. 1 (1868), pp. 23–30.

46. Allmon, "The Pre-Modern History of the Post-Modern Dinosaur," pp. 10, 12–13; Thomas Henry Huxley also considered a hopping motion for dinosaurs and during the 1860s suggested that certain dinosaurians carried an avian gait. During the late 1850s, Huxley defiantly challenged Cuvier's law of the correlation of anatomical parts, claiming that restorations of (ancestral) prehistoric animals, including dinosaurs, should not be founded in Cuvier's law, but rather on the basis of better known (possibly more recent) animals judged most related (in an evolutionary sense) to the extinct or lesser known forms. Acceptance of such philosophy, and applying the paleontological method, during the late 19th century improved the means and basis for restoring prehistoric monsters.

47. A. Debus and S. McCarthy, "A Scene from American Deep Time"; Earle E. Spamer, "The Great Extinct Lizard: *Hadrosaurus foulkii*. 'First Dinosaur' of Film and Stage," *The Mosasaur: Journal of the Delaware Valley Paleontological Society*, Vol. 7 (May 2004), pp. 109–126.

48. Paul Semonin, *American Monster: How the Nation's First Prehistoric Creature Became a Symbol of National Identity* (New York: New York University Press, 2000), pp. 400–401.

49. In 1871, British geologist John Phillips (1800–1874) suggested that *Megalosaurus* was bipedal, stating, "essentially reptilian; yet not a ground crawler like the alligator, but moving with free steps chiefly, if not solely on the hind limbs, and claiming a curious analogy, if not some degree of affinity, with the ostrich." Quoted from Phillips' *Geology of Oxford and the Valley of the Thames*, as cited in Allmon, "The Pre-Modern History of the Post-Modern Dinosaur," p. 10.

50. Louis Dollo, "Third Note on the Dinosaurs of Bernissart" (originally published in *Bulletin de Musee Royal d'Histoire Naturelle de Belgique*, Vol. 2, 1883a, pp. 85–120 and Plates 3 to 5), reprinted in *Dinosaur Papers*, ed. Weishampel and White, p. 405.

51. Ibid., p. 408.

52. Allmon, "The Pre-Modern History of the Post-Modern Dinosaur," p. 10.

53. Smit's *Megalosaurus*, printed in H.N. Hutchinson's *Extinct Monsters: A Popular Account of Some of the Larger Forms of Ancient Animal Life* (London: Chapman & Hall, LD, 1893, Plate VI facing p. 79), is bipedal yet with hind limbs poised in kangaroo fashion, with a neck that is proportionally too long. Smit also completed an early ceratosaur restoration where this dinosaur is shown as a bipedal, (non-kangaroo) creature for H.R. Knipe's *Nebula to Man* (1905). Smit's ceratosaur restoration was clearly influenced by Marsh's upright, bipedal, tail-dragging reconstruction of *Ceratosaurus*' skeleton, as published in "On the Affinities and Classification of the Dinosaurian Reptiles," *American Journal of Science*, 3rd series, Vol. 50 (1895), pp. 438–498.

54. The stegosaur clan seems to have originated in what is now China. This author has had the pleasure of sculpting miniature restorations of several of the plated varieties of dinosaurs including *Kentrosaurus*, *Lexovisaurus*, *Dacentrurus*, and Asian forms — *Tuojiangosaurus* and *Huayangosaurus* and several historical versions of *Stegosaurus*.

55. Desmond, *The Hot-Blooded Dinosaurs*, p. 148.

56. Billy de Klerk, "The First Dinosaur Fossil Discovered in South Africa: The Stegosaur *Paranthodon africanus*," posted at www.ru.ac.za./affiliates/am/paranth.html; Kenneth Carpenter and Clifford A. Miles (eds.), *The Armored Dinosaurs* (Bloomington: Indiana University Press, 2001); Allen A. Debus, "A Crystal Palace Stegosaur?" (Parts 1 to 2) *Fossil News: Journal of Avocational Paleontology*, Vol. 8, no. 11 (Nov. 2002), pp. 4–8, Dec. 2002, pp. 5–7. The Swindon Brick and Tile specimen, now residing in the British Museum collection became the holotype. In addition to 12 tail vertebrae in this specimen, bones of the left arm and hand are well preserved. Several neck and back bones were recovered, as well as hip elements (ilia and ischia) and leg bones (femur and tibia). There is only one preserved dermal spine and single plate, thought in life to have protected the neck region. But there is no known skull, and despite the fact that the 15-foot-long *Dacentrurus* is categorized as a plated dinosaur, there is no definitive basis to restore the plate geometry as aligned along the backbone, neck and tail, or to establish the tail spike number or orientation,

or to conclude whether it possessed a protective pair of shoulder spines. Such questions remain very prickly indeed! Conventionally, *Dacentrurus* is restored with a paired double row of plates. But in my original 1/27 scale sculptural restoration I speculatively adorned *Dacentrurus* with a paired, staggered row of low-ridged plates, and 4 pair of tail spikes with a pair of shoulder spines. My *Dacentrurus* blends features of two dinosaurs which are more completely known from fossils—*Stegosaurus* and the African genus, *Kentrosaurus*. At one time or another, as many as seven species have been assigned to *Dacentrurus*, and another (*Astrodon pusillus*—actually a sauropod) was later reclassified as *Dacentrurus*. Presently, however, only two species are recognized, (the second, *D. phillipsi*, may be a nomen dubium). *Omosaurus hastiger*, described by Owen in 1877 but reassigned to *D. armatus* by Peter Galton in 1990, was also found in a Kimmeridgian age clay pit in Wiltshire. Associated with the 1877 specimen were two long spines, which Owen originally interpreted as carpal spines, analogous to the defensive thumb spikes in *Iguanodon*. (By that time Owen realized that *Iguanodon's* nasal horn belonged on its hand instead.) Another genus represented by a juvenile, *Dacentrurus phillipsi* (described as *O. phillipsi* by Seeley in 1893) may be the only stegosaur known from the Oxfordian age. Following this discovery, in 1902, F.A. Lucas reestablished *Omosaurus* as *Dacentrurus*, noting that the former name had already been assigned to a crocodylian (*Omosaurus perplexus*) by Joseph Leidy in 1856. Lucas stated, "I propose the name *Dacentrurus* in allusion to the powerful spines with which the tail was armed." Another *Dacentrurus* species—*D. lennieri*, discovered in France and described by Nopsca in 1911, was at one time thought to be associated with a fossil egg—an exciting possibility! Besides France and England, *Dacentrurus* is also known from Portugal, which is where the egg in question was discovered in 1908. However, the 7-inch-long by 5-inch-wide fossil wasn't perceived as an egg until 1957, when Albert Lapparent and Georges Zbyszewski reconsidered its nature. Now, because similarly oval-shaped Late Jurassic eggs recently discovered in North America and also in Portugal (i.e. Lourinha, where an entire clutch was found) are regarded as theropod (*Preprismatoolithus*), the European eggs may not be allied to *Dacentrurus* after all. According to Galton's 1985 reassessment, "*Dacentrurus* probably represents a persistently conservative Kimmeridgian survivor of a lineage that probably diverged from the Stegosaurian stock relatively early in its phyletic history." *Dacentrurus'* closest evolutionary cousin is *Hesperosaurus*, the oldest stegosaur genus known from North America, described in 2001

by Ken Carpenter, Clifford Miles and Karen Cloward. (See their article, "New Primitive Stegosaur from the Morrison Formation, Wyoming," in *Armored Dinosaurs*, ed. Carpenter and Miles, pp. 55–75.) Armored dinosaur expert Dr. Kenneth Carpenter claims that *Hesperosaurus* had an alternating double row of plates (because no two plates are identical) and 4 tail spikes (because all known North American stegosaur species have only 4 spikes).

57. De Klerk, "The First Dinosaur Fossil Discovered in South Africa."

58. In 1909, Robert Broom (1886–1951) examined the Bushmans River material, noting the similarities between teeth in the "Cape *Iguanodon*" specimen and other herbivorous dinosaurs, particularly an armored North American ankylosaur genus then known as *Paleoscincus costatus*. Therefore, in 1912, Broom renamed the fossil "*Paleoscincus africanus*." (In 1966, Alfred Romer (1894–1973) also interpreted *Paranthodon* as an ankylosaur.) Four years later additional remains of a Cape dinosaur were found, although these were incorrectly referred to Owen's former name—"*Anthodon serrarius*"—by Prof. E.H.L. Schwarz (1873–1928). Even though these fossils had been collected in the vicinity where Cape *Iguanodon* had been found, they may not be stegosaurid. In 1929, however, F. Nopsca (1877–1933) realized that the Cape *Iguanodon* was indeed a variety of stegosaur. So Nopsca designated it *Paranthodon oweni*, although 136 years after discovery of its remains the species name was changed to P. *africanus* in 1981, following detailed study by Peter Galton and Walter Coombs. Galton and Coombs determined *Paranthodon* was decidedly a stegosaur, not an ankylosaur because, "In the Ankylosauria the outer bones of the skull are overlain by fused dermal plates and the maxillary tooth row is inset medially such that the premaxilla and maxilla form a wide almost horizontal shelf lateral to the tooth row. *Paranthodon* is therefore to be excluded from the Ankylosauria and is referred to the Stegosauridae." From their analysis of fossil tooth morphology, they concluded *Paranthodon* and African stegosaur genus *Kentrosaurus* were most closely related, although not synonymous. In 1995, to commemorate the 150th anniversary of *Paranthodon's* discovery, a five-meter long restoration was completed by Gerhard Marx. The captivating sculpture is on display in South Africa's Albany Museum, Grahamstown. Even though no dermal spines or plates have been attributed to the genus, Marx's *Paranthodon* sports a paired, double row of plates extending from the cervical area to midway along the tail, and two pairs of tail spikes. Marx based his interpretation of the poorly known genus on the anatomy of more complete plated dinosaur genera, partic-

ularly *Kentrosaurus* and Chinese genus, *Tuojiangosaurus*, as opposed to *Stegosaurus*. Why? Dr. de Klerk notes how *Paranthodon*'s teeth bear closer similarities to *Tuojiangosaurus* than to *Stegosaurus*. Furthermore, *Paranthodon* lived geographically closer to *Kentrosaurus* than *Stegosaurus*, and the back plates in *Paranthodon* and *Kentrosaurus* are both narrow and pointed, unlike those in *Stegosaurus*. Paranthodon was found in Early Cretaceous sediments dated at 135 million years. The genus, estimated to have grown to 17 feet in length, may not have been the oldest stegosaur geologically, but it is usually cited as the first to be discovered, predating Marsh's Stegosaurus by 32 years. *Stegosaurus* was not the first stegosaur discovered after all!

59. Intriguingly, in 2001, British dinosaur expert David Norman asked an analogous question as to why Owen "failed to use the discovery of another British armored dinosaur, *Scelidosaurus* [discovered in 1858, or 5 years too late for Hawkins' handiwork] to promote the intellectual rigor of his earlier vision of the dinosaurs" (D. Norman, "*Scelidosaurus*, the Earliest Complete Dinosaur," in *Armored Dinosaurs*, ed. Carpenter and Miles, pp. 3–24). "The Early Jurassic, British dinosaur *Scelidosaurus*, another armored form thought to be ancestral to the later evolving ankylosauria and stegosauria, was certainly the earliest, most complete dinosaur ever discovered. While Owen described and helped restore numerous dinosaurs based on incomplete remains, he missed a splendid opportunity to embellish his description of *Scelidosaurus* with restorations. Why? In this case, Owen was simply too preoccupied at the time to provide much more than his conventional description of the fossil slab left largely unprepared, published in 1861. However, Owen's reluctance to bring *Scelidosaurus harrisoni* to life, visually, contrasts with former enthusiasm held for his collaboration with Hawkins a decade earlier toward the restoration of Secondary saurians (i.e., Mesozoic reptiles) that were, ironically, more poorly represented as fossils than *Scelidosaurus*. The Crystal Palace prehistoric landscape display featured European dinosaurs and other extinct fauna, yet several genera were sculpted or drafted by Hawkins for the exhibit that were indigenous to two other continents, North and South America. While it's impossible to speculate how a vintage Crystal Palace stegosaur sculpted by Benjamin Waterhouse Hawkins would have looked, my guess is it would have resembled his *Hylaeosaurus*, although less spiky-looking and with plates added somewhere. A related question would be why Owen and Hawkins chose not to restore the spectacular British whale-lizard *Cetiosaurus*, described in 1841, in sculpture, perhaps posed

swimming alongside their mosasaur, plesiosaurs and ichthyosaurs."

60. George Olshevsky and Tracy L. Ford, "The Origin and Evolution of the Stegosaurs," *Gakken Mook*, no. 4 (1995), pp. 92–119. *Craterosaurus* was described by H. Seeley in 1874, based on a distinctive, incomplete dorsal vertebra found near Potton, Bedfordshire, England. Nopsca identified the holotype as stegosaurid in 1911. The holotypes of *Regnosaurus* and *Craterosaurus* are incomparable, because *Craterosaurus* is attributed to the same geological horizon as *Regnosaurus*, a circumstance leading to Olshevsky's conclusion that *Regnosaurus* and *Craterosaurus* were in fact the same, with the name *Regnosaurus* taking priority. Olshevsky believed *Regnosaurus* was dacentrurid, while Paul Barrett and Paul Upchurch later viewed *Regnosaurus* as a huayangosaur, in 1995. Barrett and Upchurch determined that *Regnosaurus* evinced evolutionary affinities to the Chinese genus, *Huayangosaurus*, a primitive, Middle Jurassic plated dinosaur thought to be ancestral to later stegosauridae. Because *Regnosaurus* (found even before South Africa's *Paranthodon*) appears to be the most primitive stegosaur of all, it may represent a relict population from Asia surviving into the Cretaceous. One wonders, if Wealden stegosaurs existed, where are their elusive yet telltale fossilized plates, dermal spines and tail-spikes? While attempting to standardize criteria for classifying British fossilized armor, recently in 2001, William Blows distinguished five characteristics evident in stegosaurian spines. While several of the British dermal spines *could* be referable to *Regnosaurus*, Blows cautiously preferred to regard them with uncertainty as Stegosauria *incertae sedis*. In contrast to Olshevsky's clade-lumping strategy, Blows viewed the holotypes of *Regnosaurus* and *Craterosaurus* as incomparable and therefore distinctive and separate. (See Blows, "Possible Stegosaur Dermal Armor from the Lower Cretaceous of Southern England," in *Armored Dinosaurs*, ed. Carpenter and Clifford, pp. 130–140.

61. Paul M. Barrett and Paul Upchurch, "*Regnosaurus northamptoni*, a Stegosaurian Dinosaur from the Lower Cretaceous of Southern England," *Geological Magazine* (Cambridge University Press), Vol. 132, no. 2 (1995), pp. 213–222.

62. O.C. Marsh, "Restoration of *Stegosaurus*," *American Journal of Science*, Series 3, Vol. 42 (1891), pp. 179–181.

63. O.C. Marsh, "A New Order of Extinct Reptilia (Stegosauria) from the Jurassic of the Rocky Mountains," *American Journal of Science*, Series 3, Vol. 14 (1877), pp. 513–514.

64. Michael F. Kohl and John S. McIntosh, *Discovering Dinosaurs in the Old West: The Field Journals of Arthur Lakes* (Washington, DC: Smithsonian Institution Press, 1997), p. 22.

65. Stephen A. Czerkas, "Discovery of Dermal Spines Reveals a New Look for Sauropod Dinosaurs," *Geology*, Vol. 20 (Dec. 1992), pp. 1068–1070.

66. Kenneth Carpenter and Peter M. Galton, "Othniel Charles Marsh and the Myth of the Eight-Spiked *Stegosaurus*," *Armored Dinosaurs*, ed. Carpenter and Clifford, pp. 76–102; Allen A. Debus and Diane E. Debus, *Paleoimagery*, pp. 73–82; my short compilation listing some of the earliest (and sometimes most quaint) stegosaur restorations and reconstructions, neglecting actual museum skeletal displays, includes the following: (1) A. Tobin (illustration), 1884. (2) Marsh (skeletal reconstruction), 1891. (3) Carl Dahlgren, (painting), 1892. (4) J. Smit (painting), 1892. (5) J. Smit (painting), 1893. (6) Charles R. Knight (painting, 1897). (7) Charles R. Knight (2nd painting), 1897. (8) Charles R. Knight (sculpture), 1899. (9) W.C. Knight and Frank Bond (painting), 1899. (10) F. John, (painting), circa 1905. (11) Lawson Wood (illustration), 1900. (12) Arthur Lakes (painting), circa 1900. (13) Charles R. Knight (painting) 1901. (14) G.E. Roberts and R. Lucas (illustration), 1901. (15) Heinrich Harder (painting) 1902. (16) Life-sized scale model displayed at National Museum of Natural History, circa 1904. (17) E. Ray Lankester (painting), 1905. (18) Richard Lydekker (illustration), circa 1900. (19) Hubert V. Zwickle (illustration), 1908. (20) G.E. Roberts (painting), 1910; (21) Richard Swann Lull (sculpture), 1910. (22) Josef Pallenberg (life-sized sculpture), circa 1910. (23) J. Smit (painting), 1911. (24) Alice B. Woodward (painting), 1912. (25) Charles Whitney Gilmore (sculpture), circa 1915. (26) Gerhard Heilmann (illustration), 1916. (27) Othenio Abel (illustration), 1920.

67. In 1868, Huxley sketched *Cetiosaurus* as a bipedal animal before its skeletal anatomy was properly interpreted; as yet the front limbs were unidentified. This cartoon was reproduced in Adrian Desmond's *Huxley: From Devil's Disciple to Evolution's High Priest* (Reading, MA: Perseus Books, 1997), Figure 21.

68. Hutchinson, *Extinct Monsters*, Plate IV facing p. 69.

69. A. Debus and D. Debus, *Paleoimagery*, pp. 226–233; A. Desmond, *The Hot-Blooded Dinosaurs*, pp. 141–190.

70. This painting was recently reproduced in Sylvia Czerkas' and Donald F. Glut's *Dinosaurs, Mammoths, and Cavemen: The Art of Charles R. Knight* (New York: E.P. Dutton, 1982), Plate 44.

71. William Ballou, "Strange Creatures of the Past," *The Century Illustrated Monthly Magazine* (New York: Century), Vol. 55, New Series, Vol. 33 (Nov. 1897 to April 1898), pp. 21–22.

72. Czerkas and Glut, *Dinosaurs, Mammoths, and Cavemen*, p. 42.

73. "Most Colossal Animal Ever on Earth Just Found Out West," *New York Journal*, Dec. 11, 1898, p. 1. For more on this, see David Rains Wallace, *The Bonehunters' Revenge: Dinosaurs, Greed, and the Greatest Scientific Feud of the Gilded Age* (Boston, New York: Houghton Mifflin, 1999), pp. 281–282.

74. Related Jurassic genus, *Allosaurus*, would soon join the popular dinosaur ranks too, especially following exhibition of an incredible American Museum display mounted in 1907. One of Knight's most brilliant, accompanying painted compositions, dated 1904, of this scene would be widely circulated in books and magazine articles.

75. Allen A. Debus, "Fin-tastic Mammals—A Quick Look at 'Naosaurus'" (parts 1–2), *Prehistoric Times*, nos. 54–55 (June/July 2002), pp. 17–19 (Aug./Sept. 2002), pp. 17–19. These articles were reprinted in A. Debus, and D. Debus, *Dinosaur Memories*, pp. 225–241.

76. H.F. Osborn, "A Great Naturalist. Edward Drinker Cope," *The Century Illustrated Monthly Magazine* (New York: Century), Vol. 55, New Series, Vol. 33 (Nov. 1897 to April 1898), pp. 10–15.

77. A. Debus and D. Debus, *Dinosaur Memories*, p. 237.

78. See note 75.

79. Another peculiar early restoration of the *Dimetrodon gigas*, by Carrie Gage, was published as figure 56 in *Hunting Dinosaurs in the Bad Lands of the Red Deer River, Alberta Canada*, by Charles Hazelius Sternberg, introduced by David A.E. Spalding (1st ed. 1917; Edmonton: NeWest Press, 1985 ed.). Among his vast collection of dinosaurabilia, Donald F. Glut owns one of Knight's as yet unpublished "Naosaurus" restorations (pencil drawing).

80. Debus, Allen A. "Get Real! Dinosaur Masquerade," *G-Fan*, no. 65 (Nov./Dec. 2003), pp. 28–34.

81. This picture is reproduced and discussed in Wallace, *The Bonehunters' Revenge*, pp. 258–261.

82. The uintathere episode was recently described in A. Debus and D. Debus, *Dinosaur Memories*, "RU-inta-'theres'?" pp. 111–127.

83. Ibid., p. 114.

84. Huxley had already demonstrated that paleontologists need not slavishly adopt Cuvier's law of the correlation of anatomical parts in making scientific restorations of extinct vertebrates. Instead, paleontologists should consider ancestral phylogeny as a more reliable guide and avenue of contemplation.

85. William Gunning, *Life History of Our Planet* (New York: Worthington, 1879); concerning Cope's 1873 sketches of *Eobasileus*, see figure 16 in H.F. Osborn, *Cope: Master Naturalist* (Princeton: Princeton University Press, 1931).

Cope wrote, "They had a proboscis I am quite sure now, and walked with the knee far below the body as elephants do. A form different enough from elephants generally, but reminding one more of the hog."

86. A. Debus and D. Debus, *Paleoimagery*, pp. 83–96.

87. O'Connor, *The Earth on Show*, pp. 91–93.

88. Ibid., p. 94; Martin J.S. Rudwick, *Worlds Before Adam: The Reconstruction of Geohistory in the Age of Reform* (Chicago: University of Chicago Press, 2008), p. 80.

89. Rudwick, *Worlds Before Adam*, p. 87.

90. While beyond the scope of the present study, Rudwick's scholarly *Worlds Before Adam* is highly recommended for in-depth study of this complex episode in the early history of paleontology.

91. Allen A. Debus, "A Pleistocene Primer: Glyptomaniac with Big Bird" (parts 1–3), *Fossil News: Journal of Avocational Paleontolog*, Vol. 8, nos. 4–6 (April 2002), pp. 4–9, (May 2002), pp. 15–17, (June 2002), pp. 8–11.

92. Desmond, *The Hot-Blooded Dinosaurs*, pp. 481–482.

93. Ronald Rainger, *An Agenda for Antiquity: Henry Fairfield Osborn and Vertebrate Paleontology at the American Museum of Natural History, 1890–1935* (Tuscaloosa: University of Alabama Press, 1991), pp. 152–181; W.D. Matthew and S.H. Chubb, *Evolution of the Horse*, 7th ed., Guide Leaflet Series no. 36 (New York: American Museum of Natural History, n.d. but circa 1930).

94. H.F. Osborn, *The Titanotheres of Ancient Wyoming, Dakota, and Nebraska*, United States Geological Survey Monograph 55 (two volumes), 1929; Donald R. Prothero, *The Eocene-Oligocene Transition: Paradise Lost* (New York: Columbia University Press, 1994).

95. Rainger, *An Agenda for Antiquity*, p. 166.

96. Ibid., pp. 166, 169.

97. A. Debus and D. Debus, *Dinosaur Memories*, pp. 128–144.

98. H.F. Osborn, "Prehistoric Quadrupeds of the Rockies," *The Century Illustrated Monthly Magazine* (New York: Century), Vol. 52 (Sept. 1896), pp. 705–715.

99. Wallace, "The Antiquity of Man in North America," p. 281.

100. James Erwin Culver, "Some Extinct Giants," *The California Illustrated Magazine*, Vol. 2 (April 1892), pp. 501–507.

101. William Ballou, "The Serpentlike Sea Saurians," *Popular Science Monthly*, Vol. 53 (June 1898), pp. 209–218.

102. Ibid., p. 209.

103. Davidson, *A History of Paleontology Illustration*, pp. 52–54.

104. H.N. Hutchinson, "Prehistoric Monsters," *Pearson's Magazine*, Vol. 10, no. 6 (Dec. 1900), pp. 628–637.

105. R.I. Geare, "Some Extinct Animals," *Outdoor Life*, Vol. 26, no. 1 (1910), pp. 6–11; contrary to Geare's crude estimations, today's Geological Time Chart, bolstered by modern radioactive age dating techniques unavailable in 1910, indicates that the Cenozoic Era (beginning of the Tertiary Period) began 65 million years ago, while the preceding Mesozoic Era commenced 245 million years ago.

106. E. Ray Lankester, *Extinct Animals* (New York: Holt, 1905).

107. H.R. Knipe, *Nebula to Man* (New York: J. M. Dent, 1905); H.R. Knipe, *Evolution in the Past* (New York: J. B. Lippincott, 1912); William Berryman Scott, *A History of Land Mammals in the Western Hemisphere* (originally published 1913; New York: Macmillan, 1937).

108. Barnum Brown, "Hunting Big Game of Other Days," *The National Geographic Magazine*, Vol. 35, no. 5 (May 1919), pp. 407–429.

109. B. Webster Smith, *The World in the Past: A Popular Account of What It Was Like and What It Contained* (London: Frederick Warne, 1931 ed.).

110. Also of interest, yet beyond scope of the present volume, are 1910s and 1920s restorations and publications by Austrian paleonbiologist, Othenio Abel (1875–1946). See chapter 14 in A. Debus and D. Debus, *Paleoimagery*, for more on Abel.

111. Harold J. Shepstone, "Big Game of Other Days," in *Wild Life of Our World* (London and Glasgow: Collins' Clear Type Press, 1934).

112. Charles R. Knight, *Before the Dawn of History* (New York: McGraw-Hill, 1935).

113. For his February 1942 *National Geographic Magazine* article (Vol. 81, no.2, pp. 141–184), "Parade of Life through the Ages," Knight offered another grand geological time tour featuring Life's history, glimpsed through 24 new paintings. This time, discounting restorations of cavemen, "Naosaurus," and fossil birds, fossil mammals — including depictions of lumbering brontotheres and sharp-horned arsinotheres — edged the dinosaurs by a tighter nine to five margin.

114. Knight, *Before the Dawn of History*, p. vii.

115. Ibid., pp. 110–111.

116. W. Maxwell Reed, and Janette M. Lucas, *Animals on the March* (New York: Harcourt, Brace, 1937). Reed was also author of another popular publication, *The Earth for Sam* (New York: Harcourt, Brace, 1930), an illustrated account of Life's history.

117. Letter dated December 14, 1932.

118. Sinclair also prepared a "Grotto of the Sinclair Exhibit" adjacent to their outside dis-

play where visitors could view additional exhibits illustrating how oil formed in the earth and how it was processed. Through its exhibition, Sinclair wanted to express that dinosaurs were *living* when the petroleum they extracted from the ground was already forming underground. According to Donald F. Glut (*The Dinosaur Scrapbook*, Secaucus, NJ: Citadel Press, 1980, p. 35), "Visitors to the Sinclair dinosaur exhibit in 1933 were transported back through time hundreds of millions of years, as they passed through the artificial rock grotto to emerge in a facsimile of the Mesozoic era."

119. Ibid., p. 36; The third display at this Fair that included three dinosaur displays based on Knight's Field Museum designs was prepared by the Century Dioramas Studios. Pictures of two of their miniature dioramas may be seen in Glut's *Dinosaur Scrapbook*, p. 37.

120. Don Glut owns two Cave Bears, an *Archaeopteryx*, wax caveman likenesses and a Pterodactyl. According to Glut (personal comm. 11/08/08), another collector bought one of the two *Archaeopteryx*es. Mr. Messmore informed Glut that many of the larger mechanical creatures were sold in the 1970s to a resort in Nagoia, Japan, and that the big Bronto and mammoth were donated to a museum devoted to circus and carnival attractions in Indianapolis. My father, Allen G. Debus, who was seven at the time of his visit, recalls seeing a Brontosaurus neck reaching over the walkway, toward foliage overhead. Current whereabouts of Sinclair's 1933–34 dinosaurs are unknown. Additionally, Glut, the consummate dinosaur collector, owns a framed painting apparently done as a design concept for the World a Million Years display showing a *Tyrannosaurus* engaging a *Triceratops* in battle, with two enthralled spectators standing behind an exhibit railing. Glut describes this rare curiosity on his collectors website as "Original watercolor painting ... by George Messmore depicting a tableau from Messmore and Damon's World a Million Years Ago."

121. Quoted term "mastodonic miracle" is from George Turner's (with Orville Goldner) *Spawn of Skull Island* (Baltimore: Luminary Press, 2002), p. 22.

122. H.G. Wells, *Mr. Blettsworthy on Rampole Island* (Garden City, NY: Doubleday, Doran, 1928).

123. By 1947, paleoartist Rudolph Zallinger completed his mesmerizing panoramic view of "The Age of Reptiles" displayed at Yale University's Peabody Museum. Whereas Knight had covered the whole of geological time for the Chicago Natural History Museum two decades, as exemplified through Zallinger, artists began to increasingly focus on the "reptile" portion of this continuum. For more on Zallinger's mid-

20th century contributions to portraying the ages of life, see chapter 16 in A. Debus and D. Debus, *Paleoimagery.*

Chapter Seven

1. For instance, although omitting Sternberg's contribution, Dennis R. Dean outlined "The Influence of Geology on American Literature and Thought," *Two Hundred Years of Geology in America: Proceedings of the New Hampshire Bicentennial Conference on the History of Geology*, ed. Cecil J. Schneer (published for the University of New Hampshire by the University Press of New England, Hanover, New Hampshire, 1979), pp. 289–303. Also see A. Debus, *Dinosaurs in Fantastic Fiction* (2006), as well as Ralph O'Connor's *The Earth on Show* (2007).

2. Charles H. Sternberg's articles and books consulted in the course of researching this chapter included the following: "Pliocene Man," *American Naturalist* (Philadelphia: Press of McCall and Stavely), Vol. 12, no. 2 (Feb. 1878), pp. 125–126; "Ancient Monsters of Kansas." *Popular Science News*, Vol. 32 (Dec. 1898), p. 268; *The Life of a Fossil Hunter* (1909; 2004 ed. with Foreword and notes by Paul F. Ciesielski, Gainesville, FL: Faulkner Press); "A New Trachodon," *Science*, Vol. 29 (1909), pp. 753–754; *A Story of the Past, or the Romance of Science* (Boston: Sherman, French, 1911); *Hunting Dinosaurs in the Bad Lands of the Red Deer River, Alberta, Canada: A Sequel to The Life of a Fossil Hunter* (1st ed., published by Charles H. Sternberg, Lawrence, Kansas, 1917); *Hunting Dinosaurs in the Bad Lands of the Red Deer River, Alberta, Canada: A Sequel to The Life of a Fossil Hunter* (3rd ed., Edmonton: NeWest Press, 1985). Only 500 self-published copies of the first edition of *Hunting Dinosaurs* (March 1917) were printed. Sternberg's own 2nd edition of 1932 added another 30 pages and 5 chapters, bringing his account up to date. While typos were corrected and some poetry including a sermonized bit of theological geology titled "The Laurentian Hills," pp. 215–219 (1st ed.), were expunged, emphasized in the additions were fossil hunting activities of the 1920s.

3. Sternberg's "pen-picture" term (*Life of a Fossil Hunter*, p. 210) is the earliest I have read used in this vein. Pen-pictures themselves represent a more vivid form of popular, speculative scientific writing which had become prevalent by the late 19th century as, for example, composed by American Museum vertebrate paleontologist Henry Fairfield Osborn (e.g., "Prehistoric Quadrupeds of the Rockies," in *Century Magazine*, Sept. 1896, Vol. LII, 89, pp. 705–715), who therein deftly imparted *animus* to the usually dry, sterile published technical descriptions of fossil mammals.

4. In 1985, David A.E. Spalding acknowledged Sternberg's science fiction story in an Introduction to the 3rd NeWest edition of *Hunting Dinosaurs*. Spalding stated that this tale "sheds interesting light on Sternberg's personality ... that it satisfied some deep personal needs" (p. xxix). Sternberg excelled at discovering important fossils, but in later life assuredly regretted his lack of formal scientific training and inability to publish scientific papers about these specimens. His technical writing lacked depth and competency sufficient for scrutiny by professional scientists— those who enthusiastically received fossils Sternberg had laboriously collected. So he fulfilled other purposes in writing — a practice he had natural talent for. Spalding noted how Sternberg was "clearly a natural storyteller" (p. xxvi). Indeed his autobiographical accounts succeed in making the reader "feel as if he is listening to him beside the campfire ... we can taste the blowing sand, feel the hot sun on our backs" (pp. xxvi, xxxviii) Spalding suggested Sternberg cherished opportunities to perform "God's work" in revealing these ancient creatures to the public. Furthermore, Sternberg may have also feared "personal oblivion" over fading memory of his considerable career accomplishments (p. xiii).

5. Sternberg later remarked, "I love creatures of other ages; and ... I want to become acquainted with them in their natural environments. They are never dead to me; my imagination breathes life into 'the valley of dry bones,' not only do the living forms of the animals stand before me, but the countries which they inhabited rise for me through the mists of these ages" (*Life of a Fossil Hunter*, p. 153).

6. It is also apparent that Sternberg believed that man was the "crowning work of the Creator's hands" (p. 164), that man wasn't coeval with the mammoths of North America (p. 194) and certainly didn't live during the Pliocene as Cope and nemesis Othniel C. Marsh (1831–1899) independently suggested in 1878 (pp. 120–124). Cope, and presumably Sternberg, espoused orthogenetic evolutionary principles (p. 202). See Sternberg and Cope ("Pliocene Man," pp. 125–126). Through Cope's sense of one-upmanship, on the basis of fossil evidence, this paper sought to substantiate on basis of fossils the largely theoretical claims of Cope's scientific rival, O.C. Marsh ("Man in the Pliocene in America," *American Naturalist*, Vol. 40, no. 9, Nov. 1877, pp. 689–690), who in 1877 had proclaimed that man had lived in North America during the Pliocene epoch. By 1909, however, Sternberg conceded that both scientists were wrong and that the human remains (attributed by Cope) to "Pliocene man" in his letter were instead deposited in recent historical times (Sternberg, *Life of a Fossil Hunter*, pp. 120–124).

7. For more on Cuvier's pervading influences during the 19th century, see Rudwick's *Scenes from Deep Time* (1992, chapter 2, pp. 27–58). According to Claudine Cohen (*The Fate of the Mammoth: Fossils, Myth, and History*, Chicago: University of Chicago Press, 2002, pp. 123–124), Cuvier's scientific writings are immensely imaginative. Cohen regards Cuvier as the "great poet" of his era, because his work "created a dreamlike world full of powerful images that still survive in popular and scientific representations of paleontology." In particular, Cuvier's *Reserches sur les ossemens fossiles de quadrupeds*, Vol. 1 (Paris, 1812), "can be read like a kind of novel whose hero is the paleontologist himself, the recreator of worlds, capable of reconstructing an animal from one of its claws or a fragment of bone." And Rudwick (*Bursting the Limits of Time*, 2005, p. 179) notes that Comte Buffon's 18th century geotheoretical systems were dismissed as "mere fantasy or novel ... or in modern terms as little more than science fiction."

8. Besides physician Conan Doyle and Sternberg, contemporaries included W.D. Matthew, Gerhard Heilmann and Vladimir Obruchev. For example, in an enlightening chapter titled "Some Fossil Birds" in *The Origin of Birds* (1926), Danish avian paleontologist/artist Heilmann (1859–1946) included elaborate pen-pictures (pp. 32–38, 51–56) of living Mesozoic birds to embellish (and otherwise lighten) his highly technical descriptions. Russian geologist Vladimir Obruchev's (1863–1956) *Plutonia* (1924), wasn't a pen-picture but rather a complete novel clearly derived from Verne's influences. Obruchev stated, "I decided to write the book after I had reread as an adult Jules Verne's *Journey to the Center of the Earth*. I was an experienced explorer by then and felt that his description of an underground voyage was not realistic; besides, many new facts on the prehistoric inhabitants of our planet have been uncovered since the novel was written" (pp. 9–10); William D. Matthew, "Scourge of the Santa Monica Mountains," *The American Museum Journal*, Vol. 16, no. 7 (1916), pp. 469–472.

9. Sternberg wrote numerous articles for scientific journals as well as the Kansas Academy of Sciences. In one of his earliest contributions, "Ancient Monsters of Kansas" (*Popular Science News*, Vol. 32, Dec. 1898, p. 268), Sternberg vividly described an imaginary combat between marine saurians including *Tylosaurus*. His pen-picture dramatically unfolds following these words, "How shall I give even a slight idea of these wonderful birds, reptiles, pterodactyls, fishes and shells? Shall we in imagination walk along the shores of that grand ocean, explore its bays and estuaries, and replace the flesh on the bony skeletons, that only remain to teach us of

their power when full of life and motion?" Clearly, over a decade prior to *Life of a Fossil Hunter*'s publication, Sternberg was already honing his fictional craft.

10. Sternberg, "A New *Trachodon*," p. 12.

11. Ibid., pp. 37–38.

12. Ibid.

13. Ibid., p. 210.

14. Science fiction writer Robert Silverberg has recently stated that all fiction can be boiled down to his "universal theory" of the story — all fiction — in which, "A sympathetic and engaging character, faced with some immensely difficult problem that it is necessary for him to solve, makes a series of attempts to overcome that problem, frequently encountering challenging sub-problems and undergoing considerable hardship and anguish, and eventually, at the darkest moment of all, calls on some insight that was not accessible to him at the beginning of the story and either succeeds in his efforts or fails in a dramatically interesting and revelatory way, therefore by arriving at new knowledge of significant kind." See Silverberg's articles titled "Toward a Theory of Story" (Parts 1 to 3) in *Asimov's Science Fiction*, April/May 2004, pp. 4–9; June 2004, pp. 4–9; and July 2004, pp. 4–9. Also see O'Connor, *The Earth on Show*, pp. 361–391, for discussion of story versus straightforward paleontological description.

15. Sternberg, *Story of the Past*, p. 4.

16. Ibid., pp. 57–58; Sternberg wrote these touching lines:
"My Master once wept at Lazarus' grave,
And bade him awake and come forth from the cave,
I believe that my darling, now free from earth's clay,
On the wings of God's love has flitted away
To the shores of the blessed, where safely she'll roam,
Redeemed through His blood; in her glorified home"

17. Sternberg, *Hunting Dinosaurs* (NeWest, 1985 edition), p. 35.

18. Here, Sternberg daringly, yet non-scientifically, used artistic license, merging paleo-environments of similar age but separated spatially to recreate his single setting. The landscape Sternberg leads us through is a composite of Late Cretaceous localities he collected fossils in before, especially the Judith River formation of Montana, now known to be 74 to 79 million years old. However, the "Trachodon" summoned for this particular segment was a former denizen of what is now Wyoming. According to Katherine Rogers (*The Sternberg Fossil Hunters: A Dinosaur Dynasty*, Missoula: Mountain Press Publishing, 1999, p. 154), Sternberg's science fiction tale received unusual impetus, following "a strange dream that continued to haunt him."

One day in the summer of 1915 he had wandered into a coal miner's tunnel in an attempt to escape the midday heat. He found the tunnel floor covered with fine dust that, he found, made a soft bed. He never knew how long he slept, but he dreamed time suddenly flashed back to the era of the Cretaceous ocean. Here — with his wife, sons, grandchildren, and especially his beloved daughter Maud — he watched the activities of an imaginary prehistoric era unfold. So vivid was his fantasy that he made sufficient notes to preserve the tempo, scene, action, and mystery until he could write it up for publication." Quote is excerpted from Sternberg, *Hunting Dinosaurs*, p. 133.

19. Niobraran deposits are now dated at 82 to 87 million years old. Overall, the Cretaceous Period — sliced into several shorter geological stages — lasted from 65 to 144 million years ago. Permian rocks are now known to be from 250 to 285 million years old.

20. Sternberg, *Hunting Dinosaurs*, Introduction by Spalding, p. xxxiv; Leo F. Laporte, *George Gaylord Simpson: Paleontologist and Evolutionist* (New York: Columbia University Press, 2000), p. 259. Sternberg's personal plight and fantasy flight from modernity may be considered in concert with a more recent and commercialized, similarly themed production — paleobiologist George Gaylord Simpson's short novel *The Dechronization of Sam Magruder* (Introduction by Arthur C. Clarke; Afterword by Stephen Jay Gould; New York: St. Martin's Griffin, 1996). Rather than a simple waking dream re-materialization, instead, founded on a modicum of disbelief-suspending quantum mechanical mumbo jumbo, in *Magruder* physicist Sam Magruder accidentally becomes temporally "de-chronized," cast abjectly into the Cretaceous Period. "In vulgar terms, he suffered a time-slip" (p. 8). While Sternberg revels in exalted prehistory, providentially assured that his journey will ultimately serve some hallowed (self-serving) purpose, Magruder, anachronistically thrust into the age of cold-blooded dinosaurs, toils much like a caveman through a lonely, miserable existence before comprehending the somber meaning of his solitary journey backward in time. Magruder quickly learns how to dodge marauding tyrannosaurids, and how to procure food and primitive clothing. Defying odds, he survives for two decades before being mortally wounded by a tyrannosaur. In both Sternberg's and Simpson's stories, prehistory settings are nostalgic, reflecting individualized views of life — yet how different are these perspectives, sullen, gloomy yet accepting in Simpson's case, while Sternberg's is vibrant, uplifting, moralistic and at times downright preachy. Strikingly, Sternberg rejoices in the splendor of a deified Prehistory, while Magruder

broods in a metaphorical Cretaceous Prison. During his life, Simpson was a much more solitary, melancholy soul than Sternberg, which is why he may have kept his *Magruder* manuscript hidden for others to find, posthumously. Had their fictionalized paths crossed in the Cretaceous, no doubt it would have been a chore for Sternberg and Maude to cheer the forlorn Sam Magruder (i.e., Simpson personified), or to view Cretaceous splendor providentially through their eyes. During their declining years, both highly accomplished paleontologists — scholarly Simpson with his hard-earned Yale Ph.D. and field-hardened Sternberg with a year of undergraduate education at the Kansas State Agricultural College — suffered from similar afflictions, anxiety as to whether future scholars would find value in their scientific contributions. Or would they be forgotten? Reflecting shared angst, both men sought solace in writing fiction near the end of their professional careers — expressive, personalized stories sharing inward philosophies, baring their souls over self-worth with readers. Did their lives really matter? From his close-knit family huddle in Cretaceous sanctuary, the loquacious, providentially minded Sternberg cries out "yes," while, imprisoned in Magruder's symbolic solitary confinement, the more agnostic evolutionist Simpson — who was more cognizant of Time's long, lonely corridors and even its imaginary nature — might hedge on this point.

21. Scottish geologist Hugh Miller (1802–1856) never wrote fiction, per se, he did refine the craft of writing geological pen-pictures. For his time, Miller was highly admired for excelling in the writing of paleo-themed descriptive passages and verbal recreations of prehistoric worlds, particularly those of the early Paleozoic Era, the age of rocks in which he collected numerous fossils. A set of his popular articles were published as a collection in 1851, titled *The Old Red Sandstone; or New Walks in an Old Field.* Concluding chapters twelve through fourteen vividly describe these geological periods as if one were witnessing them as a firsthand observer. For instance, he began chapter 12 stating, "I shall now attempt presenting it (i.e., the old red sandstone geological period), as it existed in *time* — during the succeeding periods of its formation" (pp. 212–213). Later, in chapter 14 a new geological tour commences with these words, "The curtain rises and the scene is new" (p. 243). This sort of technique is conceptually perhaps, for paleontologists, preparatory to writing science fiction. It also represents a creative fantasy style and perspective adopted, or reinvented, by Charles H. Sternberg over half a century later.

Chapter Eight

1. Donald F. Glut, *Dinosaurs: The Encyclopedia, Supplement 1* (Jefferson, NC: McFarland, 2000), pp. 359–360.

2. Christopher A. Brochu, "Osteology of *Tyrannosaurus rex*: Insights from a Nearly Complete Skeleton and High-Resolution Computed Tomographic Analysis of the Skull," *Society of Vertebrate Paleontology — Memoir 7,* Vol. 22, Supplement to Number 4 (January 14, 2003), p. 1.

3. Adrienne Mayor, *Fossil Legends of the First Americans* (Princeton: Princeton University Press, 2005), pp. 263–264; Peter Larson, *Rex Appeal* (Montpelier, VT: Invisible Cities Press, 2002) pp. 274–276.

4. Larson, *Rex Appeal,* p. 276; for details on Cope's expedition see Anthony R. Fiorillo and Edward Daeschler, "E.D. Cope's 1893 Expedition to the Dakotas Revisited," *Earth Sciences History,* Vol. 9, no. 1 (1990), pp. 57–61.

5. Larson, *Rex Appeal,* p. 277. Don't worry — *T. rex* will keep its distinctive name.

6. Jackson, Donald. *Custer's Gold: The United States Cavalry Expedition of 1874* (originally Yale University Press, 1966; Lincoln: University of Nebraska Press–Bison Books, 1972).

7. John R. Horner and Don Lessem, *The Complete T. rex* (New York: Simon & Schuster, 1993), p. 58.

8. Donald F. Glut, *Dinosaurs: The Encyclopedia,* p. 947; Horner and Lessem, *The Complete T. rex,* p. 64.

9. Glut, *Dinosaurs,* p. 947; Henry F. Osborn, "Tyrannosaurus, Upper Cretaceous Carnivorous Dinosaur. (Second Communication)," *Bulletin of the American Museum of Natural History,* Vol. 22 (1906), pp. 281–282, 295–296.

10. Glut, *Dinosaurs,* p. 947.

11. Barnum Brown, "Tyrannosaurus, a Cretaceous Carnivorous Dinosaur: The Largest Flesh-Eater That Ever Lived," *Scientific American,* Oct. 9, 1915, pp. 322–323.

12. Larson, *Rex Appeal,* p. 20.

13. Henry F. Osborn, "*Tyrannosaurus* and Other Cretaceous Carnivorous Dinosaurs," *Bulletin of the American of the American Museum of Natural History,* Vol. 21 (1905), pp. 259–265; Osborn, "*Tyrannosaurus* Upper Cretaceous Carnivorous Dinosaur." Matthew's reconstruction was the first good revelation of how this animal appeared in life.

14. William D. Matthew, "Allosaurus, A Carnivorous Dinosaur, and Its Prey," *The American Museum Journal,* Vol. 8, no. 1 (Jan. 1908), p. 5.

15. W.D. Matthew, "The Tyrannosaurus," *The American Museum Journal,* Vol. 10, no. 1 (Jan. 1910), p. 8.

16. Barnum Brown, "The Trachodon Group,"

The American Museum Journal, Vol. 8, no. 4 (April 1908), p. 51.

17. Matthew, "The Tyrannosaurus."

18. Henry F. Osborn, "Tyrannosaurus, Restoration and Model of the Skeleton," *Bulletin of the American Museum,* Vol. 22 (1913), p. 91.

19. Ibid., p. 92; but this spring action is also reminiscent of the kangaroo. (Also see note 69.)

20. Horner and Lessem, *The Complete T. rex,* p. 82.

21. Matthew, "The Tyrannosaurus."

22. Ronald Rainger, *An Agenda for Antiquity: Henry Fairfield Osborn and Vertebrate Paleontology at the American Museum of Natural History, 1890–1935* (Tuscaloosa: University of Alabama Press, 1991), p. 160.

23. Brown, "Tyrannosaurus, a Cretaceous Carnivorous Dinosaur," p. 322.

24. Joseph Augusta, *Prehistoric Animals* (London: Spring Books, 1956), Plate 34; in 1923, the bloody theme had also been captured by American Museum paleoartist E.M. Fulda as well. See Roy Chapman Andrews' *On the Trail of Ancient Man* (New York: Doubleday, Page, 1926), Plate facing p. 80.

25. Brown, "Tyrannosaurus, a Cretaceous Carnivorous Dinosaur," pp. 322–323.

26. Allen A. Debus and Diane E. Debus, *Paleoimagery: The Evolution of Dinosaurs in Art* (Jefferson, NC: McFarland, 2002), Chapter 25, "Tyrant Queen—Icon-o-saurus Rex," pp. 157–163.

27. Charles R. Knight, *Before the Dawn of History* (New York: McGraw-Hill, 1935), pp. 68–69.

28. Glut, *The Dinosaur Scrapbook,* p. 36.

29. Francis B. Messmore, "The History of the World a Million Years Ago," *Dinosaur World,* Vol. 1, no. 3 (October 1997), pp. 27–32. My father and grandparents saw this display.

30. Debus and Debus, *Paleoimagery Paleoimagery,* Chapter 31, "Building Life-Sized Dinosaurs," pp. 193–207.

31. Allen A. Debus, "Calgary's Prehistoric Zoo," *Prehistoric Times,* no. 80 (Winter 2006), pp. 52–56.

32. Bertha Morris Parker, *Animals of Yesterday* (Evanston, IL: Row, Peterson, 1941); Othenio Abel, *Geschichte und Methode der Rekonstruction Vorzeitlicher Wirbeltiere* (Jena: Gustav Fischer, 1925), p. 224; Charles R. Knight's "Parade of Life Through the Ages," *The National Geographic Magazine,* Feb. 1942, featured paintings of a *Ceratosaurus-Stegosaurus* conflict (Plate VII), and a pair of battling *T. rexes* (Plate IX).

33. Mark F. Berry, *The Dinosaur Filmography* (Jefferson, NC: McFarland, 2002), pp. 285–286. Rex versus Tops imagery was created for this film by Irving K. McGinnis, as noted in personal communication received from Mark Berry, dated 7/10/07.

34. Berry, *The Dinosaur Filmography,* p. 438.

35. Mark F. Berry, "Tyrannosaurus Flix: 80 Years of T. rex in the Cinema—Part 1," *Prehistoric Times,* no. 70 (Feb./March 2005), p. 52.

36. For more on *Hurricane Kids,* see chapter 10. For more on these and other titles, see Allen A. Debus, *Dinosaurs in Fantastic Fiction: A Thematic Survey* (Jefferson, NC: McFarland, 2006).

37. Edgar Rice Burroughs, *Out of Time's Abyss* (originally published 1918, *Blue Book Magazine*; New York: Ace Books, 1979 ed.), pp. 23–25.

38. Kenneth Robeson, *The Land of Terror* (originally published in *Doc Savage Magazine,* April 1933; New York: Bantam Books, 1965 ed.), p. 108.

39. Delos W. Lovelace, *King Kong* (originally published by Grosset & Dunlap, 1932; New York: Modern Library, 2005 ed.), p. 99.

40. Jeff Rovin, *From the Land Beyond Beyond: The Films of Willis O'Brien and Ray Harrhausen* (New York: Berkley Windhover, 1979), p. 28.

41. Z. Burian's similarly posed *Tarbosaurus* (1970) appears on page 131 of E.H. Colbert's *Dinosaurs: An Illustrated History* (Maplewood, NJ: Hammond, 1983).

42. Henry Fairfield Osborn, *The Origin and Evolution of Life on the Theory of Action, Reaction and Interaction* (New York: Scribner's, 1918 ed.), p. 225.

43. Glut, *Dinosaurs: The Encyclopedia, Supplement 1,* pp. 919–932. Before publication of Rex's formal description, Charles R. Knight had painted two magnificent *Triceratops* restorations, dated 1901 and 1904.

44. H.N. Hutchinson, *Extinct Monsters: A Popular Account of Some of the Larger Forms of Ancient Animal Life* (London: Chapman & Hall, 1893).

45. Henry R. Knipe, *Evolution in the Past* (J.B. Lippincott, circa 1910).

46. Steven M. Stanley, *Extinction* (New York: Scientific American Library, 1987). For more on the 1898 Paris Exposition *Triceratops,* see Claudine Cohen's *The Fate of the Mammoth: Fossils, Myth and History* (Editions du Seuil, 1994; Chicago: University of Chicago Press, translated by William Rodarmor, 2002 ed.), p. 171.

47. For more on *Agathaumas'* iconographic significance, see Allen A. Debus and Diane E. Debus, Chapter 13, "In Praise of Mesozoic Buffalo," in *Dinosaur Memories: Dino-Trekking for Beasts of Thunder, Fantastic Saurians, "Paleopeople," "Dinosaurabilia," and Other "Prehistoria"* (Lincoln, NE: Author's Choice Press, 2002).

48. Rovin, *From the Land Beyond Beyond,* p. 21.

49. Glut, *The Dinosaur Scrapbook,* p. 134.

50. Lovelace, *King Kong,* pp. 88–89.

51. Ibid., p. 94.

52. Donald F. Glut, *Jurassic Classics: A Col-*

lection of *Saurian Essays and Mesozoic Musings* (Jefferson, NC: McFarland, 2001), pp. 174–183; Berry, "Tyrannosaurus Flix," pp. 295–300.

53. Edgar Rice Burroughs, *Tarzan the Untamed* (originally published by the Redbook Corp., 1919, Part 1, and Frank A. Munsey Co., 1920, Part 2; New York: Ballantine Books, 1980); Burroughs, Edgar Rice. *Tarzan the Terrible* (New York Ballantine Books, 1972 ed.).

54. Edgar Rice. Burroughs, *Tarzan at the Earth's Core* (originally published in *Blue Book Magazine*, September 1929 to March 1930; New York: Ballantine Books, 1964 ed.), pp. 144–150.

55. Frank Saville, *Beyond the Great South Wall* (New York: Grosset and Dunlap, 1901). According to Glut (*Carbon Dates*, Jefferson, NC: McFarland, 1999, p. 133), in 1899, Arthur S. Coggeshall patriotically suggested naming the dinosaur later described as *Diplodocus carnegii* the Star Spangled Dinosaur.

56. Darlene Geis, *Dinosaurs and Other Prehistoric Animals* (New York: Grosset & Dunlap, 1959; illustrated by R.F. Peterson), p. 60; Jane Werner Watson, *The Giant Golden Book of Dinosaurs and Other Reptiles* (New York: Golden Press, 1965; illustrated by R.F. Zallinger); Jane Werner Watson, *Dinosaurs and Other Prehistoric Reptiles* (New York: Golden Press, 1959; illustrated by William de J. Rutherford); Willy Ley, *Worlds of the Past* (1962; illustrated by R.F. Zallinger). For more on Untermann, see A. Paul McFarland's "Ernest Untermann," *Prehistoric Times*, no. 83 (Fall 2007), pp. 36–38, 45.

57. Vincent Scully, R.F. Zallinger, Leo J. Hickey, and John H. Ostrom, *The Age of Reptiles: The Great Dinosaur Mural at Yale* (New York: Harry N. Abrams, 1990); also see Warren D. Allmon, "The Pre-Modern History of the Post-Modern Dinosaur: Phases and Causes in Post-Darwinian Dinosaur Art," *Earth Sciences History*, Vol. 25, no. 1 (2006), pp. 22, 25.

58. There are excellent summaries of the classifications and reclassifications of "Gorgosaurus" and "Albertosaurus" under individual entries for each dinosaur name in Donald F. Glut, *Dinosaurs: The Encyclopedia* (Jefferson, NC: McFarland, 1997); Glut, *Dinosaurs: The Encyclopedia, Supplement 1*; Glut, *Dinosaurs: The Encyclopedia, Supplement 2* (Jefferson, NC: McFarland, 2002); Glut, *Dinosaurs: The Encyclopedia, Supplement 3* (Jefferson, NC: McFarland, 2003); Glut, *Dinosaurs: The Encyclopedia, Supplement 4* (Jefferson, NC: McFarland, 2006); and Glut, *Dinosaurs: The Encyclopedia, Supplement 5* (Jefferson, NC: McFarland, 2008). By 1994, Field Museum artists horizontally aligned Gorgy's vertebral anatomy, posed as if eating the *Lambeosarus* corpse. This reflected current Renaissance Rex'science, and rather resembled posture and theme conveyed in the Milwaukee Public Museum's haunting, life-size Rex dio-

rama restoration (first displayed in 1983). Now our Field Museum's *Triceratops* cast skeleton fends off feasting *Daspletosaurus*, accentuating Knight's world famous "Rex vs. Tops" mural hanging on an adjacent wall. For more on the rationale for why the Field Museum's *Gorgosaurus* was referred to *Daspleosaurus*, see Thomas D. Carr, "Craniofacial ontogeny in Tyrannosauridae (Dinosauria, Coelosauria)," *Journal of Vertebrate Paleontology*, Vol. 19, no. 3 (Sept. 1990), pp. 497–520; and Carr, "FMNH PR308 — Part I: Or Analyzing an Enigmatic Tyrannosaurid Specimen," *Dinosaur World*, no. 6 (Spring/Winter 1999), pp. 16–18; and Carr, "FMNH PR308 Part II: The Cross-Dressing *Daspletosaurus*," *Dinosaur World*, no. 7 (Winter 1999/2000), pp. 21–24. For personal reminiscences about the Field Museum's "Gorgosaurus" exhibit, see the author's article, "'My' Gorgosaurus," *Fossil News*, Vol. 14, no. 4 (April 2008), pp. 12–16.

59. Glut summarized this state of affairs, accordingly, in context of a potentially new Rex specimen, "*Tyrannosaurus*, as well as other tyrannosaurids, are usually described as possessing but two fingers on each manus (a comparatively slightly vestigial third finger had already been known in the older and smaller tyrannosaurid *Gorgosaurus libratus*." (Glut, *Dinosaurs: The Encyclopedia, Supplement 4*, p. 551.)

60. Rainer Zangerl, *Dinosaurs, Predator and Prey: The Gorgosaurus and Lambeosaurus Exhibit in Chicago Natural History Museum* (1961).

61. Ibid., p. 12.

62. Geselschap's painting was recently reproduced in Debus and Debus, *Paleoimagery*, p. 81. For more on Zeman's film, see Berry, "Tyrannosaurus Flix," pp. 142–147.

63. Debus and Debus, *Paleoimagery*, p. 158.

64. Gregory M. Erickson, "Breathing Life into Tyrannosaurus Rex," in *The Scientific American Book of Dinosaurs*, Gregory S. Paul, ed. (New York: St. Martin's Press, 2000), p. 273.

65. Adrian J. Desmond, *The Hot-Blooded Dinosaurs: A Revolution in Palaeontology* (New York: Warner Books, 1975), pp. 90–91.

66. Ibid., p. 78.

67. In 1919 journalist W.H. Ballou writing in *Scientific American* had this to say of the relationship between birds and reptiles, and the extent to which the world's first known bird, *Archaeopteryx*, was reptilian. "In the sense that clothes make the man, feathers make the bird. Hence the bird is merely a flying reptile, feathered more or less, according to species. Man has little of the reptile structure, but a bird has little else. Feathers, then, merely conceal the reptile. When Robin Redbreast lifts up his head and pours out his morning song, the brain that guides it is almost identical with that of the

young alligator, which, while it can not sing, bays and roars pretty loudly. Mrs. Robin lays an egg and so does Mrs. Alligator." (Willis T. Lee, *Stories in Stone*, New York: D. Van Nostrand, p. 122). More recently, dinosaurs, including a probable *T. rex* evolutionary ancestor known as the Chinese genus, *Dilong*, have been discovered with fossilized feather impressions. Since the 1980s, Gregory S. Paul has suggested that certain theropod dinosaur lineages represent avians which became *secondarily* flightless, while still retaining feathers. (Gregory S. Paul, "Screaming Biplane Dromaeosaurs of the Air," *Prehistoric Times*, no. 60, June/July 2003, pp. 48–50). Inspired by Sara Landry's art, Paul has been illustrating furry-looking (feathered) allosauruses since the 1970s. Cladistically, there is some sense and rationale as to the possibility that Rex hatchlings sported feathers. However, few dino-monster fans would appreciate the appearance of adult feathered Rex monsters ever appearing on the motion picture screen, unless it all could be expertly done. For a summary of the feathered dinosaur phenomenon, see Debus and Debus, *Paleoimagery*, pp. 126–134, 241–242. Also see Tim Appenzeller's article, "*T. Rex* Was Fierce, Yes, but Feathered, Too," *Science*, Vol. 285 (Sept. 24, 1999), pp. 2052–2053.

68. Glut, *Dinosaurs: The Encyclopedia*, p. 951.

69. Another early pop-cultural suggestion that Rex hopped in kangaroo-like fashion appears in Willis Knapp Jones' 1927 short story, "The Beast of the Yungas," reprinted in *100 Creepy Little Creature Stories*, ed. Stefan R. Dziemianowicz, Robert Weinberg, and Martin H. Greenberg (New York: Barnes & Noble, 1994), p. 45.

70. Debus, *Dinosaurs in Fantastic Fiction*, 2006.

71. John Eric Holmes, *Mahars of Pellucidar* (New York: Ace Books, 1976), p. 204. Some other entries of relevance here would include Lin Carter's *Journey to the Underground World* (New York: Daw Books, 1979); *Zanthodon* (New York: Daw Books, 1980); and *Hurok of the Stone Age* (New York: Daw Books, 1981). For more on Carter's work, see my article "Dinosaurs in Fantastic Fiction — Extras," *Prehistoric Times*, no. 81 (Spring 2007), pp. 22–23, 56–57. There are also comics with prehistoric lost world stories settings where monstrous Rexes roam, such as *Star Spangled War Stories*, published between 1960 to 1968, *Tragg and the Sky Gods*, and *Turok — Son of Stone*. For more on *Star Spangled War Stories* see chapter ten and Rovin, *From the Land Beyond Beyond*, pp. 366–367. For more on *Tragg and the Sky Gods*, see my articles, "Traggnificent!" (Parts 1 to 2) in *Prehistoric Times* nos. 59–60, respectively (April/May 2003), pp. 18–19, and (June/July 2003), pp. 18–19.

72. Berry, "Tyrannosaurus Flix," pp. 216–

218; Berry states, "The Tyrannosaurus ... is a zoologist's nightmare. With glassy eyes, an upright stance and a humped back worthy of Lon Chaney's stand-in, it is simply too comical to take seriously. Its arms seem frozen, the robotic-looking jaw opens wider than a rattlesnake's, and it walks like ... well, like there's a man inside."

73. Allen A. Debus, "Unwittingly Western: Two Shining Knights," *G-Fan*, no. 82 (Winter 2006), pp. 54–57.

74. For considerably more on each movie title, see individual entries listed in Berry's *Dinosaur Filmography*.

75. Besides Bakker's titles listed in the Bibliography, see the following for a summary of his career. Allen A. Debus, "A Look Bakk-: Robert Bakker — Revolutionary Paleontologist," (Parts 1–3), *Prehistoric Times*, nos. 69, 70, 71 (Dec./Jan. 2005), pp. 49–51; (Feb./March 2005), pp. 49–51; (April/May 2005), pp. 50–51.

76. Debus and Debus, *Paleoimagery*, pp. 168–170, and chapters 30 to 31; Elbert H. Porter, *Dinosaur Sculptor* (Davis Printing, 1982); Sylvia Czerkas' late 1970s sculpture of a miniature *T. rex*, shown in Donald F. Glut's *The Dinosaur Scrapbook* (1980) as well as on the book jacket of his *The New Dinosaur Dictionary* (1982) (both Citadel Press) is also one of the earliest Rexes restored with a more or less horizontally aligned spinal column. Arguably, however, the Rex in the foreground of Charles R. Knight's famous Field Museum (1926–1930) mural, approaching *Triceratops*, has a spinal column that is precociously near horizontal (with its tail raised off the ground). The Rex at distance in this mural, however, is posed like AMNH-5027.

77. Desmond, *The Hot-Blooded Dinosaurs*, pp. 82–89.

78. Robert T. Bakker, *The Dinosaur Heresies: New Theories Unlocking the Mystery of the Dinosaurs and Their Extinction* (New York: William Morrow, 1986), pp. 227, 242–243, 256. Bakker's "Diracodon" is synonymous with *Stegosaurus*. From their impressive restorations it is evident that paleoartists Maurice Wilson (1914–1987) and Neave Parker, who worked with vertebrate paleontologist William E. Swinton (1900–1994) during the late 1950s and mid–1960s, envisioned *Megalosaurus* as an intermittently swift-running dinosaur. See Plates 2 (Parker's), and 12 (Wilson's), respectively, in William E. Swinton, *Dinosaurs* (London: Trustees of the British Museum, Natural History, 1969), and Swinton, *Fossil Amphibians and Reptiles* (London: Trustees of the British Museum, Natural History, 1965). Also, the stop-motion animated *T. rex* in *The Beast of Hollow Mountain* (1956) is shown sprinting after horseback riders in several sequences.

79. Robert T. Bakker, "The Return of the

Dancing Dinosaurs," *Dinosaurs Past and Present*, Vol. 1, ed. Sylvia J. Czerkas and Everett C. Olson (Seattle and London: Natural History Museum of Los Angeles County in association with University of Washington Press, 1987), pp. 38–69.

80. Bakker, *Dinosaur Heresies*, p. 416.

81. Gregory S. Paul, *Predatory Dinosaurs of the World: A Complete Illustrated Guide* (New York: Simon & Schuster, 1988); Gregory S. Paul, "The Science and Art of Restoring the Life Appearance of Dinosaurs and Their Relatives: A Rigorous How-to Guide," in *Dinosaurs Past and Present*, Vol. 2, ed. Sylvia J. Czerkas and Everett C. Olson (Seattle and London: Natural History Museum of Los Angeles County in association with University of Washington Press, 1987), pp. 5–49.

82. Paul, *Predatory Dinosaurs of the World*, p. 338.

83. Harry Adam Knight, *Carnosaur* (London: Star Books, 1984), p. 212.

84. The life-sized Rex used as a prop appeared in four movies—including two *Carnosaur* sequels, and *Dinosaur Island* (1994), which also featured a Tops prop—as discussed in Berry, *Dinosaur Filmography*. Other period novels and short stories emphasizing Rex, including David Gerrold's *Deathbeast* (1978) and David Drake's *Tyrannosaur* (1993), are outlined in latter chapters of my *Dinosaurs in Fantastic Fiction* (2006).

85. Michael Crichton, *The Lost World* (New York: Alfred A. Knopf, 1995), p. 216. Presently a number of Rex juvenile fossils have been collected, including the best specimen of all, the Burpee Museum of Natural History's teenage Rex specimen, known as "Jane," perhaps the genuine spitting image of Crichton's imagined *Lost World* monster.

86. Michael Crichton, *Jurassic Park* (New York: Alfred A. Knopf, 1990); Crichton, *The Lost World*; associated movie titles, including *Jurassic Park III* (2001), may be consulted in Berry's *Dinosaur Filmography*. Then, dramatically, in early 2007 came news of protein sequences extracted from *Tyrannosaurus* bone. Ironically, the authors of this *Science* paper chose to extract and then compare biochemical sequences from fossilized material representing America's two foremost pop-cultural monsters of the 18th–19th and 20th–21st centuries—*Tyrannosaurus* and the American mastodon! (John M. Asara, Mary H. Schweitzer, Lisa M. Freimark, Matthew Phillips, Lewis C. Cantley, "Protein sequences from Mastodon and *Tyrannosaurus Rex* revealed by Mass Spectrometry," *Science*, Vol. 316, April 13, 2007, pp. 280–285.)

87. Berry, *Dinosaur Filmography*, pp. 268–269.

88. For considerably more on the scientific issues outlined in this chapter, see references cited in note 57. More specifically, see Larson, *Rex Appeal*, p. 361; Glut, *Dinosaurs: The Encyclopedia — Supplement 4*, pp. 552–553; paleontologist Ken Carpenter has referred to Rex as the "Schwarzenegger of dinosaurs"; Mary Schweitzer referred to Rex as the "Michael Jordan of dinosaurs" due to its prevalent pathologies, despite its "champion aspect"; Gregory Erickson has referred to Rex as the "James Dean of dinosaurs" given that the Rex specimen known as Sue only lived to 29 years, cut down in its prime.

89. Debus and Debus, *Paleoimagery*, pp. 152–156.

90. Will Hubbell, *Cretaceous Sea* (New York: Ace Books, 2002), p. 224.

91. Private fine art collector John Lanzendorf did much to promote dinosaur paintings and imagery made by the celebrated current crop of paleoartists during the late 1990s. His favorite dinosaur was *T. rex*, as exemplified throughout many pages of a colorful coffee table book, *Dinosaur Imagery: The Science of Lost Worlds and Jurassic Art — The Lanzendorf Collection*, Foreword by Phillip J. Currie (San Diego: Academic Press, 2000). Nearly every issue of *Prehistoric Times* (which began publication in 1993) offers fresh and striking visuals of one kind or another dedicated to Rex, and often Rex versus Tops imagery.

92. Semonin, *American Monster*, p. 404.

93. Debus and Debus, *Paleoimagery*, p. 206.

94. Hallett 's painting was published in *ZooBooks: Dinosaurs*, Vol. 1, no. 9 (June 1985), published by Wildlife Education.

95. Artists such as Gregory S. Paul, David Peters, Robert Bakker and Bob Morales have each artistically conveyed their impressions of a skirmish between airplane-sized *Quetzalcoatlus* and select tyrannosaurids.

96. Research the references in Berry's *Dinosaur Filmography* for specific movie titles. And see my article, "At War with Dinosaurs," *G-Fan*, no. 72 (Summer 2005), pp. 8–13, as well as pp. 56–72 in my *Dinosaurs in Fantastic Fiction*.

97. Berry, *Dinosaur Filmography*, pp. 92–96. Also see note 85.

98. David W. Krauss quoted in "The Biomechanics of a Plausible Hunting Strategy for *Tyrannosaurus rex*," in *The Origin, Systematics and Paleobiology of Tyrannosauridae, September 16–18, 2005, a Symposium Hosted Jointly by Burpee Museum of Natural History and Northern Illinois University*, p. 41, as cited in Glut, *Dinosaurs: The Encyclopedia — Supplement 5*, p. 70.

Chapter Nine

1. Kenneth Carpenter, "A Dinosaur Paleontologist's View of Godzilla," pp. 102–106, in *The Official Godzilla Compendium*, ed. J.D. Lees and Marc Cerasini (New York: Random House, 1998).

2. For more on inspirations for Godzilla's cinematic origins, see Steve Ryfle's *Japan's Favorite Mon-Star: The Unauthorized Biography of "The Big G"* (Toronto: ECW Press, 1998), William Tsutsui's *Godzilla on My Mind: Fifty Years of the King of Monsters* (New York: Palgrave Macmillan, 2004).

3. For more on *The Arctic Giant*, and *The Beast from 20,000 Fathoms*, see pp. 74–78 in Allen A. Debus, *Dinosaurs in Fantastic Fiction*.

4. Mark F. Berry. *The Dinosaur Filmography*, pp. 28–37; Allen A. Debus. "Prototypical 'Mad Scientists' of the Prehistoric Monster Story," *G-Fan*, no. 83 (Spring 2008), pp. 38–43.

5. Tsutsui, *Godzilla on My Mind*, pp. 32–33; Steve Ryfle. "Godzilla's Footprint," *The Virginia Quarterly Review*, 2006, available online at vqronline.org/printmedia.php/prmMediaID/9012.

6. Ryfle, "Godzilla's Footprint."

7. For more on Bakker's career, see Don Lessem, *Kings of Creation* (New York: Simon & Schuster, 1992), and Allen A. Debus, "A Look Bakk-" (Parts 1–3), *The Prehistoric Times*, nos. 69–71 (Dec./Jan. 2005), pp. 49–51, (Feb./March 2005), pp. 49–51 (April/May 2005), pp. 50–51.

8. John H. Ostrom, *Osteology of Deinonychus antirrhopus, an Unusual Theropod from the Lower Cretaceous of Montana* (Peabody Museum of Natural History), Bulletin 30 (July 1969). R.T. Bakker's stirring life restoration of a lively, running *Deinonychus* appears as the Frontispiece to this publication, while his skeletal reconstruction adorned page 142.

9. Glut, *Classic Movie Monsters*, p. 374.

10. Edward Brock. "Godzilla Cover-age," *G-Fan*, no. 62 (May/June 2003), pp. 14–17.

11. For more on Godzilla's transformation into a radioactive superhero, see Tsutsui, *Godzilla on My Mind*, and Ryfle, "Godzilla's Footprint." for more on Godzilla's draconian status, see my article, "Prehistorical *Daikaiju* Evolution," printed in *G-Fan*, no. 84 (Summer 2008), pp. 46–51.

12. J. D. Lees and Marc Cerasini, *The Official Godzilla Compendium*, pp. 96–99.

13. Steve Ryfle, "Godzilla's Footprint," p. 27.

14. Carpenter, "A Dinosaur Paleontologist's View of Godzilla."

15. For a twist on the *Gojirasaurus* matter, see Allen A. Debus and Diane E. Debus, *Dinosaur Memories*, pp. 441–448.

16. *Velociraptor's* emergence into the most horrific dinosaur of the 1990s came on the vicious, slashing heels of the 1993 release of *Jurassic Park*, based on Michael Crichton's 1990 novel.

17. *Walking with Dinosaurs*, series produced by Tim Haines, BBC, 1999.

18. Glut, *Dinosaurs: The Encyclopedia*.

19. Giant pterosaur Rodan is a "Vernian" denizen of an inner Earth, as it is awakened inside a volcano. The American version of *Rodan* really gets off with a bang, as in deference to *Beast*, before the real story begins, we're treated to military footage of spectacular nuclear explosions and the usual assortment of somber messages implicit with unwise nuclear testing. With that table set, shortly, miners are attacked and killed by *Them!*-sized gigantic bugs. Eventually, protagonist mining engineer Shigeru identifies a picture in a file on the pterosauria presented by paleontologist Professor Kashiwagi. This prompts a scientific expedition to the mine where Shigeru saw the giant winged creature eating the huge bugs. Had the miners dug too deeply for their coal, only to awaken a bevy of prehistoric monsters? Is it the case that such creatures never really died, but that they'd only slept through eons? Next, alerted to the sensational discovery of a giant, calcareous shell fragment, the Professor now suggests that the eggshell is reptilian. The eggshell is also radiocarbon dated at twenty million years (an impossible measurement to make using carbon 14 dating, which is used for dating only much more recent artifacts). And aided by an "electronic computer," from its parabolic shape, the paleontologist contingent concludes that the winged animal must have weighed 100 tons, having a wingspan of 500 feet! Furthermore, conservatively speaking given its size, the carnivorous animal probably was fully grown when hatched. Such a creature, scientifically named *Pteranodon rodan*, would have flown at supersonic speeds, generating typhoon wind forces. In *Rodan*, the military asks two reasonable questions of their resident paleontologist. First, why are the Rodan alive if it's supposed to be extinct? Kashiwagi explains with a "theory of his own," that "millions of years ago, the egg was hermetically sealed and buried by a volcanic eruption," preserving the "germ of life." Afterward, atomic explosions fractured the Earth's crust, air and warm water percolated into the deposit, allowing eggs to finally hatch. These circumstances also presumably revivified the giant prehistoric insects. Next, the military asks where they might find the pair of Rodan, and Kashiwagi suggests a likely region between the mine and the volcano. This time, following the lack of facilitation offered by paleontologists Yamane and Tadokoro in prior cases with Godzilla and then Gigantis with Anguirus, the military doesn't even bother asking how best to kill the

monsters— that's their task. Ultimately, the Rodan meet their demise in a fiery volcano.

20. J. D. Lees, "What Is a Kaiju?" *G-Fan*, no. 78 (Winter 2007), pp. 68–72.

21. Michael Bogue, Review of "Gappa, The Triphibian Monster," *G-Fan*, no. 79 (Spring 2007), p. 41.

22. Robert Hood and Robin Pen, ed., *Daikaiju!: Giant Monster Tales* (University of Wollongong, Australia: Agog! Press, 2005), pp. vi–vii.

23. Berry, *Dinosaur Filmography*.

24. Robert Hood, "Man and Super-Monster: A History of Daikaiju Eiga and Its Metaphorical Undercurrents," *Borderlands Magazine*, no. 7 (2006; also available online at Hood's personal Web site).

25. Ibid.

26. For more on Toho's Rodan, Nikkatsu's Gappa and other prehistoric "terror-saurs" of literature and film, see both the Appendix, and— emphasizing Edgar Rice Burroughs' rhamphorynchoid Mahars— see Allen A. Debus, "Greatest 'Terror-saurs' of Sci-Fi and Horror," *Scary Monsters Magazine*, no. 48 (Sept. 2003), pp. 107–114. Also see note 31.

27. For more on *King Ghidorah* (1991), see Ryfle, "Godzilla's Footprint," pp. 265–275. For more on *Megaguirus*, see J.D. Lees, "The Unofficial Godzilla Addendum," *G-Fan*, no. 68 (Summer 2004), pp. 39–41. Beyond filmed 1950s theories for why Godzillas kept attacking Japan, expressed — respectively — by Drs. Yamane and Tadokoro in *Gojira* and *Godzilla Raids Again*, an attempt was made to re-define Godzilla's origin from exposure to atomic radiation in *Godzilla vs. King Ghidorah* (1991). In this film, Godzilla's cryptozoological progenitor is identified as a dinosaur named the Godzillasaurus whose race has survived on Lagos Island since the Mesozoic Era. When Godzillasaurus is seriously wounded in battle defending World War II Japanese forces in 1944, it is transported by Futurians who have traveled from the year 2204 to alter historical events. By moving the Godzillasaurus to another location, a decade later when the United States conducted atomic warhead testing on Lagos, altered history projects that a mutant Godzilla would never be produced. But of course this isn't the resultant, as the attempt seriously backfires. Another, earlier effort to unconvincingly reexplain Godzilla as a "real" creature in paleontological terms, as a genetic cross between a theropod dinosaur and the Jurassic dinosaur *Stegosaurus*, was filmed for the Americanized *King Kong vs. Godzilla* (1963), without sense of conviction. And a scene in the 1984 *Godzilla* remake shows the titular monster following a flock of birds, suggesting its avian-evolutionary ties, stemming from contemporary considerations of the affinity between certain theropod dinosaurs and birds.

28. D.G. Valdron. "Fossils," in Robert Hood and Robin Pen, eds., *Daikaiju!*, p. 189.

29. Marvin Livings, "Running," in Robert Hood and Robin Pen, eds., *Daikaiju!*, pp. 18, 21.

30. Debus, *Dinosaurs in Fantastic Fiction*, pp. 59–63.

31. Michael Crichton. *Jurassic Park* (New York: Ballantine Books, 1990). Also, for more on the theme of certain daikaiju as being prehistoric (especially reptilian forms), see Allen A. Debus, "Prehistorical Daikaiju Evolution," *G-Fan*, no. 84 (Summer 2008), pp. 46–51.

32. Robert Hood, "Man and Super-Monster."

33. Brian W. Aldiss, "Heresies of the Huge God," in *The Best Science Fiction Stories of Brian W. Aldiss: Man in His Time* (New York: Collier Books, 1988); J.G. Ballard, "The Drowned Giant," in *The Terminal Beach* (New York: Carroll & Graf, 1987); William Schoell, *Saurian* (New York: Book Margins, 1988); William Schoell, *Dragon* (New York: Leisure Books, 1989); Anne McCaffrey, "Dragonrider," in *The Science Fiction Hall of Fame, Vol. III — Nebula Winners 1965–1969*, Arthur C. Clarke and George W. Proctor, eds. (New York: Avon Books, 1982); Carl Dreadstone, *Creature from the Black Lagoon* (New York: Berkley Medallion, 1977); Marc Jacobson, *Gojiro* (New York: Grove Press, 1991); Joe R. Lansdale, "Godzilla's 12 Step Program," in *The Mammoth Book of Monsters* (2007).

34. David Gerrold, *Deathbeast* (New York: Popular Library, 1978).

35. Lovelace, *King Kong*; Carson Bingham, *Gorgo* (Derby, CT: Monarch Books, 1960); Dean Owen, *Reptilicus* (Derby, CT: Monarch Books, 1961); Robert K. Andreassi, *Gargantua* (New York: Tor Books, 1998).

36. The term "sauriodinosauroid" was defined by Jose Luis Sanz, in his *Starring T. Rex!*, pp. 109, 114.

37. Allen A. Debus, *Dinosaurs in Fantastic Fiction*, pp. 146–148; Allen A. Debus, "Is Saurian Kaiju?" *G-Fan*, no. 80 (Summer 2007), pp. 36–38.

38. Robert Hood, "Flesh and Bone," in Robert Hood and Robin Pen, eds., *Daikaiju! 3: Giant Monsters vs. the World Tales* (University of Wollongong, Australia: Agog! Press, 2007). Richard A. Becker, "The Return of Cthadron," in Robert Hood and Robin Pen, eds., *Daikaiju!*

39. Stephen King, *Danse Macabre* (first published by Everett House, 1981; New York: Berkley Books, 1982 ed.).

40. Robert Hood, personal E-mail communication, February 2008.

41. Brian W. Aldiss, with David Wingrove, *The Trillion Year Spree* (New York: Avon Books, 1986), p. 50.

42. Mary Shelley, *Frankenstein, or the Modern Prometheus* (1831 ed.; reprinted New York: Pyramid Books, 1964 ed.), p. 34.

43. Martin J.S. Rudwick, *Bursting the Limits of Time: The Reconstruction of Geohistory in the Age of Revolution* (Chicago: University of Chicago Press, 2005), p. 1.

44. Radu Florescu and Matei Cazacu, *In Search of Frankenstein* (Boston: New York Graphic Society, 1975); Shannon Thomas. "It's Alive—The Science behind Frankenstein's Art," Dec. 3, 1998, posted online at _sciencebehindfrankenstein.html.

45. Steve Ryfle, "Godzilla's Footprint," pp. 27, 79–81, 119; Bob Statzer, "Frankenstein: The Toho Terrors," *Castle of Frankenstein*, no. 30 (2001), pp. 41–53; Michael Bogue, "The Kyoto Kong Chronicles: Toho's Double Take on the Big Apes (Part 1 of 2)," *Scary Monsters*, no. 57 (Jan. 2006), pp. 28–37.

46. Mary Shelley, *op. cit.*; *Gojira*, produced by Tomoyuki Tanaka, Toho Motion Picture Company, 1954.

47. Shelley, *Frankenstein*, p. 8.

48. Pseudo-science behind Godzilla's regenerative powers is explored in Marc Cerasini's 1996 novel, *Godzilla Returns* (New York: Random House). I first suggested this association between the two monsters in my 2008 article, "Monster Mythology," *G-Fan*, no. 85, pp. 36–41.

49. Mark Justice. "Shinto Symbolism in Toho's Daikaiju Eiga," *G-Fan*, no. 81 (Fall, 2007), pp. 30, 37.

50. Michael Crichton, *Jurassic Park*, pp. 183, 189, 313. *Gojira* wasn't the first artistic effort to marvelously merge prehistorical/dinosaurian ideals with Frankenstein mythos. That honor goes to author and mathematician John Taine (Eric Temple Bell) who, in his 1929 novel *The Greatest Adventure*, illustrated how Godzilla-sized, pseudo-dinosaurians resulted from mad experimentation of a prehistoric race of technological beings. See my article in *G-Fan*, no. 67 (Spring 2004), pp. 12–14 for more on this. See also Allen A. Debus, *Dinosaurs in Fantastic Fiction*, pp. 49–51, chapter 7.

51. According to William Tsutsui, Osama bin Laden was inspired by TriStar Picture's 1998 film, *Godzilla*, to destroy the Brooklyn Bridge, which of course transformed instead into 9/11's parallel path of destruction. Tstusui, *Godzilla on My Mind*, p. 174.

52. Regarding its mythical status, I'm not the first to suggest that Godzilla is a modern dragon, for which Jose Luis Sanz, *Starring T. Rex*, pp. 109, 115–121, introduced the prescriptive term, "dragodinosauroid."

Chapter Ten

1. Ralph O'Connor, *The Earth on Show: Fossils and the Poetics of Popular Science, 1802–1856* (Chicago: University of Chicago Press, 2007), p. 278.

2. Ibid., pp. 273, 279.

3. Ibid., p. 273.

4. Berry, *The Dinosaur Filmography*; Donald F. Glut, *The Dinosaur Scrapbook* (Secaucus, NJ: Citadel Press, 1980); Michael Klossner, *Prehistoric Humans in Film and Television* (Jefferson, NC: McFarland, 2006).

5. Arthur Conan Doyle, *The Lost World* (originally published 1912, *Strand Magazine*; New York: A.L. Burt, c. 1925 ed.).

6. Richard Dawkins, *The Ancestor's Tale: A Pilgrimage to the Dawn of Evolution* (New York: Houghton Mifflin, Mariner Books, 2004), pp. 340, 607.

7. *The Lost World* souvenir program booklet, originally published by First National Pictures in 1925, reprinted by Image Entertainment for DVD package.

8. Rovin, *From the Land Beyond Beyond*, p. 18.

9. Berry, *Dinosaur Filmography*, pp. 242–243.

10. Ibid., p. 248.

11. Edgar Rice Burroughs, *The Land That Time Forgot* (New York: Ace Books, originally published by *Blue Book Magazine*, 1918).

12. Rovin, *From the Land Beyond Beyond*, p. 23; George E. Turner with Dr. Orville Goldner (expanded and revised by Michael H. Price with Douglas Turner), *Spawn of Skull Island* (Baltimore, MD: Luminary Press, 2002), pp. 200–203.

13. Donald F. Glut, *Jurassic Classics* (Jefferson, NC: McFarland, 2001), p. 184; in Lovelace's 1932 novel, *King Kong*, Ann Darrow proclaims "But there never was such a beast.... At least not since prehistoric times." In his *Classic Movie Monsters* (Metuchen, NJ: Scarecrow, 1978, p. 314), Glut discusses Kong's alleged symbolisms. "Intellectuals have attempted to read certain symbolisms into *King Kong*: Kong represents a Depression-stricken people lashing out against society; Kong is a Christ figure crucified on a stage before an audience of spectators; Kong symbolizes the black man and his conflict against his white oppressors; Kong on the Empire State Building depicts the most spectacular phallic symbol ever captured by film; and so on. [Merian C. Cooper] has disavowed these 'insights' into his film; he was not the kind of person to consider injecting such messages into his productions .... Whatever messages *might* lurk beneath the film's surface were entirely coincidental on Cooper's part or, at best, subliminal." In spite of Cooper's assertions, the film's symmetry and evident symbolism have since lured many admirers and writers into interpreting the real meaning of *King Kong*. In his intriguing re-

cent summary, Joseph D. Andriano, *Immortal Monster: The Mythological Evolution of the Fantastic Beast in Modern Fiction and Film* (Westport, CT: Greenwood Press, 1999, pp. 45–59), viewed Kong as a "parable of evolution anxiety." In his psychological dissection of Kong Andriano professes, "As a metaphor (Kong) is white man's image of berserk black sexuality; or he is the dragon on the island, with Ann Darrow his Andromeda; or he is simply the ape in all of us." (Quotes from p. 48.) Further consideration of the white-furred Little Kong in *Son of Kong* (1933) leads Andriano to conclude, racially-speaking, that "the beast is black when ferocious, white when tame. In both films, the giant ape is more than gorilla, clearly a reflection of humanity's collective anxiety over having evolved from other apes ... little white Kong is more civilized, more domesticated than big black Kong" (p. 58). Contrary to some suggestions, Kong is decidedly *not Gigantopithecus*.

14. Donald F. Glut, *I Was a Teenage Movie-Maker: The Book* (Jefferson, NC: McFarland, 2007), pp. 87–92; Donald F. Glut, *I Was a Teenage Movie-Maker* (DVD; Frontline Entertainment, Inc., 2006) — see Glut's *Tor, King of Beasts* filmed in 1962; Allen A. Debus, "Stop Motion Godzilla," *G-Fan*, no. 79 (Spring 2007), pp. 28–31; Perhaps, *T. rex*'s second greatest battle (besides that with a giant ape as filmed both by RKO, and Glut independently in 1962) was with *Spinosaurus*, in *Jurassic Park III* (2001).

15. Mark F. Berry, "The Top 10 Dino-Movie Scenes of All Time," *Prehistoric Times*, no. 75 (Dec./Jan. 2006), pp. 28–31, 58.

16. Ibid., p. 58.

17. Note that the commonly used term, "Skull Island," is never stated, per se, in the 1933 film. See Andriano, *Immortal Monster*, p. 59. But in Lovelace's 1932 novel, Kong's lofty lair on the island is referred to as "Skull Mountain"; Brad Strickland with John Michlig, *Kong: King of Skull Island* (created and illustrated by Joe DeVito, Milwaukie, OR: DH Press, 2004); Weta Workshop, *The World of Kong: A Natural History of Skull Island* (New York: Pocket Books, 2005); Allen A. Debus, "King Kong vs. Godzilla: Skull Island's 'True' (Alternative) Tale," Parts 1 to 2, *G-Fan*, no. 78 (Fall 2006), pp. 10–14, *G-Fan*, no. 79 (Spring 2007), pp. 23–26.

18. W. Douglas Burden quoted in Douglas J. Preston's *Dinosaurs in the Attic* (New York: Ballantine Books, 1986), p. 205.

19. Richard Milner, *The Encyclopedia of Evolution* (New York: Facts on File, 1990), p. 253.

20. George E. Turner, *Spawn of Skull Island*, p. 194.

21. I lack the space to discuss *King Kong* in further detail. Those desiring more should consult Turner's *Spawn of Skull Island* and Berry's *Dinosaur Filmography*.

22. Oskar Lebeck and Gaylord DuBois, *The Hurricane Kids on the Lost Islands* (Racine, WI: Whitman, 1941); Allen A. Debus, "Dinosaurs in Fantastic Fiction — Extras II," *Prehistoric Times*, no. 82 (Summer 2007), pp. 22–23.

23. Authors Lebeck and DuBois certainly mined all the contemporary lost world classics to produce this derivative tale. (See chapter two of my *Dinosaurs in Fantastic Fiction: A Thematic Survey*, McFarland, 2006, for more on the prehistoric lost worlds of science fiction literature.) Incidentally, in 1952 Lebeck gained notoriety as an editor of a science fiction newspaper strip titled *Twin Earths*. He was also a well known comic book editor, and for a time worked with Walt Disney Studios. In 1941 Lebeck even collaborated with DuBois on another kids' novel — *Rex, King of the Deep*, which isn't about a tyrannosaur, but instead about a World War II submarine. An on-line *ERBzine* issue (vol. 0072) states that during the 1930s and 1940s, DuBois (born in 1899), wrote numerous, highly regarded Edgar Rice Burroughs–inspired comic books featuring Tarzan. Two of these titles may have dealt with prehistoric themes, "Tarzan — the Valley of Monsters," and "Tarzan and the Beasts of Pal-ul-don." DuBois received the Thomas Alva Edison National Mass Media Award in 1955 and 1956 for Best Comic Book for Children over Eight, for scripts not involving Tarzan, although he felt his Tarzan stories were "always more fun."

24. Many of these imaginative stories were recently compiled into a book: *DC Showcase Presents: The War That Time Forgot*, Volume One (New York: DC Comics, Warner Bros. Entertainment Co., 2007).

25. A brief outline of the history of dinosaurs in comics is appropriate. Glut, "Tor of 1,000,000 Million Years Ago," pp. 165–173 in *Jurassic Classics: A Collection of Saurian Essays and Mesozoic Musings* (Jefferson, NC: McFarland, 2001); Donald F. Glut, *The Dinosaur Scrapbook*, chapter 10, "Dinosaurs in the Comics," pp. 189–235; Pat McCauslin, "Collecting Dinosaur Comics — Turok Son of Stone," *Mad Scientist*, no. 12 (Fall 2005), pp. 22–27; Allen A. Debus, "*Tragg*-nificent!" Parts 1–2, *Prehistoric Times*, nos. 59–60 (April/May 2003), pp. 18–19, and (June/July 2003), pp. 18–19; Allen A. Debus, "At War with Dinosaurs," *G-Fan*, no. 72 (Summer 2005), pp. 8–13; Allen A. Debus, "Doomsday Dinosaurs," *G-Fan*, no. 61 (March/April 2003), pp. 12–17; Don Markstein, "Toonopedia — The War That Time Forgot." As stated by dinosaur expert and science fiction writer Donald F. Glut in *The Dinosaur Scrapbook* (1980), "The mere presence of a prehistoric animal on the cover of a comic book is enough to sell the issue.... Rarely does a month pass in which at least one dinosaur or other extinct creature cannot be seen in its scaly

or shaggy majesty somewhere within the comic-book sales racks" (p. 229). In his *Scrapbook* Glut recounted exploits of several cavemen (and -women) of the comics who lived alongside dinosaurs and other dinosaur-fighting heroes, including Alley Oop, Tor, Lo-Zar, Tarzan, Ka-Zar, Kona, Turok Son of Stone, Tragg, and many others. Through the years, encounters between dinosaurs and space heroes like Flash Gordon, or superheroes wearing tights (such as Superman, Spiderman and the Fantastic Four among them) have been surprisingly common. And for the record, Glut's book appeared prior to 1980s publishing of *Calvin and Hobbes* strips involving comical vignettes of dinosaurian reverie, or Gary Larson's *The Far Side* so peppered with brain-twisting paleontological hilarity. Here, for sake of brevity, several of the more important titles and names that wouldn't necessarily seem so prominent in the comics realm today will be introduced. For instance, nearly every dinosaur enthusiast has heard of Turok, but important characters such as Alley Oop, Tor and Tragg might seem less familiar. The *Alley Oop* strip, concerning a Neanderthal man with pet dinosaur Dinny and girlfriend Oola, was created in 1934 by artist V.T. Hamlin. Alley Oop rode Dinny into many, often amusing, Stone Age adventures. Joe Kubert's and Norman Mauer's caveman *Tor,* spanning only 6 issues appeared in 1953 in 3-D splendor. Tor, primitive in age yet essentially modern in intellect and sense of righteousness, regularly battled dinosaurs and other prehistoric monsters amidst those obligatory volcanic explosions. As stated by Allan Asherman in a 1975 DC reprint of *Tor* no. 1, "He lived in a cave in the world of one million years ago, and protected himself and his friends from the perils of his day; dinosaurs, volcanoes, starvation and loneliness. The problems of a million years ago are still here, in different forms, making Tor ... relevant today" First published in 1975 (and lasting until 1977), *Tragg and the Sky Gods* was Glut's own creation, a prehistoric hero inspired by Eric von Daniken's *Chariots of the Gods?* (1968) concept (with interwoven traces of *2001: A Space Odyssey,* 1968), in which aliens visiting Earth in prehistory bio-engineer the first race of modernistic humans—one of whom becomes dinosaur-battling hero Tragg, and the other his beautiful mate Lorn.

26. Debus, "At War with Dinosaurs," pp. 8–13.

27. *DC Showcase Presents,* p. 217.

28. Ibid., pp. 76, 109.

29. Markstein, "Toonopedia—The War That Time Forgot."

30. For mere comic books, *War* tales were suffused with remarkable, page-turning *frisson.* Principally—writing and artistic talents of, re-spectively, Robert Kanigher, and Ross Andru, Mike Esposito with Joe Kubert and others contributed to these 6-time per year, low-grade science fictional (if I may) masterpieces. It's also possible to tell sometimes which paleo-artists DC comics artists borrowed dinosaur art from. For instance the (swiped) talents of Charles R. Knight, Zdenek Burian and J. Allen St. John are often recognizable. To me, some of the more memorable, if not unusual stories involved the soldier paleontologist, who is ironically platooned to the Pacific only to be amazed by the indigenous wild life which he's nearly devoured by. Then soldiers, who had been acrobats in civilian life, rely on high wire skills to, repeatedly, spare themselves from the scaly, hungry horde. There are more heartrending tales of experimental robot-soldiers who most humanely defend their human companions from the most unsettling circumstances imaginable. And in a twilight-zonish plot, several GI s are spookily plagued by identical dino-conjuring premonitions. In another tale, before dying tragically, a Kong-like, white-furred ape heroically battles a host of giant dinosaurs, rescuing stranded American soldiers from certain peril. Later, another ape tosses a Godzilla-inspired spiky-finned theropod over a cliff, recalling *King Kong vs. Godzilla's* thrilling conclusion (p. 494). Character development is not taken lightly, as in the most bizarre tale of all when a soldier actually becomes a were-dinosaur!

31. Martin J.S. Rudwick, *Worlds Before Adam: The Reconstruction of Geohistory in the Age of Reform* (Chicago: University of Chicago Press, 2008), pp. 481–498; others might disagree in claiming that Charles R. Darwin (1809–1882) was the most significant geologist of his time, instead of Lyell: they were both brilliant.

32. Rudwick's 2008 book is a scholarly ode to Lyell. Lyell meant actualistic or as recorded in human history—but see Rudwick for more.

33. Charles Lyell, *Principles of Geology, being an attempt to explain the former changes of the earth's surface by reference to causes now in operation,* 3 vols. (London, 1830–1833).

34. Lyell quoted in O'Connor, *The Earth on Show,* p. 179.

35. However, experts also recognized that the age of man was geologically preceded by a brief age of non-human primates.

36. In defining subdivisions of the Tertiary Period, Lyell perceived steady state, gradual (stochastic) addition and fluctuations of species regimes passing through time, without understanding mechanistically how species actually originated. For instance, Lyell named the Eocene, Miocene and Pliocene epochs of the Tertiary Period (named in 1760), founded upon relative proportions of fossilized marine faunal assemblages, each as compared to occur-

rences of modern species. Fossils in the oldest Eocene were merely 3 percent similar to living faunal assemblages, the more recent Pliocene contained 50 percent to 67 percent overlap, while the intermediate Miocene epoch had 17 percent commonality with modern species. Lyell aspired, but was never able to apply this approach to earlier Mesozoic Era stratigraphy. Meanwhile, between 1795 and 1879, other geologists hammered out details of the Phanerozoic rock system, now extending from the Cambrian (beginning 570 million years ago) to the Quaternary Period (ending 10,000 years ago). Historical episodes leading to identification of certain early Paleozoic periods — Cambrian, Silurian and Devonian — were charged with controversy, as leading British geologists celebrated and tenaciously defended their most ancient turf.

37. Rudwick, *Worlds Before Adam*, stated, "The successive groups of organisms summarized the directional, and indeed apparently progressive, character of the history of life … the progressiveness lay not in any greater 'perfection' in the later forms, but in an increasing diversity and the successive addition of 'higher' kinds of life. And he (Buckland) claimed that this was not the product of any gradual transmutation; somehow or other, new forms of life had been fashioned providentially to suit new environmental conditions as they developed on the ever-changing surface of a slowly cooling planet" (p. 431).

38. Quote is from Milner, Milner, *The Encyclopedia of Evolution*, p. 144; Rudwick, *Worlds Before Adam*, pp. 124–127, 545. As most geologists argued early on, any directionality denoted by the fossil record was not driven by an evolutionist (e.g., "Lamarckian") diversification of life. Conversely, close inspection of fossils dating from each of successive geological periods showed that Lyell's non-directional geotheory was left unsupported. Indeed, "Awful Changes" would have truly seemed an awful hypothesis to French savants (scientists) who, opposing Lyell, instead concluded that Earth formed out of a hot planetary nebula as a molten mass, continually, *directionally* cooling since its incandescent origin. Any heat emanating from within the earth, therefore, was residual "central" heat, indicative of long-term cooling. This conclusion was congruent with the fossil record. Still, by the mid–1840s, it remained unexplained why the planet cooled (catastrophically) during the recent "Ice Age" past, only to have then (suddenly) warmed thereafter unto the present day, that is, if the Earth had indeed steadily *cooled* throughout geological time. Steady cooling would successively cause gradual, fluctuating paleo-environmental changes. This steadily chilling trend did not, however, explain how

specific organic changes of the fossil record resulted. Nonetheless, successive fauna and flora were evidently well adapted to prevailing conditions and climate of their respective ages.

39. That is, other than mundane, stochastic changes in varieties of clams.

40. Burroughs's magnificent lost African land of "Pal-ul-don" (*Tarzan the Terrible*, 1921) with its *gryfs* and hollow-Earth Pellucidar, with its minimally evolved Mahars represent further examples of ERB's interest in evolution; Allen A. Debus, "Dinosaurs in Fantastic Fiction — Extras," *Prehistoric Times*, no. 81 (Spring 2007), pp. 22–23, 56–57.

41. In the sense that they're indistinguishable *generically* from extinct forms they're Lyellian organisms. While they may be modern day generically equivalent, evolutionary descendants (i.e., if not bioengineered from original preserved DNA), their forms may also have been conjured through a convergent evolution instead. Therefore, they may not be so genetically related after all to the original fossil varieties. Lacking a complete fossil record, we may never know for sure.

42. Christopher Golden, *King Kong* (New York: Pocket Books, 2005), p. 280.

43. Milner, *The Encyclopedia of Evolution*, p. 144.

44. See Berry, *Dinosaur Filmography*, for considerably more on these films.

45. For more on "Gaw" see references to Strickland/DeVito, *Kong: King of Skull Island*, and Debus, "King Kong vs. Godzilla: Skull Island's 'True' (Alternative) Tale," as mentioned in note 17.

46. Brian W. Aldiss with David Wingrove, *Trillion Year Spree* (New York: Avon Books, 1986), p. 81.

47. Ibid., p. 82; for the 1935 filmic production of *She*, a sabertooth cat frozen in ice — not mentioned in the novel — was added as a prop to underscore the prehistorical theme. (See Turner, *Spawn of Skull Island*, pp. 215–218.

48. Lin Carter, *Journey to the Underground World* (New York: Daw Books, 1979); Lin Carter, *Zanthodon* (New York: Daw Books, 1980); Lin Carter, *Hurok of the Stone Age* (New York: Daw Books, 1981).

49. Carter (1979), *Journey to the Underground World*, p. 36.

50. Ibid., p. 46.

51. "Zanthodon's" overlooked venue is most exemplary of an inner–Earth, primitive yet utopian ecology environment. Throughout their unlikely adventures, Carstairs and Potter do seem not only to be the luckiest folks alive in Zanthodon, but the fittest as well. At sea shortly after, Carstairs spares the noble Neandertal "Hurok" (rhymes with "Turok") from the gaping jaws of a plesiosaur. Much later, Darya

must be rescued after being abducted by an enormous thakdol, or pterodactyl — characterized as a flying stomach — that regarded her as a tasty morsel for winged hatchlings. The band of heroic adventurers often feast on *Eohippus* and *Archaeopteryx* flesh — which doesn't taste too much like chicken, dodge charging mammoths, slay a stream of sabertooth tigers and rampaging stegosaurs, and in Burroughs-like fashion resolve race and prejudicial issues between Zanthodon's various tribes, proving it is possible for men of all tribes (and species) to coexist peacefully if they're noble, instilled with chivalrous mien, and just. In Carter's sequel, *Zanthodon* (1980), Carstairs and company seek to rescue Darya once more, but instead they're soon captured by a diminutive race of humans who live inside a hollow mountain within the feared Peaks of Peril. They insidiously feed their captives to gigantic leeches which can hypnotize their victims (yes, sort of a cross between plot elements of Burroughs' Pellucidarian Mahar pterodactyls and H.G. Wells's Morlocks). Then, in Carter's satisfying *Hurok of the Stone Age* (1981), the treacherous "Xask" seeks to learn the secret of guns and explosives from Carstairs after they've become captured by the Dragonmen of Zar, ruled by a Darya look-alike goddess named Zarys. Drawing on his archaeological knowledge, Potter concludes that the Zarians are living (Minoan) descendants of the Atlanteans, The Dragonmen ride gigantic sauropod dinosaurs which are kept mesmerized, following orders (i.e. electrical brain impulses) telepathically transmitted through a mysterious metal circlet made of a strange Atlantean alloy. Heroic Hurok arrives on the scene while Potter climactically blows up the lavish Minoan citadel with gunpowder, all in the nick of time just before Carter's band of stalwart warriors become cruelly devoured by monstrous god Zorgazon — a gigantic *Tyrannosaurus*— in the spectator-filled Zarian Arena. Although vastly outnumbered, against all odds Carstairs and his loyal men are able to defeat the Zarians because "their inner strengths [had been] vitiated by the decadence of their enervating pleasures and by the soft pamperings of urban life." Carter continued Carstairs' adventures in Zanthodon's prehistoric, anachronistic environs through two additional novels, *Darya of the Bronze Age* (1981) and *Eric of Zanthodon* (1982).

52. James Gurney's insular "Dinotopia" is a curious admixture of late 19th century ideals, another lost (outer) world utopian society perhaps better suited, thematically, for a hollow earth environment. Here, the creatures commingling with humans are the same creatures that existed long ago during a variety of periods from the Paleozoic, Mesozoic and Cenozoic Eras. Time — meaning evolutionary forces —

stands still in Dinotopia. These are the same creatures— dinosaurs and other prehistoric animals alike, not all wild and some monstrous but sentient and tamed —from various ages of the prehistoric world, living together harmony without time-specific distinction.

53. As stated by Joseph Andriano, *Immortal Monster*, p. 1, "The primordial beast-monster is the sea serpent, but Herman Melville was prescient enough to realize that the ultimate monster could not be confined to the sea"; Allen A. Debus, "Dinosaurs in Fantastic Fiction — Extras III," *Prehistoric Times*, no. 83 (Fall 2007), pp. 22–23, 27; some other aquatic prehistoria of fantastic fiction include; the variety of huge sharks featured in Steve Alten's demonic *Meg* (1997) — the opening staged a sixty-foot-long *Carcharadon megalodon* shark devouring a swimming tyrannosaur; Carl Dreadstone's dino-monstrous *Creature from the Black Lagoon* (1977); and Willis Knapp Jones' "Beast of the Yungas" (1927) *Diplodocus*— written at a time when it was generally believed that sauropods were aquatic, not terrestrial, animals. There have also been numerous Loch Ness monster stories, too many to recount here.

54. H.G. Wells, *The Time Machine* (originally published in 1895), in *The Complete Science Fiction Treasury of H.G. Wells* (New York: Avenel Books, 1978), p. 66.

55. Mill's story, involving time travel through geological periods via transmigration of souls, is discussed by O'Connor, *The Earth on Show*, pp. 339–345.

56. For considerably more on the theme of time travel in literature involving menacing dinosaurs, beyond what I discussed in chapter five ("Time-relativistic Dinosaurs: Bradbury's Legacy") of my *Dinosaurs in Fantastic Fiction* (2006), please see Allen A. Debus, "Dinosaurs in Fantastic Fiction — Extras IV (Space-Time Prehistoria)," *Prehistoric Times*, no. 86 (Summer 2008), pp. 22–23, 27; Allen A. Debus, "Dinosaurs in Fantastic Fiction — Extras," *Prehistoric Times*, no. 81 (Spring 2007), pp. 22–23, 56–57; Allen A. Debus, "Time Steps Aside: Bradbury's Unsung Safari Legacy," *Prehistoric Times*, no. 80 (Winter 2006), pp. 20–22; the entertaining BBC program *Primeval*, which began airing in the United States in 2007, conjured various frightful prehistoria into modernity through a mysterious space-time warp. An interesting literary example discussed in my "Extras IV" is Richard Marsten's *Danger: Dinosaurs!* (Philadelphia: John C. Winston, 1953).

57. The topic of outer space prehistoria is addressed in chapter six of my *Dinosaurs in Fantastic Fiction* (2006). Typically, especially if they're dinosaurian in nature, such creatures inhabit Venus—formerly thought to be geologically more primitive (and swampy) than ours.

Alternatively, they exist on planets out side our solar system. Examples of the former would include Venusian tyrannosaurs described in Carey Rockwell's *The Revolt on Venus* (1954) or Venus' dinosaurian fauna, central to S.M. Stirling's *The Sky People* (2006). Dinosaurian forms of fantastic fiction evidently from much farther worlds in the cosmos would include invading alien lizards in Harry Turtledove's alternative history novel, *World War: In the Balance* (1994) and sentient dinosaurs described in L. Sprague de Camp's and Catherine Crook de Camp's *The Stones of Nomuru* (New York: Baen Publishing, 1988). Beyond fantastic fiction, memorable "alien" dinosaurs also appear in cult films, such as *Planet of Dinosaurs* (1978) and *King Dinosaur* (1955). Toho's "Ghidorah" may be another filmic example of the outer space variety.

58. Crichton, *Jurassic Park*; Crichton, *The Lost World*.

59. According to Radu Florescu (*In Search of Frankenstein*, Boston, MA: New York Graphic Society, 1975, pp. 206–207), Mary Shelley was quite familiar with Davy's scientific work. In *Frankenstein* she invoked Davy as a shining example of the modern Romantic breed of forward thinkers whose successes should be emulated by Victor Frankenstein, rather than medieval alchemy practiced by men such as Paracelsus. Four decades later, Verne structured his *Journey* novel upon Davy's volcanic theory. Without Davy's outmoded volcanic views serving to suspend disbelief, Lidenbrock's party would have burned to a crisp in the Earth's central fire. (Allen A. Debus, "Reframing the Science in Jules Verne's *Journey to the Center of the Earth*," *Science Fiction Studies*, vol. 33, no. 100, Nov. 2006, pp. 405–420.) For more on Davy, see Robert Siegfried and Robert H. Dott, Jr., *Humphry Davy on Geology: The 1805 Lectures for the General Audience* (Madison: University of Wisconsin Press, 1980); and Allen A. Debus, "Humphry Davy's *Consolations*" (Parts 1 to 2), *Fossil News: Journal of Avocational Paleontology* 12, no. 5 (May 2006), pp. 14–17; (June 2006): pp. 14–17. For more on the (general) theme of mad scientists in fiction, see Joachim Schummer, "Historical Roots of the 'Mad Scientist': Chemists in Nineteenth-Century Literature," *Ambix*, Vol. 53, no.2 (July 2006), pp. 99–127; Allen A. Debus, "Mad Scientists and Monster Men," *Erbania*, nos. 94/95 (Summer/Autumn 2007), pp. 23–27.

60. I realize that comparing books with movies is a bit like apples and oranges, but here I'm concerned with prototypical authenticity. And since I am concerned with originality, characters adapted from Verne's and Conan Doyle's classic novels for several filmed versions will not be discussed. For Verne's novel I am referring to William Butcher's recent editing of the novel, as opposed to common 1870s English translations where Lidenbrock is instead named Prof. Von Hardwigg.

61. George Sand, *Laura: A Journey into the Crystal* (originally published 1864 as *Laura, dans la voyage de Cristal*; London: Pushkin Press, 2004).

62. Gary Hoppenstand, "Dinosaur Doctors and Jurassic Geniuses: The Changing Image of the Scientist in the Lost World Adventure," *Studies in Popular Culture*, Vol. 22 (October 1999), p. 10.

63. Robert Silverberg, "Re-reading Heinlein," *Asimov's Science Fiction*, Vol. 31 (Dec. 2007), p. 9.

64. Tsutsui, *Godzilla on My Mind*.

65. Turner, *Spawn of Skull Island*, p. 89.

66. Ibid.

67. Tony Phillips, "Modern Lost Worlds," *Prehistoric Times*, no. 75 (Dec./Jan. 2006), pp. 54–56; Michael J. Wolfe, "The Season of the Dragons," *Starlog*, no. 193 (August 1993), pp. 27–31.

68. Allen A. Debus, "Unwittingly Western: Two Shining Knights," *G-Fan*, no. 82 (Winter 2008), pp. 54–57.

69. Strickland and Michlig, *Kong*.

70. Weta Workshop, *Kong: A Natural History*.

71. In Dougal Dixon's speculative *The New Dinosaurs* (Topsfield, MA: Salem House Publications, 1988), many islands and continents are populated by dinosaur descendants of Late Cretaceous forms that were not exterminated by the asteroid collision of 65 million years ago. These may be compared to Skull Island's new millennial (2005) fauna, although because mankind has evolved, the modern dinosaurs are restricted to a single island instead of being globally distributed in the absence of man. In a 2008 remake of *Journey to the Center of the Earth*, wherein a geologist played by Brendan Fraser retraces the steps of his lost brother into the bowels of the Earth, an effectively produced CGI tyrannosaur sprints after two members of the exploring party. Thus an old, classic tale is amended with requirements of the new age.

Bibliography

Several references cited in chapter footnotes have been omitted from this Bibliography, which is intended to present selected documents that should be more readily available through university or public library systems. Here, I have been highly selective about listing fanzine periodicals and society journals which ordinarily aren't archived in libraries (e.g. microfilm, microfiche, or hard copy form). That is, unless a particular fanzine article contained the germ of an idea, previously unavailable or unexpressed, of the mainstream literature presented below, I have focused here on documentation bearing *most* directly or pivotally on chapter subjects, not everything consulted in preparation of this manuscript. Of course, other information may be found under individual chapter footnotes.

Albritton, Claude C. *The Abyss of Time: Changing Conceptions of the Earth's Antiquity After the Sixteenth Century*. New York: Jeremy P. Tarcher, 1986.

Aldiss, Brian W. (with David Wingrove). *Trillion Year Spree*. New York: Avon Books, 1986.

Allmon, Warren D. "The Pre-Modern History of the Post-Modern Dinosaur: Phases and Causes in Post-Darwinian Dinosaur Art," *Earth Sciences History*, Vol. 25, no. 1, 2006, pp. 5–36.

Alvarez, L., W. Alvarez, F. Asaro, and H.V. Michel. "Extraterrestrial Cause for the Cretaceous-Tertiary Extinction: Experimental Results and Theoretical Interpretation," *Science*, Vol. 208, 1980, pp. 1095–1108.

Andriano, Joseph. *Immortal Monster: Mythological Evolution of the Fantastic Beast in Modern Fiction and Film*. Westport, CT: Greenwood Press, 1999.

Asimov, Isaac. *Biographical Encyclopedia of Science and Technology*. Garden City, NY: Doubleday, 1964

Bakker, Robert T. *The Dinosaur Heresies: New Theories Unlocking the Mystery of the Dinosaurs and Their Extinction*. New York: William Morrow, 1986.

_____. *Raptor Red*. New York: Bantam Books, 1995.

_____. "The Return of the Dancing Dinosaurs," *Dinosaurs Past and Present*. Vol. 1, ed. Sylvia J. Czerkas and Everett C. Olson. Seattle and London: Natural History Museum of Los Angeles County in association with University of Washington Press, 1987, pp. 38–69.

_____. "The Superiority of Dinosaurs," in *Discovery: Magazine of the Peabody Museum of Natural History, Yale University*, Vol. 3, no. 2, Spring 1968, pp. 11–22.

Ballou, William. "The Serpentlike Sea Saurians," *Popular Science Monthly*, Vol. 53, June 1898, pp. 209–218.

Barton, Richard. *Benchmark Papers in Geology 51: North American Geology — Early Writings*, Stroudsburg, PA: Douden, Hutchinson & Ross, 1979, Robert M. Hazen, ed., pp. 88–89.

Batory, Dana Martin, and Sarjeant, William A. S. "The Terror of Blue John Gap — A Geological and Literary Study," *ACD — The Journal of the Arthur Conan Doyle Society*, Vol. 5, 1994, pp. 108–125.

Baxter, Stephen. *The Time Ships*. New York: HarperPrism, 1995.

Bear, Greg. *Dinosaur Summer*. New York: Warner Books, 1998.

Becker, Richard A. "The Return of Cthadron," in Robert Hood and Robin Pen, eds., *Daikaiju! 3: Giant Monsters vs. the World Tales*. University of Wollongong, Australia: Agog! Press, 2007.

Berry, Mark F. *The Dinosaur Filmography*. Jefferson, NC: McFarland, 2002.

_____. "Joe DeVito: A Mighty Interview" (Parts 1–2), in *Prehistoric Times*, nos. 73–74, Aug./Sept. 2005, pp. 28–31, 47; Oct./Nov. 2005, pp. 28–31, 45.

Bingham, Carson. *Gorgo*. Derby, CT: Monarch Books, 1960.

Boyd, Julian P. "The Megalonyx, the Megatherium, and Thomas Jefferson's Lapse of Memory," *Proceedings of the American Philosophical Society*, Vol. 102, no. 5, October 1958.

Bradbury, Ray. "The Beast from 20,000 Fathoms," *The Saturday Evening Post*, June 23, 1951, pp. 28–29, 117–118.

_____. *Dinosaur Tales*. New York: Bantam Books, 1983.

Breyer, John, and William Butcher. "Nothing New Under the Earth: The Geology of Jules Verne's *Journey to the Center of the Earth*," *Earth Sciences History*, Vol. 22, no. 1, 2003, pp. 36–54.

Brochu, Christopher A. "Osteology of *Tyrannosaurus rex*: Insights from a Nearly Complete Skeleton and High-Resolution Computed Tomographic Analysis of the Skull," *Society of Vertebrate Paleontology — Memoir 7*, Vol. 22, Supplement to no. 4, January 14, 2003.

Brown, Barnum. "Hunting Big Game of Other Days." *The National Geographic Magazine*, Vol. 35, no. 5, May 1919, pp. 407–429.

_____. "The Trachodon Group," *The American Museum Journal*, Vol. 8, no. 4, April 1908, p. 51.

_____. "Tyrannosaurus, a Cretaceous Carnivorous Dinosaur: The Largest Flesh-Eater That Ever Lived," *Scientific American*, Oct. 9, 1915, pp. 322–323.

Buckland, William. *Geology and Mineralogy Considered with Reference to Natural Theology*. London: William Pickering, 1837 ed.

Burden, W. Douglas. *Dragon Lizards of Komodo: An Expedition to the Lost World of the Dutch East Indies*. New York: Putnam's, 1927.

Burroughs, Edgar Rice. *At the Earth's Core*. New York: Ace Books, 1978 ed. (orig. 1914).

_____. *Back to the Stone Age*. New York: Ace Books, 1978 ed. (orig. 1936, 1937).

_____. *The Land That Time Forgot*. New York: Ace Books, 1963 ed. (orig. 1918).

_____. *Out of Time's Abyss*. New York: Ace Books, 1979 ed. (orig. 1918).

_____. *Pellucidar*. New York: Ace Books, 1963 ed. (orig. 1923).

_____. *The People That Time Forgot*. New York: Ace Books, 1979 ed. (orig. 1918).

_____. *Savage Pellucidar*. New York: Ballantine Books, 1990 ed. (orig. 1941 to 1963).

_____. *Tanar of Pellucidar*. New York: Ace Books, 1978 ed. (orig. 1929).

_____. *Tarzan at the Earth's Core*. New York: Ballantine Books, 1964 ed. (orig. 1930).

_____. *Tarzan of the Apes*. New York: Ballantine Books, 1982 ed. (orig. 1912).

_____. *Tarzan the Terrible*. New York: Ballantine Books, 1963.

Cabot, John York. "Blitzkrieg in the Past," *Amazing Stories*, Vol. 16, no. 7, July 1942, pp. 8–47.

Capek, Karel. *War with the Newts*, new translation by Ewald Osers, Highland Park, NJ: Catbird Press, 1985 (orig. London: George Allen & Unwin, 1936).

Carey, Diane, and James I. Kirkland. *Star Trek: First Frontier*. New York: Pocket Books, 1995.

Carrington, Richard. *Mermaids and Mastodons: A Book of Natural and Unnatural History*. London: Chatto and Windus, 1957.

Carpenter, Kenneth. "A Dinosaur Paleontologist's View of Godzilla," pp. 102–106 in *The Official Godzilla Compendium*, New York: Random House, 1998, J. D. Lees and Marc Cerasini, eds.

Carter, Lin. *Hurok of the Stone Age* New York: Daw Books, 1981.

_____. *Journey to the Underground World*. New York: Daw Books, 1979.

_____. *Zanthodon*. New York: Daw Books, 1980.

Castor, Cartelle, and Gerardo De Iuliis. "*Eremotherium laurillardi*: The Panamerican Late Pleistocene megatheriid Sloth," *Journal of Vertebrate Paleontology*, Vol. 15, no. 4, December 1995, pp. 830–841.

Clarke, Arthur C. *2001: A Space Odyssey*. New York: A Signet Book, 1968.

Cohen, Claudine. *The Fate of the Mammoth: Fossils, Myth and History*. Chicago: University of Chicago Press, 2002.

Cohen, I. Bernard. *Science and the Founding Fathers*. New York: W.W. Norton, 1995.

Cohen, Jeffrey, Jerome. (ed.) *Monster Theory: Reading Culture*. Minneapolis: University of Minnesota Press, 1996.

Collinson, Peter. "Of Some Very Large Fossil Teeth Found in North America, and Described by Peter Collinson," *Philosophical Transactions* (Anno 1767), London, p. 477.

Conybeare, W.D. "On the Discovery of an Almost Perfect Skeleton of the *Plesiosaurus*" (originally published in *Transactions of the Geological Society of London*, 2nd series, Vol. 1, 1824, pp. 381–390), reprinted in *The Dinosaur Papers (1676–1906)*, ed. David B. Weishampel and Nadine M. White, Washington DC: Smithsonian Inst., 2003.

Cope, Edward D. "The Fossil Reptiles of New Jersey" (Part 1), *The American Naturalist*, Vol. 1, 1868, pp. 23–30.

Costello, Peter. *Jules Verne: Inventor of Science Fiction*. New York: Scribner's, 1978.

Crichton, Michael. *Jurassic Park*. New York: Ballantine Books, 1990.

_____. *The Lost World*. New York: Alfred A. Knopf, 1995.

Culver, James Erwin. "Some Extinct Giants," *The California Illustrated Magazine*, Vol. 2, April 1892, pp. 501–507.

Czerkas, S.J., and E.C. Olson (eds.). *Dinosaurs Past and Present— Volume 1*, published by the Natural History of Los Angeles County in association with University of Washington Press, 1987.

Czerkas, Sylvia M., and Donald F. Glut. *Dinosaurs, Mammoths, and Cavemen: The Art of Charles R. Knight*. New York: E.P. Dutton, 1982.

Darnton, John. *Neanderthal*. New York: St. Martin's Press, 1996.

Davidson, Jane P. *A History of Paleontology Illustration*. Bloomington & Indianapolis: Indiana University Press, 2008.

Davy, Humphry, Robert Siegfried and Robert H. Dott, Jr. (eds.). *Humphry Davy on Geology: The 1805 Lectures for the General Audience*. Madison: University of Wisconsin Press, 1980.

Darwin, Charles. *The Origin of Species: By Means of Natural Selection, or the Preservation of Favoured Races in the Struggle for Life*, with a Foreword by George Gaylord Simpson. New York: Collier Books, 6th ed., 1872, reprint.

_____. *The Voyage of the Beagle*. New York: Mentor, 1988 ed.

Dean, Dennis R. *Gideon Mantell and the Discovery of Dinosaurs.* Cambridge: Cambridge University Press, 1999.

_____. "The Influence of Geology on American Thought," in *Two Hundred Years of Geology in America: Proceedings of the New Hampshire Bicentennial Conference on the History of Geology.* Cecil J. Schneer, ed., Published for the University of New Hampshire by the University Press of New England: Hanover, New Hampshire, 1979.

Debus, Allen A. "Decade of the Dinosaur," *Fossil News: Journal of Avocational Paleontology*, Vol. 11, no. 5, May 2005, pp. 7–16.

_____. *Dinosaurs in Fantastic Fiction: A Thematic Survey.* Jefferson, NC: McFarland, 2006.

_____. "Greatest 'Prehistoric' Monsters of All-Time! (Parts 1, 3–4)" *Scary Monsters Magazine*, nos. 57–59; Jan. 2006, pp. 68–73; April 2006, pp.7–14; June 2006, pp. 31–39.

_____. "Greatest 'Prehistoric' Monsters of All-Time! (Part 2)" *Scary Monsters Memories — 2006*, Yearbook no. 14, March 2006, pp. 98–103.

_____. "Humphry Davy's 'Consolations' Prize (Parts 1–2)," *Fossil News: Journal of Avocational Paleontology*, Vol. 12, no. 5, May, 2006, pp.14–17; no. 6, June 2006, pp. 14–17.

_____. "Reframing Verne's Paleontological *Journey*," *Science Fiction Studies*, Vol. 33, no. 100, Part 3, November 2007, pp. 405–420.

_____. "Sorting Fossil Vertebrate Iconography in Paleoart," *Bulletin of the South Texas Geological Society*, Vol. 44, no. 1, Sept. 2003, pp. 5, 10–24.

Debus, Allen A., and Diane E. Debus. *Dinosaur Memories: Dino-trekking for Beasts of Thunder, Fantastic Saurians, "Paleo-people," "Dinosaurabilia," and Other "Prehistoria."* Lincoln, NE: Authors Choice Books, 2002.

_____, and _____. *Paleoimagery: The Evolution of Dinosaurs in Art.* Jefferson, NC: McFarland, 2002.

Debus, Allen A., and McCarthy, Steve. "A Scene from American Deep Time: New York's Palaeozoic Museum — Revisited," *The Mosasaur — The Journal of the Delaware Valley Paleontological Society*, Vol. 6, May 1999, pp. 105–115.

Debus, Allen G. *Man and Nature in the Renaissance.* Cambridge: University of Cambridge Press, 1978.

De Camp, L. Sprague. *Rivers of Time.* Riverdale, NY: Baen Books, 1993.

De Camp, L. Sprague, and Catherine Crook De Camp. *The Stones of Nomuru.* New York: Baen Fantasy, 1988.

De Noux, O'Neil. "Tyrannous and Strong," *Asimov's Science Fiction*, Vol. 24, no. 2, Feb. 2000, pp. 66–81.

DePaolo, Charles. *Human Prehistory in Fiction.* Jefferson, NC: McFarland, 2003.

Desalle, Rob, and David Lindley. *The Science of Jurassic Park.* New York: Basic Books, 1997.

Desmond, Adrian J. *The Hot-Blooded Dinosaurs: A Revolution in Palaeontology.* New York: Warner Books, 1975.

_____. *Huxley: From Devil's Disciple to Evolution's High Priest.* Reading, MA: Perseus Books, 1997.

DeVito, Joe, and Brad Strickland. *Merian C. Cooper's King Kong.* New York: St. Martin's Griffin, 2005

Dingus, Lowell, and Timothy Rowe. *The Mistaken Extinction: Dinosaur Evolution and the Origin of Birds.* New York: W.H. Freeman, 1998.

Dixon, Dougal. *The New Dinosaurs: An Alternative Evolution.* Topsfield, MA: Salem House, 1988.

Doyle, Arthur Conan. "The Lost World," *The Strand Magazine*, Vol. 43 to Vol. 45, April to November 1912.

_____. *The Lost World of Arthur Conan Doyle* (Collector's Anniversary Edition by John R. Lavas). New Zealand, 2002.

_____. "The Terror of Blue John Gap," *The Strand Magazine*, Vol. 40, September 1910.

Drake, David. *Tyrannosaur.* New York: Tor Books, 1993.

Edwards, W.N. *The Early History of Palaeontology.* London: Trustees of the British Museum (Natural History), 1967.

Endor, Guy. *The Werewolf of Paris.* 1933.

Evans, Arthur B. "The Illustrators of Jules Verne's *Voyages Extraordinaires*," *Science Fiction Studies*, July 1998, pp. 241–270.

_____. "Literary Intertexts in Jules Verne's *Voyages Extraordinaires*," *Science Fiction Studies*, Vol. 23, no. 2, July 1996, pp. 172–173.

Farina, R.A., and R.E. Blanco. "*Megatherium*, the Stabber," *Proceedings of the Royal Society of London*, B, Vol. 263, 1996, pp. 1725–1729.

Farlow, James O., Michael Brett-Surman, and Robert Walters (eds.). *The Complete Dinosaur.* Bloomington: Indiana University Press, 1997.

Field, Henry. *Prehistoric Man — Hall of the Stone Age of the Old World*, Leaflet 31, Chicago: Field Museum of Natural History, 1933.

Figuier, Louis. *The World Before the Deluge.* London: Cassell, 1867.

Fiorillo, Anthony R., and Daeschler, Edward. "E.D. Cope's 1893 Expedition to the Dakotas Revisited," *Earth Sciences History*, Vol. 9, no. 1, 1990, pp. 57–61.

Fitting, Peter (ed.). *Subterranean Worlds: A Critical Anthology.* Middletown, CT: Wesleyan University Press, 2004.

Flammarion, Camille. *Le monde avant la creation de l'homme. Origines de la terre. Origines de la vie. Origines de l'humanité.* Paris: C. Marpon and E. Flammarion, 1886.

Florescu, Radu. *In Search of Frankenstein.* Boston, MA: New York Graphic Society, 1975.

Foster, Alan Dean. "He," *Fantasy and Science Fiction*, June 1976, pp. 47–64.

Foster, Elizabeth S. "Melville and Geology," *American Literature*, Vol. 17, no. 1, March 1945, pp. 50–65.

Franco, Barbara. *The Cardiff Giant: A Hundred Year Old Hoax.* Cooperstown, NY: New York State Historical Association, 1990.

Franklin, Benjamin. "Conjectures Concerning the Formation of the Earth," *Am. Philos. Soc. Trans.*, Vol. 3, 1793, pp. 1–5.

Gerrold, David. *Deathbeast.* New York: Popular Library, 1978.

Glut, Donald F. *Classic Movie Monsters.* Metuchen, NJ: Scarecrow, 1978.

_____. *The Dinosaur Scrapbook.* Secaucus, NJ: Citadel Press, 1980.

_____. *Dinosaurs: The Encyclopedia.* Jefferson, NC: McFarland, 1997.

_____. *Dinosaurs: The Encyclopedia, Supplement 1.* Jefferson, NC: McFarland, 2000.

_____. *Dinosaurs: The Encyclopedia, Supplement 2.* Jefferson, NC: McFarland, 2002.

_____. *Dinosaurs: The Encyclopedia, Supplement 3.* Jefferson, NC: McFarland, 2003.

_____. *Dinosaurs: The Encyclopedia, Supplement 4.* Jefferson, NC: McFarland, 2006.

_____. *Dinosaurs: The Encyclopedia, Supplement 5.* Jefferson, NC: McFarland, 2008.

_____. *Frankenstein in the Lost World.* Castle of Frankenstein (Tome 6). Highwood, IL: Druktenis, 2002 (orig. 1971).

_____. *Jurassic Classics: A Collection of Saurian Essays and Mesozoic Musings.* Jefferson, NC: McFarland, 2001.

Goetzmann, William. *New Lands, New Men.* New York: Viking Penguin, 1986.

Golden, Christopher. *King Kong.* New York: Pocket Star Books, 2005.

Gould, Charles. *Mythical Monsters.* New York: Crescent Books, 1889 (originally published 1886).

Gould, Stephen J. *Bully for Brontosaurus.* New York: W.W. Norton, 1991.

_____. *Dinosaur in a Haystack: Reflections in Natural History.* New York: Harmony Books, 1996.

_____. *The Lying Stones of Marrakech: Penultimate Reflections in Natural History.* New York: Harmony Books, 2000.

Green, Jonathan. *Pax Brittania: Unnatural History.* Oxford, UK: Abaddon Books, 2007.

Greene, John C. *American Science in the Age of Jefferson.* Ames: Iowa State University Press, 1984.

_____. *The Death of Adam.* Ames: Iowa State University Press, 1959.

Gunning, William. *Life History of Our Planet.* New York: Worthington, 1879.

Gurney, James. *Dinotopia: A Land Apart from Time.* Atlanta: Turner, 1992.

_____. *Dinotopia: The World Beneath.* Atlanta: Turner, 1995.

Hachya, Michihiko. *Hiroshima Diary.* New York: Avon Books, 1955.

Hamilton, Edith. *Mythology.* New York: New York American Library, 1942.

Harrison, Harry. *Return to Eden.* New York: Bantam Books, 1988.

_____. *West of Eden.* New York: Bantam Books, 1984.

_____. *Winter in Eden.* New York: Bantam Books, 1986.

Hawkins, Benjamin Waterhouse. "On Visual Education as Applied to Geology" (originally published in *Journal of the Society of Arts*, Vol. 2, 1853–54, pp. 444–449), reprinted in *The Dinosaur Papers (1676–1906)*, ed. David B. Weishampel and Nadine M. White, Washington DC: Smithsonian Institution, 2003.

Heilmann, Gerhard. *The Origin of Birds.* London: Witherby, 1926.

Hodgson, William B. *Memoir on the Megatherium and Other Extinct Gigantic Quadrupeds of the Coast of Georgia with Observations on Its Geologic Features.* New York: Bartlett & Welford, 1846.

Holiday, F. W. *The Great Orm of Loch Ness.* New York: Avon Books, 1969.

Holmes, John Eric. *Mahars of Pellucidar.* New York: Ace Books, 1976.

Holtzmark, Erling B. *Edgar Rice Burroughs.* Boston: Twayne, 1986.

Hood, Robert. "Flesh and Bone," in Robert Hood and Robin Pen, eds., *Daikaiju! 3: Giant Monsters vs. the World Tales.* University of Wollongong, Australia: Agog! Press, 2007.

Hood, Robert, and Robin Pen, eds. *Daikaiju!: Giant Monster Tales.* University of Wollongong, Australia: Agog! Press, 2005.

Hooke, Robert. *Micrographia, or Some Physiological Descriptions of Minute Bodies Made by Magnifying Glasses.* London: Martyn and Allestry, 1665; Dover Press, 1961.

Hopp, Thomas P. *Dinosaur Wars.* Lincoln, NE: Authors Choice Press, 2000.

_____. *Dinosaur Wars: Counterattack.* iUniverse, 2002.

Hoppenstand, Gary. "Dinosaur Doctors and Jurassic Geniuses: The Changing Image of the Scientist in the Lost World Adventure," *Studies in Popular Culture*, Vol. 22, October 1999, pp. 1–14.

Horner, John R. and Lessem, Don. *The Complete T. rex.* New York: Simon & Schuster, 1993.

Horrocks, Thomas. "Thomas Jefferson and the Great Claw," *Virginia Calvacade*, Vol. 35, no. 2, Fall 1985, pp. 70–79.

Howard, Robert West. *The Dawnseekers: The First History of American Paleontology.* New York: Harcourt, Brace Jovanovich, 1975.

Hubbell, Will. *Cretaceous Sea.* New York: Ace Books, 2002.

Hulke, Malcolm. *Dr. Who and the Cave-Monsters.* London: A Target Book, 1983.

_____. *Dr. Who and the Dinosaur Invasion.* London: A Target Book, 1976.

Hunter, William. "Observations on the Bones, Commonly Supposed to Be Elephant Bones, Which Have Been Placed Near the River Ohio in America," *Philosophical Transactions* (Anno 1768), London, p. 505.

Hutchinson, H.N. *Extinct Monsters: A Popular Account of Some of the Larger Forms of Ancient Animal Life*, London: Chapman & Hall, LD, 1893.

Inwood, Stephen. *The Forgotten Genius: The Biography of Robert Hooke, 1635–1703.* San Francisco: MacAdam/Cage, 2002.

Jackson, Donald. *Custer's Gold: The United States Cavalry Expedition of 1874* (originally published by Yale University Press, 1966). Lincoln: University of Nebraska Press– Bison Books, 1972.

Jefferson, Thomas. "A Memoir on the Discovery of Certain Bones of a Quadruped of the Clawed Kind in the Western Parts of Virginia" (originally published in *Am. Philos. Soc. Trans.*, Vol. 4, pp. 246–260, 1799, pp. 246–258), in *Benchmark Papers in Geology/151 North American Geology: Early Writings*, ed. Robert M. Hazen. Stroudsburg, PA: Dowden, Hutchington & Ross, 1979, pp. 74–86.

_____. *Notes on the State of Virginia*. Philadelphia, 1782.

Johanson, Donald, and Maitland Edey. *Lucy: The Beginnings of Mankind*. New York: Warner Books, 1981.

Johnson, Craig. "Preservation of Biomolecules in Cancellous Bone of *Tyrannosaurus rex*," *Journal of Vertebrate Paleontology*, Vol. 17, no. 2, June 1997, pp. 330–348.

Keynes, Richard Darwin. *Fossils, Finches and Fuegians*. New York: Oxford University Press, 2003.

Kircher, Athanasius. *Mundus Subterraneus*. Rome, 1665.

Knight, Charles R. *Before the Dawn of History*. New York: McGraw-Hill, 1935.

Knight, Harry Adam. *Carnosaur*. London: A Star Book, 1984.

Knipe, Henry Robert. *Nebula to Man*. New York: J.M. Dent, 1905.

Koch, Albert C. *Description of the Missourium, or Missouri Leviathan; Together with Its Supposed Habits and Indian Traditions Concerning the Location from Whence It Was Exhumed, Also Comparison of the Whale, Crocodile and Missourium with the Leviathan as Described in 41st Chapter of the Book of Job* (2nd edition, Enlarged, 1841), published by Prentice and Weissinger of Louisville, Kentucky.

_____. *Journey Through a Part of the United States of North America in the Years 1844 to 1846*, translated by Ernst A. Stadler in 1972. Carbondale: Southern Illinois University Press, 1972.

Kolbl-Ebert, Martina. "Female British Geologists in the Early Nineteenth Century," *Earth Sciences History*, Vol. 21, no. 1, 2002, pp. 3–25.

Kreps, Penelope Banka. *Carnivores*. New York: Kensington Publishing, 1993.

Lankester, E. Ray. *Extinct Animals*. London: Archibald Constable, 1905.

Laporte, Leo F. *George Gaylord Simpson: Paleontologist and Evolutionist*. New York: Columbia University Press, 2000.

Larson, Peter. *Rex Appeal*. Montpelier, VT: Invisible Cities Press, 2002.

Laudan, Rachel. *From Mineralogy to Geology: The Foundations of a Science 1650–1830*. Chicago: University of Chicago Press, 1987.

Lees, J.D. "What Is a Kaiju?" *G-Fan*, no. 78, Winter 2007, pp. 68–72.

Lees, J.D., and Marc Cerasini (eds.). *The Official Godzilla Compendium*. New York: Random House, 1998.

Leidy, Joseph. "Notices of Remains of Extinct Reptiles and Fishes, Discovered by Dr. F.V. Hayden in the Bad Lands of the Judith River, Nebraska Territory" (originally published in *Proceedings of the Academy of Natural Sciences of Philadelphia*, Vol. 8, March 1856, pp. 72–73), reprinted in *Dinosaur Papers (1676–1906)*, ed. David B. Weishampel and Nadine M. White, Washington DC: Smithsonian Institution, 2003.

Levine, Joseph M. *Dr. Woodward's Shield: History, Science, and Satire in Augustan England*. Berkeley: University of California Press, 1977.

Livings, Marvin. "Running," in Robert Hood and Robin Pen, eds. *Daikaiju!: Giant Monster Tales*. University of Wollongong, Australia: Agog! Press, 2005.

London, Jack. *Before Adam*. New York: Macmillan, 1907.

Longyear, Barry B. *The Homecoming*. New York: Walker, 1989.

Lopez Pinero, Jose. "Juan Bautista Bru (1740–1799) and the Description of the Genus *Megatherium*," *Journal of the History of Biology*, 1988, Vol. 21, pp. 146–163.

_____. *Juan Bautista Bru de Ramon: El Atlas zoologoico, el megaterio y las tecnicas de pesca valencianas 1742–1799.* Valencia: Ayuntamiento de Valencia, 1996, pp. 420–435.

Lourie, Eugene. *My Work in Films.* San Diego: Harcourt, Brace, Jovanovich, 1985.

Lovelace, Delos W. *King Kong.* New York: Grossett and Dunlap, 1932.

Lyell, Charles. *Charles Lyell on North American Geology* (reprinted from *Am. Journal of Science*, vol. 2, no. 3, 1847), New York: Arno Press, 1978), pp. 36–37, 267–269, 322–323.

_____. *The Geological Evidence of the Antiquity of Man* (originally published as *The Geological Evidences of the Antiquity of Man: With Remarks on Theories of the Origin of Species by Variation* by J. Murray, 1863), New York: Dover, 2004, p. 161.

Mackal, Roy P. *A Living Dinosaur?* Leiden, The Netherlands: E.J. Brill, 1987.

Mantell, Gideon. "On the Structure of the Jaws and Teeth of the *Iguanodon*" (originally published in *Philosophical Transactions of the Royal Society of London*, Vol. 138, part 1, 1848, pp. 183–202), reprinted in *Dinosaur Papers (1676–1906)*, ed. David B. Weishampel and Nadine M. White. Washington DC: Smithsonian Institution, 2003.

Marche, Jordan D., II. "Edward Hitchcock's Poem: The Sandstone Bird (1836)," *Earth Sciences History*, Vol. 10, no. 1, 1991, pp. 5–8.

Marsh, O.C. "A New Order of Extinct Reptilia (Stegosauria) from the Jurassic of the Rocky Mountains," *American Journal of Science*, Series 3, Vol. 14, 1877, pp. 513–514.

_____. "Restoration of *Stegosaurus*," *American Journal of Science*, Series 3, Vol. 42, 1891, pp. 179–181.

Mathews, Cornelius. *Behemoth: A Legend of the Mound-Builders* (1839). Reprinted, Whitefish, MT: Kessinger, 2007.

Matthew, William D. "Allosaurus, A Carnivorous Dinosaur, and Its Prey," *The American Museum Journal*, Vol. 8, no. 1, Jan. 1908, p. 5.

_____. "Scourge of the Santa Monica Mountains," *The American Museum Journal*, Vol. 16, no. 7, 1916, pp. 469–472.

_____. "The Tyrannosaurus," *The American Museum Journal*, Vol. 10, no. 1, Jan. 1910, p. 8.

Mayor, Adrienne. *The First Fossil Hunters.* Princeton: Princeton University Press, 2000.

_____. *The First Fossil Hunters: Paleontology in Greek and Roman Times.* Princeton: Princeton University Press, 2000.

_____. *Fossil Legends of the First Americans.* Princeton: Princeton University Press, 2005.

McCarthy, Steve, and Mick Gilbert. *The Crystal Palace Dinosaurs: The Story of the World's First Prehistoric Sculptures.* Croydon: Crystal Palace Foundation, 1994.

McConnell, Frank. *The Science Fiction of H.G. Wells*, New York: Oxford University Press,1981, pp. 88–106.

McGowan, Christopher. *The Dragon Seekers: How an Extraordinary Circle of Fossilists Discovered the Dinosaurs and Paved the Way for Darwin.* Cambridge, MA: Perseus, 2001.

Melville, Herman. *Mardi: And a Voyage Thither ... 1849* (1970 ed.). Evanston, IL, and Chicago: Northwestern University Press and the Newberry Library, 1970.

_____. *Moby-Dick* (1851). New York: Signet Classics, 1961 ed.

Menzies, Gavin. *1421: The Year That China Discovered America.* New York: 1st Perennial ed., 2004.

Meritt, Abraham. *The Face in the Abyss.* New York: Liveright, 1931.

Miller, Hugh. *The Old Red Sandstone; or New Walks in an Old Field.* Boston: Gould and Lincoln, 1851.

Miller, Perry. *The Raven and the Whale: Poe, Melville and the New York Literary Scene.* Baltimore: Johns Hopkins University Press, 1997.

Milner, Richard. *The Encyclopedia of Evolution: Humanity's Search for Its Origins.* New York: Facts on File, 1990.

Mitchell, W.J.T. *The Last Dinosaur Book: The Life and Times of a Cultural Icon.* Chicago: University of Chicago Press, 1998.

Moore, James R. (ed.). *History, Humanity and Evolution: Essays for John C. Greene.* Cambridge: Cambridge University Press, 1989.

Moser, Stephanie. *Ancestral Images: The Iconography of Human Origins.* Ithaca, NY: Cornell University Press, 1998.

Moskowitz, Sam (ed.). *Science Fiction by Gaslight.* Cleveland and New York: World Publishing, 1968.

Nahin, Paul J. *Time Travel: A Writer's Guide to the Real Science of Plausible Time Travel,* Cincinnati, OH: Writer's Digest Books, 1997.

Nye, Nicholas. *Return to the Lost World.* Upton upon Severn, UK: Self-Publishing Association, 1991.

Obruchev, Vladimir. *Plutonia* (orig. 1924). Moscow: Raduga, 1988.

Osborn, H.F. *Cope: Master Naturalist.* Princeton: Princeton University Press, 1931.

_____. *The Hall of the Age of Man (Revised by William K. Gregory and George Pinkley),* New York: American Museum of Natural History, 1938.

_____. "Prehistoric Quadrupeds of the Rockies," *The Century Illustrated Monthly Magazine,* Vol. 52, Sept. 1896, New York: Century, pp. 705–715.

_____. "Thomas Jefferson as a Paleontologist," *Science,* Vol. 82, 1935, pp. 533–538.

_____. *The Titanotheres of Ancient Wyoming, Dakota, and Nebraska.* United States Geological Survey Monograph 55 (two volumes), 1929.

O'Connor, Ralph. *The Earth on Show: Fossils and the Poetics of Popular Science, 1802–1856.* Chicago: University of Chicago Press, 2007.

Osborn, Henry F. "*Tyrannosaurus* and Other Cretaceous Carnivorous Dinosaurs," *Bulletin of the American of the American Museum of Natural History,* Vol. 21, 1905, pp. 259–265.

_____. "Tyrannosaurus, Restoration and Model of the Skeleton," *Bulletin of the American Museum,* Vol. 22, 1913, p. 91.

_____. "Tyrannosaurus, Upper Cretaceous Carnivorous Dinosaur (Second Communication), *Bulletin of the American Museum of Natural History,* Vol. 22, 1906, pp. 281–296.

Osborn, Henry Fairfield. *The Origin and Evolution of Life on the Theory of Action, Reaction and Interaction.* New York: Scribner's, 1918 ed.

Ostrom, John H. "Osteology of *Deinonychus antirrhopus,* an Unusual Theropod from the Lower Cretaceous of Montana," *Bulletin 30* (Peabody Museum of Natural History), July 1969.

_____. "The Supporting Chain," *Discovery — Magazine of the Peabody Museum of Natural History,* Vol. 5, no. 1, Fall 1969.

Owen, Dean. *Reptilicus.* Derby, CT: Monarch Books, 1961.

Owen, Richard. "An Excerpt from the Report on British Fossil Reptiles" (originally published in *Report of the Eleventh Meeting of the British Association for the Advancement of Science,* Plymouth, England, July 1841, John Murray, publisher, London, 1842, pp. 60–204), in *Dinosaur Papers (1676–1906),* ed. David B. Weishampel and Nadine M. White, Washington, DC: Smithsonian Institution, 2003.

Paul, Gregory S. *Predatory Dinosaurs of the World: A Complete Illustrated Guide.* New York: Simon & Schuster, 1988.

_____. "The Science and Art of Restoring the Life Appearance of Dinosaurs and their Relatives: A Rigorous How-to Guide," *Dinosaurs Past and Present,* Vol. 2, ed. Sylvia J. Czerkas and Everett C. Olson. Seattle and London: Natural History Museum of Los Angeles County in association with University of Washington Press, 1987, pp. 5–49.

Peale, Rembrandt. *Historical Disquisition on the Mammoth or, Great American Incogni-*

tum, an Extinct, Immense, Carnivorous Animal Whose Fossil Remains Have Been Found in North America (London, 1803), reprinted in *Natural Sciences in America: Selected Works in Nineteenth Century North American Paleontology.* Introduction by Keir B. Sterling. New York: Arno Press, 1974.

Pellegrino, Charles R. "Dinosaur Capsule," *Omni*, Vol. 7, no. 4, January 1985, pp. 38–40, 114–115.

Petticolas, Arthur. "Dinosaur Destroyer," *Amazing Stories*, Vol. 23, January 1949, pp. 8–71.

Phillips, Tony. "Modern Lost Worlds," *Prehistoric Times*, no. 75, Dec./Jan. 2006, pp. 54–56.

Piers, Anthony. *Balook*. New York: Ace Books, 1990.

Pohl, Frederick. *Alternating Currents*. New York: Ballantine Books, 1956.

Poinar, George and Hess, R. "Ultrastructure of 40-million-year-old Insect Tissue," *Science*, Vol. 215, March 5, 1982, pp. 1241–1242.

Pope, Gustavius, W. *Romances of the Planets: N. 1, Journey to Mars*. New York: G.W. Dillingham, 1894.

Popescu, Petru. *Almost Adam*. New York: Avon Books, 1996.

Preston, Douglas. *Tyrannosaur Canyon*. New York: Forge, 2005.

Price, Robert M. (ed.) *The Antarktos Cycle: At the Mountains of Madness and Other Chilling Tales*. Oakland: A Chaosium Book, 1999.

Prothero, Donald R. *The Eocene-Oligocene Transition: Paradise Lost*. New York: Columbia University Press, 1994.

Rainger, Ronald. *An Agenda for Antiquity: Henry Fairfield Osborn and Vertebrate Paleontology at the American Museum of Natural History, 1890–1935*. Tuscaloosa: University of Alabama Press, 1991.

Reid, Constance. *The Search for E.T. Bell: Also Known as John Taine*. Washington, DC: Mathematical Association of America, 1993.

Resnick, Michael and Martin H. Greenberg (eds.). *Dinosaur Fantastic*. New York: Daw Books, 1993.

_____ and _____ (eds.). *Return of the Dinosaurs*. New York: Daw Books, 1997.

Rivkin, J.F. *Age of Dinosaurs no.1: Tyrannosaurus rex*. New York: Roc, 1992.

Roberts, Charles G.D. *In the Morning of Time*. London: Hutchinson, 1919.

Robertson, Garcia y. "The Virgin and the Dinosaur," *Asimov's Science Fiction*, Vol. 16, February 1992, pp. 132–168.

Robeson, Kenneth. *The Land of Terror* (orig. 1933). New York: Bantam Books, 1965 ed.

Rovin, Jeff. *The Encyclopedia of Monsters*. New York: Facts on File, 1989.

_____. *Fatalis*. New York: St. Martin's Paperbacks, 2000.

Rudwick, Martin J.S. *Bursting the Limits of Time: The Reconstruction of Geohistory in the Age of Revolution*. Chicago: University of Chicago Press, 2005.

_____. "Caricature as a Source for the History of Science: De la Beche's Anti-Lyellian Sketches of 1831," *Isis*, Vol. 66, 1975, pp. 534–560.

_____. "The Emergence of a Visual Language for Geological Science," *History of Science 1760–1840*, Vol. 14, 1976, pp. 149–195.

_____. *Georges Cuvier, Fossil Bones, and Geological Catastrophes*. Chicago: University Press, 1997.

_____. *The Meaning of Fossils: Episodes in the History of Palaeontology*. Chicago: University of Chicago Press, 2nd ed., 1985.

_____. *Scenes from Deep Time: Early Pictorial Representations of the Prehistoric World*. Chicago: University of Chicago Press, 1992.

_____. *Worlds Before Adam: The Reconstruction of Geohistory in the Age of Reform*. Chicago: University of Chicago Press, 2008.

Rush, Joseph Harold. *The Dawn of Life*. Garden City, NY: Hanover House, 1957.

Russett, Cynthia Eagle. *Darwin in America: The Intellectual Response 1865–1912.* San Francisco: W. H. Freeman, 1976.

Ryfle, Steve. *Japan's Favorite Mon-Star: The Unauthorized Biography of "The Big G."* Toronto: ECW Press, 1998.

Sagan, Carl. *The Dragons of Eden.* New York: Ballantine, 1977.

Sand, George. *Laura: A Journey into the Crystal.* London: Pushkin Press, 2004 ed.

Sanz, José Luis. *Starring T. Rex!: Dinosaur Mythology and Popular Culture.* Bloomington: Indiana University Press, 2002.

Sarjeant, William A.S. "Crystal Palace," *Encyclopedia of Dinosaurs,* Philip J. Currie and Kevin Padian, eds. San Diego: Academic Press, 1997, pp. 161–164.

Saville, Frank. *Beyond the Great South Wall.* New York: Grosset and Dunlap, 1901.

Schoell, William. *Saurian.* New York: Book Margins, 1988.

Schweitzer, Mary H. "Soft-tissue Vessels and Cellular Preservation in *Tyrannosaurus rex*," *Science,* Vol. 307, March 25, 2005, pp. 1952–1955.

Scott, William Berryman. *A History of Land Mammals in the Western Hemisphere* (originally published 1913). New York: Macmillan, 1937.

Scully, Vincent, R.F. Zallinger, Leo J. Hickey, and John H. Ostrom. *The Age of Reptiles: The Great Dinosaur Mural at Yale.* New York: Harry N. Abrams, 1990.

Semonin, Paul. *American Monster: How the Nation's First Prehistoric Creature Became a Symbol of National Identity.* New York: New York University Press, 2000.

Shelley, Mary. *Frankenstein.* New York: Pyramid Books, Fifth printing of the 1831 ed., *Frankenstein, or the Modern Prometheus,* 1964)

Shortland, Michael. "Darkness Visible: Underground Culture in the Golden Age of Geology," *History of Science,* Vol. 32, Part 1, no. 95, March 1994.

Silverberg, Robert. *Scientists and Scoundrels: A Book of Hoaxes.* New York: Thomas Y. Crowell, 1965.

Simpson, George Gaylord. "The Beginnings of Vertebrate Paleontology in North America," *Proceedings of the American Philosophical Society.* Vol. 86, no. 1, Sept. 1942, pp. 130–188.

_____. *The Dechronization of Sam Magruder.* New York: St. Martin's Griffin, 1996.

_____. *Discoverers of the Lost World.* New Haven: Yale University Press, 1984.

Silverberg, Robert. "Heart of Stone," *Asimov's Science Fiction,* Vol. 27, no. 12, Dec. 2004, pp. 4–9.

_____. "Toward a Theory of Story" (Parts 1 to 3), *Asimov's Science Fiction,* April/May 2004, pp. 4–9; June 2004, pp. 4–9; and July 2004, pp. 4–9.

Sommer, Marianne. "The Romantic Cave? The Scientific and Poetic Quests for Subterranean Spaces in Britain," *Earth Sciences History,* Vol. 22, no. 2, 2003, pp. 172–208.

Spamer, Earle E. "The Great Extinct Lizard: *Hadrosaurus foulkii.* 'First Dinosaur' of Film and Stage," *The Mosasaur: Journal of the Delaware Valley Paleontological Society,* Vol. 7, May 2004, pp. 109–126.

Standish, David. *Hollow Earth: The Long and Curious History of Imagining Strange Lands, Fantastical Creatures, Advanced Civilizations, and Marvelous Machines Below the Earth's Surface.* Cambridge, MA: De Capo Press, 2006.

Sternberg, Charles, H. "Ancient Monsters of Kansas." *Popular Science News,* Vol. 32, Dec. 1898, p. 268.

_____. *Hunting Dinosaurs in the Bad Lands of the Red Deer River, Alberta, Canada: A Sequel to The Life of a Fossil Hunter* (1st ed.). Published by Charles H. Sternberg. Lawrence, Kansas, 1917.

_____. *The Life of a Fossil Hunter* (1909). With Foreword and notes by Paul F. Ciesielski. Gainesville, FL: Faulkner Press, 2004.

_____. "A New *Trachodon.*" *Science,* Vol. 29, 1909, pp. 753–754.

_____. "Pliocene Man," *American Naturalist*, Vol. 12, no. 2, Feb. 1878 (Philadelphia: Press of McCall and Stavely), pp. 125–126.

_____. *A Story of the Past, or the Romance of Science*. Boston: Sherman, French, 1911.

Sternberg, Charles Hazelius. *Hunting Dinosaurs: In the Bad Lands of the Red Deer River, Alberta Canada* (1st ed. 1917). With Introduction by David A. E. Spalding. Edmonton: NeWest Press, 1985.

Strickland, Brad, with John Michlig. *Kong: King of Skull Island*. Created and Illustrated by Joe DeVito. Milwaukie, OR: DH Press, 2004.

Taine, John. *Before the Dawn* (Williams & Wilkins Co. 1934), rep. by Arno Press, 1974.

_____. *The Iron Star* (orig. 1930). Westport, CT: Hyperion Press, 1976 ed.

Trinkhaus, Erik and Shipman, Pat. *The Neandertals: Changing the Image of Mankind*. New York: Alfred A. Knopf, 1993.

Tsutsui, William. *Godzilla on My Mind*. New York: Palgrave Macmillan, 2004.

Turner, George, with Goldner, Orville. *Spawn of Skull Island*. Baltimore: Luminary Press, 2002.

Valdron, D.G. "Fossils," in Robert Hood and Robin Pen, eds. *Daikaiju!: Giant Monster Tales*. University of Wollongong, Australia: Agog! Press, 2005.

Verne, Jules. *Journey to the Center of the Earth* (orig. 1867). A new translation by William Butcher. Oxford: Oxford University Press, 1998.

_____. *The Village in the Treetops* (orig. 1902). New York: Ace Books, 1964.

Walsh, John Evangelist. *Unraveling Piltdown: The Science Fraud of the Century and Its Solution*. New York: Random House, 1996.

Ward, Henry A. *Catalogue of Casts of Fossils from the Principal Museums of Europe and America, with Short Descriptions and Illustrations*. Rochester, NY: Benton & Andrews, 1866.

Warren, Bill. *Keep Watching the Skies!: American Science Fiction Movies of the Fifties, Vol. II, 1958–1962*. Jefferson, NC: McFarland, 1986.

Wells, H.G. *The Island of Dr. Moreau* in *The Complete Science Fiction Treasury of H G. Wells with a Preface by the Author*. New York: Avenel Books, 1978.

_____. *Mr. Blettsworthy on Rampole Island*. Garden City, NY: Doubleday, Doran, 1928.

_____. *The Time Machine*, in *The Science Fiction Hall of Fame, Volume IIA*. Ben Bova (ed.). New York: Avon Books, 1973.

Wells, H.G., Julian Huxley, and G.P. Wells. *The Science of Life*. Garden City, NY: Doubleday, Doran, 1935 ed.

Wendt, Herbert. *Before the Deluge*. New York: Doubleday, 1968.

Weta Workshop. *The World of Kong: A Natural History of Skull Island*. New York: Pocket Books, 2005.

Wilford, John Noble. *The Riddle of the Dinosaur*. New York: Alfred A. Knopf, 1985.

Williams, Robert Moore. "The Lost Warship," *Amazing Stories*, Vol. 17, no. 1, January 1943, pp. 8–56.

Winslow, J.H., and A. Meyer. "The Perpetrators of Piltdown," *Science*, Vol. 83, September 1983, pp. 32–43.

Womack, Todd. "Plentifully Charged with Fossils: The 1822 Discovery of the *Eremotherium* at Skidaway," *Fossil News: Journal of Avocational Paleontology*, Vol. 6, no. 7, July 2000, pp. 14–16.

Woodward, John. *Essay Toward a Natural History of the Earth*. London, 1723 ed.

Yochelson, Ellis S. "Peale's 1799 Theory of the Earth," *Earth Sciences History*, Vol. 10, no. 1, 1991, pp. 51–55.

Zangerl, Rainer. *Dinosaurs, Predator and Prey: The Gorgosaurus and Lambeosaurus Exhibit in Chicago Natural History Museum*. 1961.

Zimmer, Carl. *At the Water's Edge: Macroevolution and the Transformation of Life*. New York: Free Press, 1998.

Index

Numbers in **bold italics** indicate pages with illustrations.